30

MICROBIOLOGICAL RISK
ASSESSMENT SERIES

Interventions for the Control of Non-typhoidal *Salmonella* spp. in Beef and Pork

MEETING REPORT AND SYSTEMATIC REVIEW

Food and Agriculture Organization of the United Nations
World Health Organization

Rome, 2016

WHO Library Cataloguing-in-Publication Data:
Interventions for the control of non-typhoidal Salmonella spp. in beef and pork: meeting report and systematic review

(Microbiological risk assessment series, 30)

1.Food Contamination - prevention and control. 2.Animal Husbandry - methods. 3.Salmonella. 4.Salmonella Infections, Animal. 5.Meat Products. 6.Swine Diseases. 7.Cattle Diseases. 8.Review. 9.Meeting Abstracts. I.World Health Organization. II.Food and Agriculture Organization of the United Nations.

ISBN 978 92 4 156524 0 (WHO) (NLM classification: WC 268)
ISBN 978-92-5-109049-7 (FAO)
ISSN 1726-5274

Recommended citation:
FAO/WHO [Food and Agriculture Organization of the United Nations/World Health Organization]. 2016. Interventions for the control of non-typhoidal Salmonella spp. in beef and pork: Meeting report and systematic review. Microbiological Risk Assessment Series No. 30. Rome. 276pp.

Contents

ANNEXES

Acknowledgements

The Food and Agriculture Organization of the United Nations and the World Health Organization would like to express their appreciation to all those who contributed to the preparation of this document through the provision of their time, expertise and other relevant information both before, during and after the meeting. Special appreciation is extended to Dr Ian Young and Dr Barbara Wilhelm for leading the Systematic Review of the Efficacy of Interventions to Control *Salmonella* in Beef and Pork. The FAO/WHO JEMRA secretariat would also like to express their appreciation to the Codex working group on Non-typhoidal *Salmonella* spp. in beef and pork and in particular the lead countries of that working group for their availabity to address queries and provide support in the design and implementation of the review and the expert meeting. Appreciation is also extended to all those who responded to the call for data that was issued by FAO and WHO and provided information that was not readily available in the peer reviewed literature. All contributors are listed on the following pages.

FAO and WHO would also like to acknowledge the financial resources provided by the United States of America to support this work.

Final editing for language was by Thorgeir Lawrence and layout by Joanne Morgante.

Contributors

EXPERTS

Lis Alban, Chief Scientist, Food Safety and Veterinary Issues, Risk Assessment Group, Danish Agriculture and Food Council, Århus, Denmark.

Declan Bolton, Principal Research Officer, Teagasc, Food Research Centre, Dublin, Ireland.

Sava Buncic, Professor, Department of Veterinary Medicine, Faculty of Agriculture, University of Novi Sad, Serbia.

Todd R. Callaway, Research Microbiologist, Food and Feed Safety Research Unit, Agricultural Research Service, United States Department of Agriculture, College Station, USA.

Marisa Ribeiro de Itapema Cardoso, Universidade Federal do Rio Grande do Sul, Faculdade de Veterinária, Porto Alegre, Brazil.

Jan Dahl, Danish Agriculture and Food Council, Copenhagen, Denmark.

James S. Dickson, Professor, Department of Animal Science, Inter-Departmental Program in Microbiology, Iowa State University, USA.

Glen Edmunds, Senior Scientist, Residues and Microbiological Policy, Exports Division, Australian Government Department of Agriculture, Brisbane, Australia.

Moses Gathura Gichia, Deputy Director, Directorate of Veterinary Services, Kangemi, Nairobi, Kenya.

Renata Ivanek-Miojevic, Associate Professor of Epidemiology, Department of Population Medicine and Diagnostic Sciences, College of Veterinary Medicine, Cornell University, USA.

Ian Jenson, Manager, Market Access Science and Technology, Meat & Livestock Australia, Sydney, Australia.

Ákos Jóźwiak, Vice-director, National Food Chain Safety Office, System Management and Supervision Directorate, Budapest, Hungary.

Andreas Kiermeier, Director, Statistical Process Improvement Consulting and Training Pty Ltd, Australia.

Tareq Osaili, Associate Professor of Food Safety, Department of Nutrition and Food Technology, Jordan University of Science and Technology, Irbid, Jordan.

Blaise Ouattara, Chief, Risk Prioritization, Food Safety, Canadian Food Inspection Agency, Ottawa, Canada.

Héctor Argüello Rodríguez, Research Officer, Food Safety Department, Teagasc, Food Research Centre, Dublin, Ireland.

Javier Sanchez, Associate Professor of Epidemiology, Health Management Department, Atlantic Veterinary College, University of Prince Edward Island, Canada.

Hajime Toyofuku, Professor, Joint Faculty of Veterinary Medicine, Yamaguchi University, Yamaguchi, Japan.

RESOURCE PERSONS

Terrance M. Arthur, Microbiologist, United States Meat Animal Research Center, Agricultural Research Service, United States Department of Agriculture, USA.

Annamaria Bruno, Senior Food Standards Officer, FAO/WHO Food Standards Programme, Rome, Italy.

Verna Carolissen-Mackay, Food Standards Officer, FAO/WHO Food Standards Programme, Rome, Italy.

Jose Emilio Esteban, Chair of Codex Committee on Food Hygiene, USDA, Food Safety Inspection Service, Athens, USA

Neal J. Golden, Senior Risk Analyst, USDA, Food Safety and Inspection Service, Office of Public Health Science, Risk Assessment and Analytics Division, Washington, USA.

Gillian Mylrea, Deputy Head, Department of International Trade, World Organisation for Animal Health, Paris, France.

Gudrun Sandø, Special Veterinary Advisor, Ministry of Environment and Food, The Danish Veterinary and Food Administration, Glostrup, Denmark.

Takako Yano, Food Standards Officer, FAO/WHO Food Standards Programme, Rome, Italy.

JEMRA SECRETARIAT

Sarah Cahill, Food Safety Officer, Food Safety and Quality Unit, Agriculture and Consumer Protection Department, FAO, Rome, Italy.

Patricia Desmarchelier, Director, Food Safety Principles, Pullenvale, Queensland, Australia.

Rei Nakagawa, Technical Officer, Department of Food Safety and Zoonoses, WHO, Geneva, Switzerland.

Andrijana Rajić, Food Safety Officer, Food Safety and Quality Unit, Agriculture and Consumer Protection Department, FAO, Rome, Italy.

Barbara Wilhelm, Ontario Veterinary College, University of Guelph, Guelph, Canada.

Ian Young, Food Safety Expert, Food Safety and Quality Unit, Agriculture and Consumer Protection Department, FAO, Rome, Italy.

Declarations of Interest

All participants completed a Declaration of Interests form in advance of the meeting. Six of the experts declared an interest in the topics under consideration:

- Lis Alban and Jan Dahl are employees of the Danish Agriculture and Food Council, which represents the meat industry.
- Todd R. Callaway performs scientific research in live cattle which is financially supported by the National Cattlemen's Beef Association.
- James Dickson performs consultancy work with the meat industry and conducts scientific research which is financially supported by the United States National Pork Board.
- Ian Jenson is an employee of Meat and Livestock Australia, which represents the meat industry.
- Andreas Kiermeier performs consultancy work for Australian Pork Limited, which represents the meat industry.

Upon detailed review of these declarations, it was considered that the activities of Lis Alban, Jan Dahl, James Dickson, Ian Jenson, and Andreas Kiermeier represent potential conflicts of interest. However, it was also considered that these experts provide a unique and very practical expertise on meat processing and the application of interventions in a range of real world settings. Therefore, these experts were invited to the meeting but did not participate in the final adoption of the conclusions and recommendations of the meeting. The activities of Todd Callaway were also considered to represent a potential conflict of interest in the evaluation of farm-level interventions. However, similarly his expertise was considered necessary for the meeting discussions and therefore, he was invited but did not participate in the final adoption of the conclusions and recommendations of the meeting that related to farm-level interventions.

All of the declarations, together with any updates, were made known and available to all the participants at the beginning of the meeting. All the experts participated in their individual capacity and not as representatives of their country, government or organization.

Abbreviations used in this report

ACR	assumed control risk
ADG	average daily gain
APC	aerobic plate count
CAC	Codex Alimentarius Commission
CI	confidence interval
CCFH	Codex Committee on Food Hygiene
CFU	Colony-Forming Unit(s)
DTLN	deep tissue lymph node
ECP	experimental chlorate preparation
FAO	Food and Agriculture Organization of the United Nations
FC	feed conversion
GFP	Good Farming Practice
GHP	Good Hygiene Practice
GRADE	Grades of Recommendation, Assessment, Development and Evaluation
HACCP	Hazard Analysis Critical Control Point
MA	meta-analysis average [estimate from random-effects model]
OIE	World Organisation for Animal Health
OR	odds ratio
PFU	plaque-forming unit
PWMS	post-weaning multi-systemic wasting syndrome
pWG	Physical Working Group
QMRA	quantitative microbial risk assessment
RR	risk ratio
RSR	Rapid Systematic Review
SC	Summary Card
WHO	World Health Organization

Executive Summary

Non-typhoidal *Salmonella* spp. (hereafter simply referred to as *Salmonella*) are estimated to cause 93.8 million cases of acute gastroenteritis and 155 000 deaths globally each year, approximately 85% of which are estimated to be foodborne. As a result, *Salmonella* have a significant public health and economic impact on society. Pork products are among the top food-borne sources of *Salmonella* globally. While beef products are a less significant source of salmonellosis in many countries, they have been implicated in several large outbreaks in recent years. Contamination of beef and pork with *Salmonella* can also have a negative impact on the agri-food and trade sectors due to costly recalls of products and by limiting market access.

The 45th Session of the Codex Committee on Food Hygiene (CCFH) agreed to develop "Guidelines for the Control of Non-typhoidal *Salmonella* spp. in Beef and Pork Meat". To facilitate this work, CCFH (November 2014) requested FAO and WHO to provide scientific advice on this matter, by conducting a systematic literature review of the publicly available scientific literature to ensure that any relevant measures for the control of *Salmonella* in beef and pork are identified, and by convening an expert meeting to review the technical basis of the interventions proposed by CCFH. The experts were specifically asked to: advise on the most appropriate point(s) of application of specific interventions and decontamination treatments; verify, based on the available data, the efficacy of the interventions in terms of reduction of *Salmonella*; and advise, with some level of confidence, to the extent possible, on the quantifiable level of reduction that interventions achieve, and whether these are appropriate to include in the Codex guideline.

The FAO/WHO systematic review covered all possible interventions from primary production, to the end of processing to control *Salmonella* in beef and pork. FAO and WHO also issued to all Member Countries a public "call for data" on control measures for *Salmonella* in beef and pork. Several replies were received in response to the call, a couple of which included previously unavailable information. All this information was then considered by the expert meeting, which was convened 28 September to 2 October 2015 at FAO Headquarters, Rome, Italy.

During the meeting, the experts considered any intervention for which there was available evidence, and which could be applied to prevent, reduce or control *Salmonella* in the production and processing of fresh beef or pork. These were referred to as hazard-based interventions. While the focus was on identified hazard-based interventions, the experts emphasized that these interventions must not be considered in isolation, but rather as an integral part of an overall meat hygiene programme. It was noted that there are a range of contextual factors that will guide decisions on whether a particular intervention is implemented, and that efficacy will also vary according to the conditions at the point of implementation. It was agreed that all interventions should be verified at the local establishment of implementation.

In developing conclusions and recommendations, the experts considered the efficacy of each intervention to reduce *Salmonella* prevalence or concentration, and the most appropriate point(s) of application to achieve the intended reduction. They then agreed on a recommendation for each intervention as to whether or not it should be considered by CCFH for inclusion in the Codex guideline, and by regulators, food safety risk managers or establishment operators as a hazard-based intervention or good hygiene practice (GHP). There are some unique animal, production and processing characteristics that can affect the opportunity for contamination and interventions for the control of *Salmonella* along the beef and pork chains. The experts therefore considered the two meat types separately for most stages of the chain.

With regard to beef, no specific hazard-based interventions were identified in primary production, although the experts agreed that biosecurity could contribute to general on-farm control of *Salmonella* and other zoonotic foodborne infections.

Decontamination treatments of cattle hides using chemical washes, including organic acids and other chemicals, were recommended for consideration as potential hazard-based interventions for the control of *Salmonella* when applied post-exsanguination and before de-hiding. However, decontamination of the hides of live animals was not recommended for consideration due to a lack of confidence in supporting evidence and concerns for animal welfare.

Carcass decontamination treatments were recommended for consideration as potential hazard-based interventions for the control of *Salmonella* after hide removal and before chilling. Decontamination treatments recommended by the experts for consideration included hot-water washes and steam pasteurization that achieve a carcass surface temperature of at least 70°C, and chemical washes (including organic acids and other chemicals with proven efficacy). Additionally, chemical washes with proven efficacy were recommended for consideration as potential hazard-based interventions for the control of *Salmonella* in fabricated beef.

For pork, the experts agreed that biosecurity is an important good farming practice (GFP) that can help to prevent the introduction of *Salmonella* to *Salmonella*-negative farms and to reduce the prevalence in finisher pigs in *Salmonella*-positive farms. If a regulator, risk manager or farmer wishes to reduce the pre-harvest *Salmonella* prevalence, then potential on-farm hazard-based interventions for the control of *Salmonella* could be considered, including feed management, such as feeding meal vs pellets, and acidification of feed or water using organic acids. Vaccination could also be considered as a potential hazard-based intervention for the control of *Salmonella* on-farm; however, the experts also identified a number of factors that need to be considered if vaccination is used as a food safety measure. Moreover, if measures are taken only pre-harvest, then there may be a limited effect on the reduction of *Salmonella* on carcasses.

Scalding and singeing are process steps that were recommended for consideration as potential hazard-based interventions for the control of *Salmonella* due to the extensive evidence for reductions in prevalence on pork carcasses. Carcass decontamination treatments with proven efficacy were recommended for consideration as potential hazard-based interventions before chilling. These included hot water washes and steam pasteurization achieving a carcass surface temperature of at least 70°C during treatment, and organic acid washes.

For both beef and pork it was acknowledged that other steps during production and processing are also important for reduction of *Salmonella*; however, a lack of consistent and credible evidence, and insufficient evidence of efficacy specifically for *Salmonella,* meant that they could not be considered as potential hazard-specific interventions. Instead, several of these were considered as important GHP measures, including: hygiene during transport to slaughter and in lairage to limit the spread of *Salmonella*; hygiene during carcass dressing to minimize contamination; bunging to reduce faecal spillage during processing; carcass trimming and steam vacuuming to remove visible contamination; chilling to prevent growth of *Salmonella*; and practices to prevent carcass cross-contamination in the chilling room. In addition, the following were considered as important GHP measures in the pork chain: feed withdrawal to reduce rupturing of the intestines and intestinal spillage during carcass dressing; hygiene during de-hairing and polishing to reduce cross- and re-contamination of carcasses; and full carcass steam vacuuming as a potential alternative to hot water washes in small establishments with limited resources.

During packaging, the experts recommended that irradiation should be considered as a potential hazard-based intervention for the control of *Salmonella* in beef and pork products. In terms of post-packaging interventions, it was noted that there were a number of interventions that could be applied from product distribution to consumption, but these varied widely and limited information was available for their consideration. However, the experts highlighted some key areas in terms of *Salmonella* control, including the importance of cold chain management and application of hazard analysis critical control point (HACCP)-based principles and hygiene prerequisites.

The experts also discussed some limitations and caveats regarding available data which should be considered when interpreting the meeting results.

1

Background and Rationale

Globally, non-typhoidal salmonellosis is estimated to cause 93.8 million cases of acute gastroenteritis and 155 000 deaths each year (Majowicz *et al.*, 2010). It is estimated that 80.3 million of these cases are food-borne (Majowicz *et al.*, 2010). Non-typhoidal salmonellosis has also been associated with various chronic sequelae, such as reactive arthritis and irritable bowel syndrome (Keithlin *et al.*, 2015). As a result, non-typhoidal *Salmonella* spp. (hereafter simply referred to as *Salmonella*) can have a significant public health and economic impact on society.

Non-typhoidal salmonellosis cases and outbreaks are most commonly attributed to exposure to contaminated food, and along with eggs and poultry, pork products are among the major food-borne sources globally (Pires *et al.*, 2014). While beef products are not one of the main sources of salmonellosis for most countries, they have been implicated in several large outbreaks of *Salmonella* infections in the United States of America during the past 15 years, sometimes due to multidrug resistant strains (Laufer *et al.*, 2015). These outbreaks are increasingly associated with ground beef rather than with intact beef cuts (Laufer *et al.*, 2015). Non-typhoidal salmonellosis outbreaks resulting from the consumption of beef products contaminated with multidrug resistant strains have also been recently reported in several European countries (Friesema *et al.*, 2012; Mindlin *et al.*, 2013; Raguenaud *et al.*, 2012).

Salmonella reside in the intestinal tract of pigs, cattle and other food-producing animals, as well as wildlife and pests. Food-producing animals are frequently asymptomatic carriers, who shed the bacterium without showing any signs of disease (Buncic and Sofos, 2012). *Salmonella* are spread between animals on the farm, during transport, in lairage, and by contact with these environments (Baer, Miller and Dilger, 2013; Buncic and Sofos, 2012). *Salmonella* from contaminated and infected incoming animals and from environmental contamination in the slaughterhouse can be transferred to carcasses during processing, and the contamination level can potentially be amplified during storage and distribution if not controlled (Arthur *et al.*, 2008; Duggan *et al.*, 2010; Mannion *et al.*, 2012; Rivera-Betancourt *et al.*, 2004). Implementation of good production and hygiene practices are needed along the food chain continuum, along with appropriate controls and interventions that minimize *Salmonella* contamination and prevent bacterial growth, as part of an integrated approach to reduce *Salmonella* contamination of beef and pork and the associated burden of food-borne illness in humans.

Such an approach is important not only because of the health significance of food-borne *Salmonella*, but also because the agri-food and trade sectors are negatively affected by this organism. Contamination of beef and pork with *Salmonella* can lead to costly recalls of products (Dey *et al.*, 2013), including destruction of such products, and can hinder economic development by limiting market access. In addition, the global demand for food of animal origin is increasing. With the growing global population, it is expected that by 2050 we will need to be producing 70% more livestock products to meet this demand (FAO, 2009). This has many implications, of which food safety is among them.

In this context, there is an increasing need for global guidance and standards on the control of *Salmonella* in beef and pork. In November 2013, the 45th Session of the Codex Committee on Food Hygiene (CCFH) agreed to develop "Guidelines for the Control of Non-typhoidal *Salmonella* spp. in Beef and Pork Meat". To facilitate this work, the 46th Session of the CCFH (CCFH46), held in November 2014, requested the FAO and WHO to provide the CCFH with scientific advice on this matter, by

- conducting a systematic literature review to ensure that any relevant measures for the control of *Salmonella* in beef and pork are identified; and
- convening an expert meeting to review the technical basis of the mitigation/intervention measures proposed by a CCFH physical working group (pWG) prior to the 47th Session in November, 2015.

A CCFH pWG meeting was held in Brussels, Belgium, 6–9 May 2015. The pWG meeting resulted in proposed draft guidelines based on the draft prepared by the United States of America and Denmark, and taking into account comments previously submitted to CCFH46 and the interim report of the systematic review prepared by FAO/WHO. The pWG at the Brussels meeting refined the request to the FAO/WHO expert meeting to provide advice on *Salmonella* control measures that might be included in the draft guidelines, and for each of these interventions to specifically:

- advise on the most the appropriate point(s) of application of specific interventions and decontamination treatments;
- verify, based on the available data, the efficacy of the intervention in terms of reduction of *Salmonella*; and
- advise with some level of confidence, to the extent possible, on the quantifiable level of reduction that the interventions achieve, and whether these are appropriate to include in the Codex guideline.

To address *Salmonella* control measures in pigs and cattle at the pre-harvest level of the food chain, the World Organisation for Animal Health (OIE) has convened an *ad hoc* group to develop draft chapters for their Terrestrial Animal Health Code. Relevant sections of the code will be referenced in the CCFH guideline for pre-harvest measures. Therefore, the focus of the expert meeting guidance was requested to be on control measures at processing, while also recognizing that pre-harvest measures are important considerations affecting the management and control of *Salmonella* across the food chain continuum.

2

Expert Meeting

The expert meeting was held from 28 September to 2 October 2015, at FAO Headquarters in Rome, to address the requests from the CCFH and the pWG.

2.1 SCOPE AND OBJECTIVES

The experts agreed that the scope of the meeting included the consideration of any treatment or action for which there was available evidence that could be applied to prevent, reduce or control *Salmonella* in the production and processing of fresh beef or pork along the food chain, from farm to consumer. Manufactured beef and pork products (e.g. sausages and salamis) were not considered[1].

The objectives of the meeting were threefold:

- To review the technical and scientific basis of potential measures for the control of *Salmonella* in beef and pork, with a focus on the processing level of the food chain.
- To make recommendations to CCFH on the effectiveness of specific control measures that should be considered for inclusion in the "Proposed Draft Guidelines for the Control of Non-typhoidal *Salmonella* spp. in Beef and Pork Meat", including how that information should be presented and quantified.
- To give expert guidance to FAO/WHO Member Countries on contextual factors to consider when planning and implementing control measures for *Salmonella* in beef and pork.

[1] Codex Code of Hygienic Practice for Meat (CAC/RCP 58-2005)

2.2 APPROACH

FAO and WHO conducted a systematic review of the publicly available scientific literature on measures to control *Salmonella* in beef and pork. The review covered all possible interventions from primary production to the end of processing. Post-processing interventions were initially considered in the review, but were later excluded, as the search was not appropriately optimized to capture such studies. The full review document, including a detailed description of the methodology used, is presented in Annex 1, *Rapid Systematic Review of the Efficacy of Interventions to Control Salmonella in Beef and Pork*, with a brief description below. Limitations and caveats of the review methodology and available data are noted in Chapter 8.

The systematic review considered all primary research study designs as well as previously published systematic reviews, risk assessments and stochastic models. The review followed standard methods for systematic reviews (Higgins and Green, 2011; Rajić and Young, 2013), but some steps were streamlined using a rapid approach to ensure that preliminary findings would be available within approximately two months to inform the CCFH pWG meeting in May 2015, with updates and finalization over the following four months to inform this expert meeting in September, 2015. Results were presented as six "Summary Cards", presenting the evidence stratified by commodity of focus (pork or beef) and stage in the food chain (farm, transport-lairage and processing).

Whenever intervention efficacy results from identified research studies were summarized quantitatively in the review, the level of confidence in the intervention efficacy estimates was assessed using a modified version of the Cochrane Collaboration's Grades of Recommendation, Assessment, Development and Evaluation (GRADE) approach (Guyatt *et al.*, 2011; WHO, 2014). The GRADE approach classifies the confidence in intervention efficacy estimates into one of four levels: very low; low; moderate; or high. Due to inherent differences in strength of evidence for intervention efficacy by study design (Guyatt *et al.*, 2011; Sargeant, Kelton and O'Connor, 2014), controlled trials started at a rating of "high" and other study designs at "moderate". Five downgrading and three upgrading criteria were then assessed for each estimate, which could lead to reducing or increasing, respectively, the pre-defined GRADE ratings. The downgrading criteria included: risks of bias; heterogeneity in effects across studies; overall precision of the included studies; the degree to which the studies are reflective of the topic of interest and target conditions (directness of evidence); and publication bias. The potential upgrading criteria were: presence of a dose-response relationship; evidence for underestimation of the intervention effect (if one exists); or presence of a very large intervention effect. The corresponding GRADE rating represents the level of confidence that one can place in the measured intervention effect estimates in terms of how closely they would be expected to match the actual intervention effect in practise or future studies conducted under similar conditions.

FAO and WHO also issued a public "call for data" to all Member Countries on control measures for *Salmonella* in beef and pork. Several replies were received in response to the call, two of which included previously unavailable documents that were relevant to the meeting. One was an executive summary of another systematic review lead by a team from Iowa State University, and funded by the United States National Pork Board, that summarized research on the magnitude of change in the prevalence and quantity of *Salmonella* after administration of pathogen reduction treatments on pork carcasses. This review was conducted concurrently to but independently from the FAO/WHO review. The other document reported on preliminary results from the United States Department of Agriculture (USDA)'s Nationwide Beef and Veal Carcass Baseline Survey. These data described differences in *Salmonella* prevalence and concentration from paired samples taken at different points in commercial slaughter plants: immediately after hide removal (pre-evisceration) and at pre-chill (after all decontamination interventions).

Prior to the meeting all experts were sent a copy of the most recent version of the CCFH "Proposed Draft Guidelines for the Control of Non-typhoidal *Salmonella* spp. in Beef and Pork Meat" (CX/FH 15/47/5, August 2015), as well as a report of the FAO/WHO systematic review and the two relevant documents received in the "call for data".

The meeting was attended by 18 internationally recognized experts from 11 countries, covering all global regions, together with eight resource persons and six individuals from the FAO/WHO secretariat (see list of Contributors in the front matter). The meeting was chaired by Dr Glen Edmunds.

During the meeting, the experts identified interventions from primary production to consumption and assessed them for relevance, processing parameters, and evidence for efficacy. As there was no Codex, FAO or WHO definition of "intervention", the experts interpreted these according to the Oxford English Dictionary as any "action of intervening, 'stepping in', or interfering in any affair, so as to affect its course or issue", within the context of controlling *Salmonella* in beef and pork. Interventions were recommended as "hazard-based" if evidence was available specific to their effect on reducing *Salmonella* prevalence and/or concentration. If such evidence was not available for a potential intervention, but the experts agreed that it was important for general meat hygiene, and as a result could potentially have an effect on *Salmonella*, then experts considered the intervention as a good hygiene practice (GHP).

To assess the interventions, the experts worked in two groups, one for beef and the other for pork. The beef group was chaired by Dr Sava Buncic, with Dr Renata Ivanek-Miojevic as Rapporteur, and the pork group was chaired by Dr Javier Sanchez, with Dr Héctor Argüello Rodríguez as Rapporteur. Within their groups, the experts made reference to the FAO/WHO systematic review, their knowledge and experience, more recent publications and other sets of data in their deliberations. The experts assessed the interventions in sequence according the steps of the food chain as outlined in the CCFH "Proposed Draft Guidelines for the Control of Non-typhoidal *Salmonella* spp. in Beef and Pork Meat" (Annex 2). The experts grouped interventions into categories depending on where in the food chain they would mostly likely be applied, and in their assessments made reference to the appropriate steps in the CCFH draft guideline.

For each intervention, the assessment began with a review of the key findings from the systematic review, taking into consideration the GRADE ratings of the intervention efficacy estimates. Experts then discussed the practical applicability of the review results to commercial conditions and appropriately qualified and, where deemed necessary, added to the evidence by drawing on their expertise, experience and any additional data or other resources. They then agreed on a recommendation for each intervention as to whether or not it should be considered by CCFH for inclusion in the Codex guideline and by regulators, food safety risk managers or establishment operators as a hazard-based intervention or GHP. For hazard-based interventions recommended for consideration, the experts also discussed and agreed on how to most appropriately quantify, if at all, the estimates of intervention efficacy.

3

Salmonella-specific interventions in the context of meat hygiene

3.1 RISK MANAGEMENT CONSIDERATIONS

The Codex "Code of Hygienic Practice for Meat" (CAC/RCP58-2005) states that the principles of food safety risk management should be incorporated wherever appropriate in the design and implementation of meat hygiene programmes. Regulators, export markets, customers and establishment operators may set targets for *Salmonella* in beef or pork. To achieve these targets, each food safety manager can develop a food safety system which may include good practices for farming, veterinary medicine, hygiene, manufacturing and production. Where these practices are not sufficient to meet the proposed target, then additional risk management options may be used, including specific process interventions whether GHP- or hazard-based.

In the slaughter process, GHP is essential to ensure the prevention or minimization of contamination with *Salmonella*. Where GHP does not achieve this alone, application of additional interventions may be considered. However, these interventions should not replace the application of GHPs.

It is important to remember that the establishment operators bear primary responsibility for the production of hygienic and safe meat, and they should develop a food safety system that is documented and identifies how they will meet the hazard-control objective. Where the establishment operator has used interventions, they will need to be fully developed and validated for the specific conditions of the food business and the individual product. It is up to the establishment operator to decide on which interventions to use, and their decision may be influenced by regulatory factors, cost, feasibility, technical requirements, in-house expertise, consumer expectation and other contextual factors. While experts recognized some interventions, including chemical treatments and irradiation, may need approval by the competent authority, this aspect was not considered in detail in the contextual assessment in the meeting.

Any intervention used should be applied at the most effective or appropriate step to obtain the maximum benefit in reducing public health risk. Hazard-based control is applied through the entire food chain and the positive effect of any intervention could be undone at any later stage. It is important to remember that the application of any intervention may be country, region, establishment or production-system specific. The effectiveness of interventions may also be dependent on *Salmonella* serovar and levels of contamination, point of application, background microbiota, presence and amount of soil or organic matter, and season, among other factors. For these reasons, it is critical that any intervention is appropriately validated in the local establishment of application. Once implemented, the intervention should be monitored and verified. Where the intervention is not attaining the stated hazard-control objective, it should be reviewed, any required corrective action should be taken, and the intervention should be revalidated. Interventions should also be reviewed in a timely manner to ensure the most appropriate intervention is being used considering new knowledge and issues as they arise.

At a national or regional level, risk managers require a profile of *Salmonella* in the beef and pork food chains. This can be obtained from surveillance and monitoring programmes and includes data such as prevalence of *Salmonella* in meat production animals, meat and foods, and human incidence of salmonellosis, together with characterization of *Salmonella* isolates, such as by serovar, genotype or antimicrobial resistance profile. Surveys of the prevalence of *Salmonella* on beef and pork carcasses and on meat provide baselines of current production performance and interventions in place, as well as a point of reference for evaluating the effect of any interventions introduced. These systems, if not already in place, will incur significant cost for government and industry.

Practical implementation factors should be considered by the establishment manager when choosing an intervention. These vary with the establishment, intervention type and point of application. Some broad areas to consider, in no particular order, include:

- Ability to contribute to targeted hazard control objectives.
- Cost-effectiveness.

- Reliable and timely supplies of materials or other resources, e.g. chemicals, power.
- Adequate quantity and reliable supply of potable water.
- Impact of chemicals on equipment, and effect of accumulation in the establishment environment.
- Development of resistance in *Salmonella* strains with long-term use of chemicals and biocides.
- Occupational and safety risks to workers.
- Acceptance of intervention agents as food additives by regulators in domestic and export markets; and need for labelling.
- Technical complexity and ease of use.
- Cost and availability of infrastructure, with ongoing maintenance.
- Impact on meat quality.
- Consumer acceptance.
- Environmental impact, e.g. waste disposal and pollution.

Multiple interventions (multiple-hurdle strategy) can be placed in a single step or in consecutive steps in a processing line. Where multiple interventions are applied, the level of control expected should be greater than any single intervention, but is not likely to be the sum of the individual measures of efficacy of each intervention. The Codex "Code of Hygienic Practice for Meat" (CAC/RCP 58-2005) noted:

> *A contemporary risk-based approach to meat hygiene requires that hygiene measures should be applied at those points in the food chain where they will be of greatest value in reducing food-borne risks to consumers. This should be reflected in application of specific measures based on science and risk assessment, with a greater emphasis on prevention and control of contamination during all aspects of production of meat and its further processing. Application of HACCP principles is an essential element. The measure of success of contemporary programmes is an objective demonstration of levels of hazard control in food that are correlated with required levels of consumer protection, rather than by concentrating on detailed and prescriptive measures that give an unknown outcome.*

This report aims to provide information that builds on the above approach to meat hygiene, by focusing on particular steps of the beef and pork chains and on interventions where there is a potentially demonstrable effect on *Salmonella*.

3.2 DIFFERENCES IN THE NATURE OF BEEF AND PORK PRODUCTION

There are some unique animal, production and processing characteristics that can have an impact on the opportunity for contamination and the use of interventions for the control of *Salmonella* along the beef and pork chains. Some of these are outlined below.

3.2.1 Beef

Cattle can be raised for meat or milk production, with dairy operations also sending bull calves and aged dairy cows for meat processing. Less is known of the epidemiology of *Salmonella* in cattle compared with farmed pigs. *Salmonella* can cause clinical salmonellosis in calves and to a much lesser extent in adult cattle. All age groups can be asymptomatic *Salmonella* carriers, and with varying prevalence in age groups at a location over time. Cattle can carry a wide range of serovars, some are commonly associated with bovine disease and a small number may predominate in a region at a point in time.

Production systems can be extensive and/or intensive, can include pasture and/or feedlot beef herds or dairy herds, and cattle can be part of mixed-species enterprises, all of which can be associated with multiple and differing risk factors for *Salmonella*. The nature of cattle production systems, the scale of some systems, the movement of cattle between these systems, and the multitude of risk factors for *Salmonella,* make on-farm control more challenging and less practical than for pigs. In comparison with pigs, hazard-based measures for reducing the *Salmonella* contamination of cattle entering slaughter have been implemented closer to the point of slaughter (e.g. hide decontamination).

Cattle hides can be heavily contaminated with *Salmonella* when entering the processing establishment. Protecting the newly exposed carcass surface from this source of contamination during de-hiding and from spillage of intestinal contents during evisceration is essential. As this is not always achieved, decontamination after hide removal has become a common hazard-based food safety measure in beef processing in some regions.

3.2.2 Pork

Pig production has a single end focus: the production of pork, and is often conducted in intensive farming systems. This commonality limits the variability in factors affecting *Salmonella* carriage in production systems. Carriage of *Salmonella* in farmed pigs is common worldwide but is not usually associated with clinical disease.

Salmonella serovars found in pigs include the more host-adapted, as well as many other broad-host-range serovars. *S.* Typhimurium is a common serovar found in pigs. *Salmonella* can be spread between pigs through the faecal-oral route, nose-to-nose contact within and between groups, and possible fomite and aerosol transmission. The epidemiology of *Salmonella* varies during the stages of production. Of particular relevance to food safety is the infection status among finishers, as this increases the number of shedding pigs at slaughter. Pigs can rapidly develop new or resurgent infections at this late stage before processing.

Processing of pigs has some unique aspects relevant to *Salmonella* control. Scalding and singeing are process steps where heat is applied to the hide to facilitate hair and hoof removal. These steps, if properly applied, provide an additional opportunity for reducing the prevalence of *Salmonella*. Pigs are frequently processed with the hide intact so that the carcass surface is not directly exposed to hide contamination. However, there are also process steps that can increase the risk of contamination, such as during de-hairing and polishing.

In some regions it has been possible to apply control strategies to declare and maintain *Salmonella*-free herds. This may be in part because pigs are monogastric animals and cattle ruminants, and therefore this may play a role in consideration of interventions in the animal host. In general, it may be that multiple on-farm controls are more feasible and practical in pig production relative to beef production.

More detailed information can be found in the OIE publication: "A review of the scientific literature on the control of *Salmonella* spp. in food-producing animals other than poultry" (Belluco *et al.*, 2015).

4

Beef chain intervention assessment, from primary production to processing

4.1 PRIMARY PRODUCTION (STEPS 1 AND 2)[2]

The primary source of *Salmonella* for cattle occurs at the farm level. Therefore, on-farm control of *Salmonella* may assist in reducing *Salmonella* carriage and potential transmission during cattle production, thereby minimizing the *Salmonella* contamination of cattle destined for beef processing. On-farm control of *Salmonella* may thus contribute to the whole food chain continuum of measures for reducing the food safety risk of *Salmonella* in beef.

Limited evidence was identified in the systematic review for the effect of farm-level interventions on the control of *Salmonella* in the population of greatest interest (i.e. beef or culled dairy cattle). The evidence was summarized across five main categories of interventions: non-therapeutic use of antimicrobials; biosecurity; feed management and feed additives; manipulation of microbiota; and vaccination. Some of the specific interventions investigated were also considered good farming practices (GFPs), with additional benefits beyond their specific effects on *Salmonella* prevalence or concentration. The key findings of the review and conclusions of the experts are summarized below.

---

[2] See Annex 2 for steps in the beef chain, as outlined in the CCFH Draft Guidelines

4.1.1 Non-therapeutic use of antimicrobials

Antimicrobials may be used in cattle for therapeutic purposes (to treat clinical disease), and also for non-therapeutic purposes (to prevent clinical disease, or to enhance physiological performance, in accordance with the specific regulatory environment of the farmer or operator).

The systematic review identified a small number of controlled trials that found inconsistent or non-statistically significant effects for the non-therapeutic use of antimicrobials (in-feed sub-therapeutic levels of Tylosin™, a macrolide antimicrobial, or in-feed use of the ionophore monensin) on *Salmonella* prevalence or concentration in cattle (Annex 1, SC4.3).

While antimicrobial drugs and medications may be warranted to treat clinical disease in cattle, the experts agreed that they should not be used in the treatment of sub-clinical salmonellosis nor should they be used as an intervention in cattle to control *Salmonella* as a food safety hazard, due to the serious potential for development of antimicrobial resistance.

- **Non-therapeutic use of antimicrobial drugs and medications was not recommended for consideration as a food safety intervention for the control of *Salmonella* in cattle on-farm, due to inconsistent evidence for efficacy and the serious potential for the development of antimicrobial resistance.**

4.1.2 Biosecurity

Biosecurity refers to a package of measures implemented on-farm to control infectious disease in general, and as such, is needed for many reasons besides *Salmonella* control. Numerous observational studies support the use of various biosecurity measures to reduce the likelihood of isolating *Salmonella* in cattle on-farm (Annex 1, SC4.4).

- **Biosecurity was recommended for consideration as an important GFP to control *Salmonella* in cattle on-farm.**

4.1.3 Feed management and feed additives

Feed management may include multiple factors, such as diet composition, feed withdrawal, use of partitioning agents, or energy supplements. Limited and inconsistent evidence was identified for the effect of feed management strategies to control *Salmonella* in cattle (Annex 1, SC4.5–4.6). Controlled trials and observational studies showed that some dietary components may, in some settings, potentially increase the prevalence of faecal *Salmonella* shedding (for example, feeding of distiller's grains to cattle, or non-pasteurized milk to dairy calves). The experts noted that considerable care should be taken when using feed management for the control of *Salmonella* in cattle, as the impact may be inconsistent or may even increase the prevalence of *Salmonella*.

- **Feed management was not recommended for consideration as a hazard-based intervention for the control of *Salmonella* in cattle on-farm, while it was noted that considerable care should be taken to ensure feed management and selection does not increase faecal *Salmonella* shedding.**

4.1.4 Manipulation of gut microbiota

Probiotics are living microorganisms that are fed to animals to colonize the gut environment and improve the balance of favourable microorganisms. A variety of probiotics (e.g. *Lactobacillus* spp., *Propionobacterium freudenreichii*) have been investigated in a small number of studies (Annex 1, SC4.7). A limited impact of probiotics was found on *Salmonella* prevalence in adult cattle in controlled trials; some probiotics decreased *Salmonella* prevalence and concentration in calves in challenge trials.

The effectiveness of the manipulation of gut microbiota on *Salmonella* reduction was considered by the experts to be dependent on multiple factors, such as cattle age, stage of rumen development, and diet.

- **Manipulation of microbiota using probiotics was not recommended for consideration as a hazard-based intervention for the control of *Salmonella* in cattle on-farm due to the limited available evidence for efficacy.**

4.1.5 Vaccination

Vaccination programmes in cattle have targeted various *Salmonella* antigens. Vaccine efficacy depends on various factors (e.g. vaccine type, *Salmonella* serovars targeted, age of cattle at vaccination, type of cattle production system).

Controlled trials have shown limited reductions in the prevalence of *Salmonella* in faecal samples of dairy cattle after vaccination (Annex 1, SC4.8); one study found no effect of vaccination on prevalence in faecal samples in feedlot cattle. The experts indicated that the magnitude of effect of vaccination in the studies evaluated was not large enough to be of practical importance. The experts also noted there was no published data on the effect of vaccines on the presence of *Salmonella* located in lymph nodes.

- **Vaccination for *Salmonella* was not recommended for consideration as an intervention for *Salmonella* in cattle on-farm due to insufficient evidence of its efficacy.**

4.1.6 Transport duration and hygiene

Cattle ready for slaughter are transported a variable distance from the production unit to the abattoir. Limited evidence from observational studies was available to estimate the potential impact of transport-based interventions on *Salmonella* control (Annex 1, SC5.3).

The experts recognized that it is not always practical to minimize the duration of transport for cattle in commercial settings, and this was therefore not considered as an intervention. However, the experts suggested that GHPs, including appropriate cleaning and disinfection protocols for transport vehicles, were expected to reduce *Salmonella* contamination. They noted that adequate cleaning and disinfection of transport vehicles is affected by many factors (e.g. water temperature and pressure, detergents and disinfectants used, their concentration, contact time, contact surfaces, personnel training).

- **Interventions during transport were not recommended for consideration as hazard-based interventions for the control of *Salmonella* in cattle due to lack of evidence and impracticality.**
- **Hygienic maintenance of transport vehicles was recommended as a GHP-based control measure.**

4.2 PROCESSING (STEPS 3–19)

As it is not possible to completely control *Salmonella* in cattle at the primary production stage and during transport, interventions can be applied during processing to reduce the level of *Salmonella* prevalence and concentration on beef carcasses and meat products and to prevent *Salmonella* growth and cross-contamination.

4.2.1 Lairage (Step 4)

At this step after unloading and before slaughter, cattle may be exposed to further *Salmonella* contamination within and between herds, and endure stress that may result in an increased load of *Salmonella* in animals entering slaughter establishments. No evidence for specific interventions applied at the lairage stage in cattle was identified in the systematic review. However, observational and molecular studies suggest the potential for lairage to be area of amplification and transmission of *Salmonella* among cattle (Annex 1, SC5.4).

The experts recognized that it is not always practical to minimize the duration in lairage for cattle in commercial settings, and this was therefore not considered as an intervention.

GHPs were considered necessary in hygienic maintenance of holding pens. The experts noted that the effect of cleaning and disinfection in the lairage area varies with the protocol applied, and could also be affected by design of the lairage area.

- **Interventions during lairage were not recommended for consideration as hazard-based interventions for the control of *Salmonella* in cattle due to lack of evidence and impracticality.**
- **Hygiene in lairage was recommended for consideration as a GHP-based control measure.**

4.2.2 Hide decontamination interventions (between Step 4 and Step 8)

Hide decontamination interventions could include washes using ambient or hot water, organic acids and other chemicals, as well as hide de-hairing. These interventions could be applied to the hides of live animals in lairage, or could be applied to animal hides post-exsanguination (pre-hide removal). The key findings of the systematic review and conclusions of the experts regarding these interventions are summarized below.

4.2.3 Live animal hide washes

The systematic review found limited and inconsistent evidence for the effect of ambient water, organic acid and other chemical washes applied to the hides of live animals in lairage (Annex 1, SC6.3).

The experts agreed that animal welfare has to be considered for hide decontamination processes applied to live animals.

- **Hide interventions applied to live animals were not recommended for consideration as a hazard-based intervention for the control of *Salmonella* due to a lack of consistent evidence and concerns for animal welfare.**

4.2.4 Hide de-hairing

Hide de-hairing refers to the process of applying successive washes with water and chemical solutions containing, for example, sodium hydroxide and sodium sulfide, to remove hair from and improve visible cleanliness of cattle hides. The process might also remove or inactivate *Salmonella*.

The systematic review identified two challenge trials conducted under laboratory and pilot plant conditions that indicated that de-hairing can reduce *Salmonella* concentrations on hides (Annex 1, SC6.3). However, it was the opinion of the experts that this intervention is not practical in a commercial setting due to difficulty in disposal of waste products.

- **Although it may be effective, hide de-hairing was not recommended for consideration as a hazard-based intervention for the control of *Salmonella* due to the practical difficulty of disposing of the waste materials accumulated.**

4.2.5 Post-exsanguination hide washes

Cattle hides may be washed with ambient or hot water, organic acids (e.g. lactic acid) or other chemical solutions (e.g. chlorine; hydrogen bromide) post-exsanguination to reduce microbial contamination of cattle hides prior to de-hiding.

The systematic review identified four quasi-experimental studies conducted under commercial conditions that found that washes containing chemicals other than organic acids (including hydrogen bromide, chlorine, or sodium hydroxide) significantly reduced the prevalence of *Salmonella* on hides from 62.3% to 18.1–35.8% (Annex 1, SC6.3). One study found a similar effect for an organic acid wash, reducing the prevalence of *Salmonella* on hides from 74.0% to 30.1–69.9%. Challenge trials conducted under laboratory and pilot plant conditions also found that organic acid and other chemical washes were effective in reducing *Salmonella* concentrations on hides compared with ambient water washes. No evidence was found of the effect of water washes at any temperature to reduce the prevalence of *Salmonella* on hides.

The experts noted that the efficacy of all hide decontamination treatments will depend on the type of hide wash, how it is applied (e.g. chemical concentration, and time of application), and the nature of hide contamination, including the amount, type of soiling and its distribution over the animal's hide. If organic acid or other chemical washes are used, they should be applied at appropriate combinations of concentration, time, duration of application and temperature to achieve the intended *Salmonella* reduction.

The experts noted that hide treatments potentially damage the commercial value of the hide. The washes can be water-intensive processes and there are concerns about the negative environmental impact of waste products.

- **Washing hides with ambient or hot water post-exsanguination was not recommended for consideration as a hazard-based intervention for the control of *Salmonella* due to a lack of evidence of efficacy.**
- **Organic acid and other chemical washes applied to hides post-exsanguination with proven efficacy were recommended for consideration as hazard-based interventions to control *Salmonella*.**

4.3 CARCASS DRESSING AND BUNGING (STEPS 8–14)

Carcass dressing refers to the slaughter processes where the appendages and viscera are removed. Bunging is a slaughter process where a cut is made around the rectum (i.e. terminal portion of the large intestine) to free it from the carcass, and then it may be tied off and/or bagged to prevent spillage of faecal material.

The systematic review identified only one study related to the effects of bunging on control of *Salmonella* in beef (Annex 1, SC6.4). The study was a controlled trial that found that bunging conducted prior to evisceration washes may provide greater *Salmonella* reductions compared with not bunging prior to evisceration. No studies were identified related to the effects of other carcass dressing procedures on reducing *Salmonella* in beef.

Given a lack of evidence for the effect of hygiene during carcass dressing or bunging specifically to control *Salmonella*, the experts agreed to recommend these processes as GHPs, noting that their efficacy is affected by the hygienic practices used.

- **Hygiene during carcass dressing and bunging were recommended for consideration as GHP-based control measures in meat processing.**

Post-mortem inspection (Step 15), which refers to the process before chilling of conducting a detailed inspection of carcasses, was not included in the scope of the systematic review and was not explicitly considered by the experts as an intervention to control *Salmonella*. However, it was noted that the manner in which it is implemented may increase the risk of bacterial cross-contamination of carcasses.

4.4 PRE-CHILL CARCASS INTERVENTIONS (BETWEEN STEP 8 AND STEP 16)

Some level of contamination of carcasses with faecally contaminated material is possible when the hide is removed and if leakage of gut contents is not prevented. Decontamination of carcasses can be used to physically remove and/or inactivate *Salmonella*, preferably as soon after the bacteria contact the carcass surface as possible.

Pre-chill decontamination interventions could include ambient water washes, hot water washes and steam pasteurization, organic acid and other chemical washes, and carcass trimming to remove visible contamination. Although the systematic review identified studies that investigated other pre-chill carcass interventions such as electricity and natural extracts, these were laboratory-based challenge studies which showed promise, but were not evaluated by experts as they were not commercially applicable. The key findings of the systematic review and conclusions of the experts regarding these interventions are summarized below.

4.4.1 Ambient water washes

Washing carcasses using ambient temperature water may be a routine activity in-plant that may remove superficial visible soiling. However, the systematic review found inconsistent evidence for removal of *Salmonella* from carcasses by washing with water at ambient temperature (Annex 1, SC6.5).

- **Ambient water washes were not recommended for consideration as a hazard-based intervention to control *Salmonella* due to inconsistent evidence.**

4.4.2 Hot water washes and steam pasteurization

The systematic review identified one controlled trial study conducted under commercial conditions that found that hot water applied at 74–87.8°C for 18–39 sec reduced the prevalence of *Salmonella* on beef carcasses from 30.3% to 2.2% (Annex 1, SC6.5). In addition, several challenge trials conducted under laboratory and pilot plant conditions were identified that found wide-ranging reductions of *Salmonella* concentrations on pre-chill carcasses for these interventions. The experts agreed that realistic reductions due to hot water washes and steam pasteurization under commercial conditions could be up to 1–2 $\log_{10}$ CFU/cm^2.

The experts noted that these interventions should be applied uniformly over the carcass surface at a time-temperature combination to achieve the intended reduction. They also noted that the efficacy of water washes and steam pasteurization depends on the carcass surface temperature achieved, the duration of application and the coverage of the carcass. The experts agreed that the carcass surface temperature should reach at least 70°C to effectively reduce *Salmonella* and that the time-temperature combinations required to achieve a targeted *Salmonella* reduction were specific to an individual processing plant.

- **Hot water washes that achieve a carcass surface temperature of at least 70°C were recommended for consideration as a hazard-based intervention for the control of *Salmonella*, and the experts considered realistic reductions could be up to 1–2 log_{10} CFU/cm^2.**

4.4.3 Organic acid and other chemical washes

The systematic review identified several challenge studies conducted under laboratory and pilot plant conditions that investigated the effect of using organic acid (e.g. lactic or acetic acid) and other chemical washes (e.g. peroxyacetic acid and acidified sodium chlorite) (Annex 1, SC6.5). The studies reported a wide-range of reductions in *Salmonella* levels compared with ambient water washing (from almost no reduction up to 3 log_{10} CFU/cm^2 for organic acids and up to 2.6 log_{10} CFU/cm^2 for other chemicals). However, based on the majority of studies, the experts agreed that realistic reductions under commercial conditions could be up to 1 log_{10} CFU/cm^2.

The experts noted that, if used, organic acid and other chemical washes should be applied uniformly over the carcass at appropriate combinations of concentration, time, duration of contact time, pH, and temperature to achieve the intended reduction with uniform coverage.

In general, the experts noted that the effect of organic acid and other chemical washes is enhanced by increasing the concentrations, but the upper limit of the concentration that can be practically used depends on retaining acceptable sensory quality of the meat. At higher concentrations of chemicals, water rinsing may be needed following their application. The temperature of the organic acid washes will affect their efficacy in reducing *Salmonella*. Additionally, the experts noted that the application of organic acid and other chemical washes might affect consumer acceptability of the meat.

- **Organic acid and other chemical washes were recommended for consideration as hazard-based interventions for the control of *Salmonella*, and realistic reductions could be up to 1 log_{10} CFU/cm^2.**

4.4.4 Trimming

Trimming of beef carcasses by slaughterhouse operators can remove visible surface contamination. The systematic review identified limited evidence showing that trimming can reduce *Salmonella* prevalence and concentration on carcasses (Annex 1, SC6.5). The experts noted that the efficacy of trimming to reduce *Salmonella* is affected by the hygienic practices used by operators and by equipment hygiene, and as a result, a specific hazard-based reduction estimate could not be determined.

- **The use of carcass trimming was recommended for consideration as a GHP-based control measure in beef processing.**

4.4.5 Removal of lymph nodes

While the experts noted that the carcass surface decontamination treatments assessed above are not effective in inactivating *Salmonella* located in lymph nodes, at this time, they agreed that there is insufficient evidence to assess the practice of removing bovine lymph nodes from carcasses as a *Salmonella* intervention in beef.

4.5 CHILLING (STEP 16)

The primary *Salmonella* control measure associated with carcass chilling is inhibition of microbial growth. Carcass chilling may include air, water or chemical spray chilling that causes carcass surface temperature reductions together with carcass surface drying or bacterial inactivation by chemicals, depending on the method used. The systematic review identified limited and inconsistent evidence on the potential effect of chilling to reduce *Salmonella* contamination of beef carcasses (Annex 1, SC6.6), which was in agreement with the opinion of the experts. The experts had some concerns about the overestimation of the reported lethal effect of chilling reported on *Salmonella*, as the poor recovery of *Salmonella* cells from an injured state induced by the effects of the chilling process could influence the results and interpretation of efficacy.

- **Chilling was recommended for consideration as a GHP-based control measure to prevent the growth of *Salmonella* on carcasses.**
- **GHP-based control measures should also be considered to avoid carcass cross-contamination in the chilling room.**

4.6 POST-CHILLING INTERVENTIONS (BETWEEN STEPS 16 AND 17)

Adequate and hygienic chilling should prevent cross-contamination of carcasses and prevent growth of *Salmonella*. Excessively high contamination levels and/or poor chilling and poor hygiene practices could result in cross-contamination and in growth of *Salmonella* remaining on carcasses following processing.

Chilled carcasses may be washed with water or chemical solutions (including organic acids and other chemicals such as peroxyacetic acid). Steam vacuuming can also be applied to remove visible contaminating material on a carcass surface. The systematic review identified a limited number of studies investigating the effects of post-chilling interventions, which indicated low or limited reductions of *Salmonella* on beef carcasses (Annex 1, SC6.7). The experts noted that interpretation of the efficacy of interventions at this stage is limited due to the low prevalence of *Salmonella* in available quasi-experimental studies.

The experts noted that if chemical washes are used, they should be applied at combinations of concentration, time, duration of application, and temperature to achieve the intended reduction. They also noted that intervention processes that will significantly increase the temperature of chilled carcasses (e.g. hot water washes) could have a negative impact on the prior effect of carcass chilling.

- **Post-chilling interventions were not recommended for consideration as hazard-based interventions for the control of *Salmonella* due to a lack of available evidence for efficacy.**

4.7 POST-FABRICATION INTERVENTIONS (BETWEEN STEPS 17 AND 19; STEP 9)

Post-fabrication decontamination interventions primarily involve treatments of small meat cuts and trimmings. Because small meat cuts and trimmings are irregularly shaped, very non-homogeneously contaminated and originate from different carcasses, post-fabrication interventions (e.g. spray washes with liquid solutions of ambient water, hot water, organic acids or other chemicals) were treated differently from carcasses, and assessed together by experts. Cheek meat, which is derived from trimming of heads at Step 9, was included with these steps because the interventions were expected to be applied similarly.

The systematic review identified several studies that investigated the effects of post-fabrication interventions; however, all were challenge trials conducted under laboratory and pilot plant conditions (Annex 1, SC6.8). Interventions considered included ambient water washes, heat-based treatments (e.g. hot water), and various chemical washes (including organic acids and other chemicals such as peroxyacetic acid and acidified sodium chlorite). There were limited data for the efficacy of heat-based treatments. The studies reported wide-ranging reductions in *Salmonella* levels for organic acid and other chemical washes compared with ambient water washes (from almost no reduction to up to 4 $\log_{10}$ CFU/g). However, the experts agreed that based on the majority of studies, reductions under commercial conditions from these interventions could be up to 1 $\log_{10}$ CFU/g.

The experts noted that, if used, post-fabrication interventions should be applied at combinations of concentration, time, duration of application and temperature to achieve the intended reduction. In general, the effect of chemical treatments is enhanced by increased concentration, but the upper limit of concentration that can be practically used depends on retaining acceptable sensory quality. At higher concentrations, water rinsing after application of chemical washes may be needed. The experts cautioned that application of heat-based treatments could increase the temperature of the product and therefore increase the risk of *Salmonella* growth.

The experts recognized that there are emerging technologies (e.g. high pressure processing) that could be applicable as a post-fabrication intervention to control *Salmonella* in beef, but there was insufficient evidence to consider their efficacy.

- **Chemical treatments with proven efficacy were recommended for consideration as hazard-based interventions for the control of *Salmonella* in fabricated beef.**

5

Pork-chain intervention assessment, from primary production to processing

5.1 PRIMARY PRODUCTION (STEPS 1 AND 2)[3]

The primary source of *Salmonella* for pigs is at the farm level. Therefore, on-farm control of *Salmonella* may assist in reducing *Salmonella* carriage and potential transmission during pig production, thereby minimizing *Salmonella* contamination of pigs destined for pork processing. On-farm control of *Salmonella* may thus contribute to the whole food chain continuum of measures for reducing the food safety risk of *Salmonella* in pork.

The systematic review identified a large amount of literature investigating farm-level interventions to control *Salmonella* in pigs, but these frequently did not measure similar sample types in the population of greatest interest (i.e. finisher pigs). The evidence was summarized across six main categories of interventions: non-therapeutic antimicrobials and bacteriophage therapy; biosecurity; feed and water acidification; feed management and feed additives; manipulation of microbiota; and vaccination. Some of the specific interventions investigated were also considered GFPs, with additional benefits beyond their specific effects on control of *Salmonella*. The key findings of the systematic review and conclusions of the experts are summarized below.

[3] See Annex 2 for the list of steps as outlined in the CCFH Draft Guidelines.

5.1.1 Non-therapeutic antimicrobials and bacteriophage therapy

Antimicrobials may be used in pigs for therapeutic purposes (to treat clinical disease), and also for non-therapeutic purposes (to prevent clinical disease, or to enhance physiological performance, in accordance with the specific regulatory environment of the farmer or operator). Evidence regarding the effects of non-therapeutic use of antimicrobials in pigs on *Salmonella* prevalence or concentration was inconsistent (Annex 1, SC1.3).

While antimicrobial drugs and medications may be warranted to treat clinical disease in pigs, the experts agreed that they should not be used in the treatment of sub-clinical salmonellosis nor should they be used as an intervention in pigs to control *Salmonella* as a food safety hazard, due to the serious potential for development of antimicrobial resistance.

Bacteriophages are typically applied orally, as a cocktail of several bacteriophages, and are applicable in varying production conditions. Bacteriophages are inexpensive, easy to apply, widely available in the environment, and acceptable to the public. However, current evidence reports inconsistent effects of bacteriophage therapy on prevalence of *Salmonella* shedding, intestinal concentrations, or lymph node colonization in pigs.

- **Non-therapeutic use of antimicrobial drugs and medications was not recommended for consideration as a food safety intervention for the control of *Salmonella* in pigs on-farm, due to inconsistent evidence for efficacy and the serious potential for the development of antimicrobial resistance.**
- **Bacteriophage therapy was not recommended for consideration as a hazard-based intervention for the control of *Salmonella* in pigs on-farm given the limited data currently available.**

5.1.2 Biosecurity

Biosecurity refers to a package of measures implemented on-farm to control infectious disease in general, and as such, is needed for many reasons besides *Salmonella* control. External biosecurity, i.e. measures to prevent the introduction of pathogens on-farm, is important to reduce the likelihood of introduction of *Salmonella* into *Salmonella*-free herds. Numerous observational studies support the use of external biosecurity measures (such as rodent control programmes, use of a hygienic lock, restriction of farm visitors, and managing the *Salmonella* status of replacement gilts and boars) to reduce the likelihood of introducing *Salmonella* on-farm (Annex 1, SC1.4).

Observational studies and a small number of controlled trials support the use of internal biosecurity measures (e.g. pig flow, cleaning and disinfection between batches, strategic movement) to control the spread of *Salmonella* within the farm (Annex 1, SC1.4). Internal biosecurity measures could reduce the prevalence of *Salmonella* in finishers in infected herds, but are unlikely to eliminate it in the absence of other measures applied concurrently (e.g. feeding of organic acids; vaccination).

- **Biosecurity was recommended for consideration as an important GFP to control *Salmonella* in pigs on-farm.**

5.1.3 Feed and water acidification

Feed and water acidification refers to the use of organic acids (e.g. lactic or formic acid), administered at conditions which provide a reduction in *Salmonella* prevalence, by taking into account the type of acid, its concentration, the duration of administration, and the age of pigs (e.g. weaners vs finishers) receiving treatment. Organic acids have a direct effect of reducing the level of *Salmonella* in feed.

Adding organic acids in water or feed has been shown in controlled trials to yield varying degrees of reduction of *Salmonella* prevalence in an infected herd (Annex 1, SC1.5).

The experts noted that the effect of organic acids depends on other factors (e.g. *Salmonella* contamination, disease status, type of feed, other management factors). As with other interventions, the use of acids alone might have a limited effect on reducing *Salmonella* prevalence in the finishers leaving the farm, in an infected herd.

- **Acidification of feed or water using organic acids was recommended for consideration as a hazard-based intervention for the control of *Salmonella* in infected herds.**

5.1.4. Feed management and feed additives

Feed management may include several strategies that manipulate feed form, such as: feeding meal versus pellets; varying the degree of feed coarseness; or feeding fermented liquid feed. Controlled trials and observational studies support that feeding meal (vs pellets), coarse feed (vs fine feed) and fermented liquid feed (vs dry feed) each have a varying magnitude of effect in reducing *Salmonella* prevalence on an infected farm (Annex 1, SC1.6).

A variety of feed additives have been investigated for reduction of *Salmonella* prevalence and concentration in pigs, one of the most studied of which is sodium chlorate. Limited published information based on experimental data reported a potential for reduction in *Salmonella* prevalence measured by faecal shedding, with no effect on lymph nodes, from feeding sodium chlorate (Annex 1, SC1.6).

- **Feed management strategies with proven efficacy were recommended for consideration as hazard-based interventions for the control of *Salmonella* in infected herds.**
- **Sodium chlorate addition to feed was not recommended for consideration as a hazard-based intervention for the control of *Salmonella* in pig herds, based on the limited available evidence.**

5.1.5 Manipulation of gut microbiota

Probiotics are living microorganisms that are fed to animals to colonize the gut environment and improve the balance of favourable microorganisms. A prebiotic may be defined as a non-digestible food ingredient that selectively stimulates the favourable growth and activity of one or a limited number of bacteria in the colon. A variety of probiotics (e.g. *Lactobacillus* spp.) and prebiotics (e.g. galactomannan) have been reported to be used in finisher pigs.

A small number of published controlled trials show the potential for prebiotics and/or probiotics to reduce *Salmonella* prevalence and concentration in pigs, although with unclear magnitude of effect (Annex 1, SC1.7).

Variability in the efficacy and use of probiotics and prebiotics depends on various factors, including microbial species involved (probiotics), type of nutrient (prebiotics), combinations of pro- and pre-biotics used concurrently, dosages used, and age at administration. Although probiotics and/or prebiotics can reduce *Salmonella* colonization when administered, there may also be potential re-infections later in the productive life of the animal.

- **Manipulation of microbiota using pre- or pro-biotics was not recommend for consideration as a hazard-based intervention for the control of *Salmonella* in pigs due to the limited available evidence.**

5.1.6 Vaccination

Vaccination programmes targeted various *Salmonella* antigens to reduce prevalence. Vaccination was shown in controlled trials to reduce the prevalence of *Sal-*

monella from 26.7% to 0.7–22.1% for individual faecal prevalence in finisher pigs, 47.5% to 12.2–43.8% for lymph node prevalence at slaughter, and 46.3% to 13.4–43.4% for caecal prevalence at slaughter (Annex 1, SC1.8). The effect of vaccination when measured at the pen level was inconsistent.

The experts noted that the effect of vaccination is variable due to factors such as vaccine type, age of pigs at vaccination application, and *Salmonella* serovars targeted. Vaccination could interfere with herd monitoring for *Salmonella* where serological control programmes are used. The cost-effectiveness of a vaccination programme is situation dependent. Because the aim of vaccination is to reduce contamination at slaughter, its effect should ideally be assessed in pigs as close to the time of slaughter as possible.

- **Vaccination was recommended for consideration as a hazard-based intervention for the control of *Salmonella* in pigs on-farm; however, only where the context is carefully described and considered.**

5.1.7 Transport duration and hygiene

Market pigs are transported for a variable distance and time from the production unit to the abattoir. Sufficient evidence was not identified to estimate the effect of duration of transport on *Salmonella* control (Annex 1, SC2.3). The experts agreed that transport duration depends on other related factors (e.g. logistics, animal welfare). The experts also agreed that duration of transport should be kept as short as possible.

Sufficient evidence was not identified to estimate the effect of cleaning and disinfection of transport trucks on *Salmonella* control, although numerous observational studies have reported that current disinfection protocols for cleaning transports between batches can be limited in removal of bacteria, including *Salmonella*, which are present after hauling animals to slaughter (Annex 1, SC2.3). However, the experts agreed that cleaning and disinfection should be considered between transport loads as a GHP. The experts noted that adequate cleaning and disinfection of transport vehicles is influenced by many factors (e.g. water temperature and pressure, detergents and disinfectants used, their concentration, contact time, contact surfaces, personnel training).

- **Interventions during transport were not recommended for consideration as hazard-based interventions for the control of *Salmonella* in pigs due to a lack of evidence, and impracticality.**
- **Hygienic maintenance of transport vehicles was recommended for consideration as a GHP-based control measure.**

5.2 PROCESSING (STEPS 3–21)

Salmonella contamination of carcasses and meat products may occur throughout processing due to cross-contamination (e.g. from inadequate carcass dressing and environmental contamination), and growth of *Salmonella* has to be controlled. *Salmonella* contamination can be controlled by applying GHPs (e.g. regular sanitation of equipment). As well, a reduction in *Salmonella* contamination may be accomplished by applying hazard-based interventions at various points in the processing chain. However, these interventions should not be considered substitutes for GHPs.

5.2.1 Lairage (Step 4)

At this step after unloading and before slaughter, pigs may be exposed to further *Salmonella* contamination within and between herds and they may endure stress that could result in increased *Salmonella* contamination on animals entering slaughter establishments. Limited evidence was identified in the systematic review for the effect of lairage-based interventions on the control of *Salmonella* in pigs. The key findings of the review and conclusions of the experts are summarized below.

5.2.2 Feed withdrawal

Feed may be withdrawn from the animals for a period, varying with the duration of the interval elapsed between leaving the farm, and slaughter. Feed withdrawal serves to reduce the volume of intestinal content, and the subsequent risk of intestinal spillage at evisceration and contamination of the carcass. However, the length of feed withdrawal should be taken into consideration with animal welfare concerns.

Contradictory evidence exists from controlled trials and challenge trials regarding the effect of feed withdrawal on *Salmonella* shedding in finisher pigs (Annex 1, SC2.3).

- **The use of feed withdrawal was recommended for consideration as a GHP-based control measure to reduce rupturing of the intestines and intestinal spillage during carcass dressing.**

5.2.3 Duration and hygienic practices in lairage

The duration spent in lairage varies, and increasing duration could increase the *Salmonella* contamination of pigs due to increased exposure to contamination and animal stress. There was insufficient evidence, derived from observational studies, to estimate the effect that duration in lairage could have on *Salmonella* prevalence and concentration in slaughter pigs (Annex 1, SC2.4). The experts agreed that lairage duration involves other, related, factors (e.g. logistics, meat quality, animal welfare). The experts also agreed that duration of lairage should be kept as short as possible.

Similarly, there was insufficient evidence, derived from controlled trials, to estimate the effect that hygienic practices in lairage pens could have on *Salmonella*. However, the experts suggested that as a GHP, pen hygiene could reduce the *Salmonella* contamination in slaughter pigs, and that cleaning and disinfection should be performed between batches delivered, although it was recognized that it is not always practical in commercial settings. The experts noted that the effect of cleaning and disinfection of holding pens in lairage is influenced by many factors (e.g. water temperature and pressure, detergents and disinfectants used, their concentration, contact time, contact surfaces, personnel training).

- **Interventions during lairage were not recommended for consideration as hazard-based interventions for the control of *Salmonella* in pigs due to a lack of evidence and impracticality.**
- **Hygienic practices in lairage were recommended for consideration as GHP-based control measures.**

5.2.4 Logistic slaughter

Logistic slaughter refers to the strategic scheduling of pigs for slaughter based on pre-harvest information on *Salmonella* status. This intervention, therefore, must also involve the segregated transport, and holding in lairage, of pigs from herds with different risks of *Salmonella* infection. Across experimental and observational studies, no consistent evidence was identified to estimate the effect of logistic slaughter (Annex 1, SC2.5). However, theoretically, the experts agreed it would reduce *Salmonella* contamination in post-farm stages.

The success of logistic slaughter depends on many factors (e.g. it requires a pre-harvest *Salmonella* surveillance programme with high accuracy, and good hygiene in post-farm stages). This should not preclude special handling of pigs highly infected relative to the general population.

- **Logistic slaughter of pigs was not recommended for consideration as a hazard-based intervention for the control of *Salmonella* due to insufficient evidence; however, this should not preclude special handling of pigs originating from herds that are known to have a high prevalence of *Salmonella* infection relative to the general population.**

5.3 SCALDING AND SINGEING (STEPS 7 AND 10)

Scalding and singeing are routine process steps in the slaughter line, performed primarily for removal of hair. Some processors might skin and do not de-hair, in which case scalding and singeing are not applied. The systematic review identified evidence on scalding and singeing from a range of study designs (Annex 1), the key findings of which are summarized below.

One previous systematic review of primary research studies describing the prevalence of *Salmonella* in pork from slaughter to chilling found that both scalding and singeing steps were associated with a decrease in prevalence. Quasi-experimental studies conducted under commercial conditions found that both scalding and singeing significantly reduced *Salmonella* prevalence on pork carcasses. These effects are supported by observational studies, which found associations between scalding and singeing practices and reduced *Salmonella* contamination of carcasses.

5.3.1 Scalding

Scalding involves the use of hot water at temperatures and contact times to facilitate subsequent removal of hair and hooves. This will also result in inactivation of *Salmonella* with adequate parameters of temperature and time, and with uniform coverage of the carcass. There was extensive evidence from quasi-experiments of reduction of *Salmonella* prevalence on carcasses during scalding, from 35% to 1.5% (Annex 1, SC3.3).

Scalding at inappropriate temperatures, or in the presence of organic matter in the water, can be a source of *Salmonella* contamination of carcasses. This may be a particular concern with contamination of the pig's pharynx, as subsequent carcass decontamination steps would not address this internal contamination. The experts recommended scald conditions of temperature and time (or equivalent combinations) effective for inactivation of *Salmonella* could include 61°C for 8 min or 70°C for 2–3 min.

- **Scalding was recommended for consideration as a hazard-based intervention for the control of *Salmonella* when applied at appropriate combinations of temperature and time.**

5.3.2 Singeing

Singeing involves the use of direct-fire bursts on the animal surface to improve hair removal. This will also result in inactivation of *Salmonella*, given adequate parameters for temperature and time, and with uniform coverage of the carcass. There was extensive evidence from quasi-experiments of reduction of *Salmonella* prevalence on carcasses during singeing from 18.1% to 5.4% (Annex 1, SC3.3). A 2 $\log_{10}$ CFU/cm^2 reduction in *Salmonella* concentration on carcasses after singeing was estimated in a single challenge trial.

Singeing temperature should be homogeneous across the carcass, as areas such as the base of the ears might not reach the required temperature to inactivate *Salmonella*. In some circumstances, a second singeing step might be considered after polishing.

- **Singeing was recommended for consideration as a hazard-based intervention for the control of *Salmonella*.**

5.3.3 De-hairing and polishing

Although scalding and singeing can reduce carcass contamination, there is a risk of recontamination in de-hairing and polishing steps if GHPs are not used.

- **GHP measures were recommended during de-hairing and polishing to reduce cross-contamination of carcasses.**

5.4 CARCASS DRESSING AND BUNGING (STEPS 12–16)

Carcass dressing refers to the slaughter processes where the appendages and viscera are removed. Bunging is a slaughter process where a cut is made around the rectum (i.e. terminal portion of the large intestine) to free it from the carcass, and then it may be tied off and/or bagged to prevent spillage of faecal material.

The systematic review did not capture any studies investigating the effects of hygiene during carcass dressing or bunging to reduce the prevalence or concentration of *Salmonella* in pork. Given a lack of evidence for their effect specifically to control *Salmonella*, the experts agreed to recommend these measures as GHPs, noting that the efficacy of the processes is affected by the hygienic practices used.

- **Hygiene during carcass dressing and bunging were recommended for consideration as GHP-based control measures in meat processing.**

5.5 POST-MORTEM INSPECTION (STEP 17)

This refers to the process before chilling of conducting a detailed inspection of carcasses, but was not included in the scope of the systematic review and was not explicitly considered by the experts as an intervention to control *Salmonella*. However, it was noted that the manner in which it is implemented may increase the risk of bacterial cross-contamination of carcasses.

5.6 PRE-CHILL CARCASS INTERVENTIONS (BETWEEN STEPS 16 AND 18)

Decontamination of carcasses based on washing interventions (e.g. hot water, organic acids or chemicals) are specifically intended to control microbiological contamination such as *Salmonella*. There are a number of different treatments that can be applied on carcasses, based on the washing of its surface. The systematic review identified several relevant studies investigating the following types of pre-chill interventions of pork carcasses: ambient water washes; hot-water washes and steam treatments; organic acid washes; other chemical washes; and steam vacuuming. The key findings of the systematic review and conclusions of the experts are summarized below.

5.6.1 Ambient water washes

Washing with ambient water is a routine step in some establishments for aesthetic reasons (to remove bone dust). There was limited evidence identified in the systematic review of its effect on reduction of *Salmonella* prevalence or concentration of pork carcasses (Annex 1, SC3.4).

- **Washing with ambient water was not recommended for consideration as a hazard-based intervention for the control of *Salmonella* due to lack of evidence for its efficacy.**

5.6.2 Hot-water washes and steam pasteurization

The systematic review identified several controlled trials that found that hot-water washes and steam pasteurization at different process parameters (e.g. hot water at 74–81°C for 5–15 sec, or steam at 82–85°C for 60 sec) could reduce the prevalence of *Salmonella* on pork carcasses from 4.3% to 1.6% (Annex 1, SC3.4), and unpublished data reported in a European Commission cost-benefit analysis found that hot-water washes could reduce the prevalence of *Salmonella* from 13% to 1% (Food Control Consultants Consortium, 2013). Challenge studies conducted under laboratory and pilot plant conditions have shown that hot water washes can achieve up to a 2 $\log_{10}$ CFU/cm^2 reduction, which was in agreement with the expert opinion of a realistic reduction.

Hot water should be applied at a time×temperature combination appropriate to achieve the intended reduction, and should be uniformly applied over the carcass. The experts agreed the time×temperature combinations required to achieve a specific reduction were processing plant-specific. It was generally accepted that the carcass surface temperature should reach at least 70°C.

- **Hot water washes that achieve a carcass surface temperature of at least 70°C were recommended for consideration as a hazard-based intervention for the control of *Salmonella*, and the experts considered realistic reductions could be up to 2 $\log_{10}$ CFU/cm^2.**

5.6.3 Organic acid washes

Washes containing organic acids (e.g. lactic or acetic acid) have been shown in controlled trials to reduce the prevalence of *Salmonella* on carcasses from 7% to 2% (Annex 1, SC3.4). In challenge studies conducted under laboratory and pilot plant conditions, organic acid washes have shown a wide range of reductions in *Salmonella* levels compared with ambient water washing (from almost no reduction up to 1.8 $\log_{10}$ CFU/cm^2). The experts concluded that a realistic reduction under commercial conditions could be up to 0.5 to 1 $\log_{10}$ CFU/cm^2.

The experts noted that, if used, organic acid washes should be applied uniformly over the carcass at combinations of concentration, time, duration of contact time, and temperature to achieve the intended reduction. Washing concentrations need to be measured at the site of application. Concentrations required to achieve a specific reduction are processing plant specific, and vary between acids. Contact time of washes may need to be considered, especially if followed by a rinse step.

- **Organic acid washes were recommended for consideration as a hazard-based intervention for the control of *Salmonella*, and the experts considered realistic reductions could be up to 0.5 to 1 $\log_{10}$ CFU/cm^2.**

5.6.4 Other chemical washes

Examples of other chemical washes include acidified sodium chloride, chlorine and hydrogen peroxide. There was limited evidence identified in the systematic review for reduction of *Salmonella* on carcasses using these washes (Annex 1, SC3.4).

- **Other chemical washes were not recommended for consideration as a hazard-based intervention for the control of *Salmonella* due to insufficient data.**

5.6.5 Full-carcass steam vacuum

Steam vacuuming may be applied, targeting the whole carcass. There was limited evidence on its efficacy in published studies (Annex 1, SC3.4). However, the experts noted that full carcass steam vacuuming can be a valuable intervention for small establishments, because it is relatively inexpensive and flexible as an alternative to whole carcass hot water washes. However, its efficacy to reduce *Salmonella* can be highly variable and depends on application and operator. Given the lack of published evidence supporting its effect on *Salmonella*, it was not considered as a hazard-based intervention.

- **Full carcass steam vacuuming was recommended for consideration as a GHP-based control measure in small establishments with limited resources.**

5.6.6 Manual removal of visible contamination

Removal of visible contamination may be performed by knife trimming of carcasses or spot steam vacuuming. Steam vacuuming and knife trimming are useful in removing visible contamination; however, the efficacy depends on various factors, including the operator and hygienic application. The experts noted that the efficacy of these approaches in reducing *Salmonella* can be highly variable, with very limited published evidence supporting their use to reduce *Salmonella*.

- **Knife trimming and/or spot-vacuum were recommended for consideration as GHP-based control measures in pork processing.**

5.7 CHILLING (STEP 18)

The primary microbial control measure associated with carcass chilling is inhibition of microbial growth, which may be affected by carcass spacing, air flow, and cooling capacity of the chiller. Carcass chilling may include air, water or chemical spray chilling that cause carcass surface temperature reductions together with carcass surface drying or bacterial inactivation by chemicals depending on the method used.

The systematic review identified two previous meta-analyses that measured the effect of chilling to reduce the prevalence of *Salmonella* on pork carcasses (Annex 1, SC3.5). The results indicate a significant effect on reduction of *Salmonella* prevalence due to chilling, but with wide variation in results across studies. The study authors found that the total sample size, number of batches sampled in an abattoir, and size of the carcass swab area had a significant impact on estimates of the effect of chilling, indicating that the sampling methods contributed to the heterogeneity among studies. However, the experts had some concerns about the overestimation of the effect of chilling reported, because experimental factors related to the poor recovery of *Salmonella* cells from an injured state induced by the effects of the chilling process could influence the results.

The experts were of the opinion that blast chilling may yield larger *Salmonella* reductions on pork carcasses compared with conventional chilling; however, there was not enough published evidence to confirm this conjecture.

- **Chilling was recommended for consideration as a GHP-based control measure to prevent growth of *Salmonella* on pork carcasses.**
- **GHP-based control measures should be considered to avoid carcass cross-contamination in the chilling room.**

6

Assessment of interventions from packaging to consumption for beef and pork

Processing steps for beef and pork from packaging to consumer were considered together by the experts, as food safety control measures are very similar for both meats.

6.1 PACKAGING (STEP 19 FOR BEEF; STEP 21 FOR PORK)[4]

Packaging protects finished products from contamination post-processing. The addition of preservatives or application of preservation technologies can be used to extend shelf life and may have an effect on the presence and growth of pathogens such as *Salmonella*. Irradiation is summarized in this section as it is typically applied at the packaging step; however, it could also be applied earlier in the processing chain (e.g. post-fabrication). The key findings of the systematic review and conclusions of the experts are summarized below.

6.1.1 Packaging and emerging technologies

Packaging-based interventions can include modifying the package environment (e.g. modified atmosphere, vacuum packaging), the addition of microbial inhibitors such as chemicals, biological extracts and lactic acid bacteria, and the ap-

[4] See Annex 2 for list of steps as outlined in the CCFH Draft Guidelines.

plication of non-thermal technologies. The systematic review identified several challenge trials conducted under laboratory conditions that reported limited and/ or conflicting results for different packaging-based interventions for *Salmonella* control in beef and pork meat (Annex 1, SC3.6 and SC6.9).

The experts recognized that packaging-based interventions to control *Salmonella* in beef and pork were subject to many factors, such as the native microbiota, temperature, pH, storage time, and packaging materials which confound the ability to achieve a consistent reduction in *Salmonella*. Vacuum packaging and modified atmosphere packaging were considered very useful to extend the shelf-life of beef and pork products, but there was no evidence of effect on *Salmonella*.

The experts recognized that there are emerging non-thermal technologies (e.g. high pressure processing) that could be applicable as an intervention to control *Salmonella* in fresh beef and pork, but there was insufficient evidence to consider the efficacy of these at this time. Similarly, while the systematic review identified limited evidence from laboratory-based studies on the effect of non-thermal and non-chemical technologies other than irradiation (e.g. cold plasma), they were not assessed by the experts as they were not yet commercially applicable.

- **Packaging interventions were not recommended for consideration as hazard-based interventions for the control of *Salmonella* due to conflicting evidence.**

6.1.2 Irradiation

Irradiation consists of the use of ionizing radiation to eliminate or reduce *Salmonella*. The systematic review identified a limited number of challenge trials that evaluated irradiation to reduce *Salmonella* concentrations in ground beef, showing potential for large reductions (Annex 1, SC6.8). A previous review found that irradiation of minced pork meat has resulted in D_{10} values (kGy) of 0.403–0.860 for *S.* Typhimurium, and irradiation of ground beef has resulted in D_{10} values (kGy) of 0.618–0.661 for *Salmonella* spp., with differences in serovar ranging from 0.55 for *S.* Typhimurium to 0.78 for *S.* Stanley (Farkas, 1998). The experts were confident about the efficacy of irradiation to mitigate or eliminate the *Salmonella* contamination in beef and pork.

Consumer acceptance may be a factor in determining the uptake of irradiation as a hazard-based intervention for the control of *Salmonella*.

- **Irradiation was recommended for consideration as a hazard-based intervention for the control of *Salmonella* in beef and pork products.**

6.2 DISTRIBUTION TO CONSUMPTION (STEPS 20–26 FOR BEEF; STEPS 22–25 FOR PORK)

The systematic review scope did not include interventions in these steps. Nevertheless, the experts discussed these and agreed that control of *Salmonella* in beef and pork from distribution to consumption was reliant on strict maintenance of the cold chain. *Salmonella* has a minimum growth temperature of 5°C and temperatures of meat must be maintained below this temperature to prevent growth of *Salmonella* (EFSA BIOHAZ Panel, 2014).

- **Maintenance of the cold chain was recommended for consideration as a GHP-based control measure to reduce the risk of *Salmonella* growth and recontamination of beef and pork.**

The experts recognized that various approaches can be taken from product distribution to consumption to control and mitigate the risk of food-borne illness due to *Salmonella* and other pathogens. However, given the wide variability and diversity in these steps across different food chains, regions and countries, it was not feasible to consider these measures in detail during the meeting.

The experts acknowledged that hazard-based interventions for the control of *Salmonella* and other pathogens can be applied between distribution and retail; however, the diversity of retail outlets and food service businesses was too extensive, variable and not feasible to assess at this meeting. Nevertheless, the experts noted that food safety in the retail and catering sectors should be based on the effective application of programmes based on HACCP and its prerequisites.

Consumers should follow the WHO "Five Keys to Safer Food" to prevent food-borne illness (WHO, 2006), including cooking beef and pork to a temperature that is sufficient to inactivate *Salmonella*.

7

Summarized response to CCFH request

In response to the request made to FAO/WHO, this chapter summarizes considerations agreed by the invited experts during the Joint FAO/WHO Technical Meeting on *Salmonella* in beef and pork.

The experts were specifically asked to address the following aspects of hazard-based interventions for the control of *Salmonella* in beef and pork, as outlined by the CCFH pWG meeting held in Brussels, Belgium, on May 6-9, 2015:

- Advise on the most appropriate point(s) of application of specific interventions and decontamination treatments.
- Verify, based on the available data, the efficacy of the intervention in terms of reduction of *Salmonella*.
- Advise, with some level of confidence, to the extent possible, on the quantifiable level of reduction that the intervention achieves, and whether these are appropriate to include in the Codex guideline.

The following is a summary of recommendations resulting from the expert meeting. The quantitative reduction values that are presented in this section were taken from the scientific literature available at the time of the expert meeting, or from the opinion of the experts. Any reference to the efficacy of an intervention to reduce *Salmonella* prevalence or concentration must consider the specific circumstances of the studies referenced and the conditions under which they were conducted.

7.1 HAZARD-BASED INTERVENTIONS FOR BEEF UP TO STEP 18

7.1.1 Pre-harvest interventions (Step 1)

No specific hazard-based interventions were identified in primary production although the experts agreed biosecurity could contribute to general on-farm control of *Salmonella* and other zoonotic foodborne infections.

7.1.2 Hide decontamination (Step 4)

Decontamination treatments of hides were recommended for consideration as potential hazard-based interventions for the control of *Salmonella*; however, the experts concluded that the step at which hide intervention is performed is important. Decontamination of the hides of live animals was not recommended due to lack of confidence in supporting evidence, and concerns for animal welfare.

Decontamination treatments of hides post-exsanguination and before de-hiding (Step 8) using chemical washes including organic acids and other chemicals were recommended for consideration as a potential hazard-based intervention for the control of *Salmonella*.

The efficacy of washes would depend on the amount, type and distribution of soil on the hide. The expected levels of *Salmonella* reduction under commercial conditions were:

- Organic acid washes could be expected to reduce the prevalence of *Salmonella* on hides from 74.0% to 30.1–69.9%.
- Other chemical washes, such as chlorine and hydrogen bromide washes, could be expected to reduce the prevalence of *Salmonella* on hides from 62.3% to 18.1–35.8%.

7.1.3 Pre-chill carcass decontamination (between Step 8 and Step 16)

Carcass decontamination treatments with proven efficacy were recommended for consideration as potential hazard-based interventions for the control of *Salmonella* after hide removal and before chilling.

The efficacy of carcass decontamination depends on application of specific process criteria and the uniform coverage of the treatment over the carcass surface and should be validated at an individual establishment. The first treatment should be applied as soon as possible after hide removal to counteract bacterial attachment. One or more treatments could be applied depending on the performance required;

however, the overall effect would be greater than a single treatment, although not the sum of individual treatments, and the effect of a sequential treatment should not invalidate the effect of a prior treatment.

Decontamination treatments recommended by the experts were as follows:

Hot water washes and steam pasteurization

Experts agreed the carcass surface temperature should reach at least 70°C to effectively reduce *Salmonella* and the exposure time×temperature combinations required to achieve a targeted *Salmonella* reduction were specific to an individual processing plant.

In one controlled trial conducted under commercial conditions, hot water at 74–87.8°C for 18–39 sec reduced the prevalence of *Salmonella* from 30.3% to 2.2%. The experts considered that realistic reductions in concentration could be up to 1–2 $\log_{10}$ CFU/cm^2.

Chemical washes

Efficacy of washes depends on combinations of concentration, time, duration of contact, temperature, and uniform carcass surface coverage to achieve the intended objective. Examples included:

- Organic acid washes (e.g. lactic or acetic acid) resulted in a wide-range of reductions of *Salmonella* from almost 0 to 3 $\log_{10}$ CFU/cm^2 in challenge studies under laboratory and pilot plant conditions.
- Other chemical washes (e.g. peroxyacetic acid and acidified sodium chlorite) resulted in a wide-range of reductions in *Salmonella* from almost 0 to 2.6 $\log_{10}$ CFU/cm^2 in similar studies.

Based on the majority of studies, the experts considered that realistic reductions for chemical washes could be up to 1 $\log_{10}$ CFU/cm^2.

7.1.4 Post-fabrication decontamination (after Step 17)

Chemical washes with proven efficacy were recommended for consideration as potential hazard-based interventions for the control of *Salmonella*.

The types of washes, types of studies and factors on which efficacy depends were as outlined for pre-chill carcass washes.

Organic acid and other chemical washes resulted in a wide range of reductions in *Salmonella* from almost 0 to 4 $\log_{10}$ CFU/g. Based on the majority of studies, the experts considered that realistic reductions could be up to 1 $\log_{10}$ CFU/g.

7.2 GHP-BASED CONTROL MEASURES FOR BEEF UP TO STEP 18

The following processes or interventions were recommended for consideration by the experts as GHPs and not hazard-based interventions, due to combinations of lack of credible studies, methodological difficulties in confidently measuring efficacy, and insufficient evidence of efficacy specifically for *Salmonella*:

- Hygiene during transport to slaughter and in lairage to limit the spread of *Salmonella*.
- Hygiene during carcass dressing to minimize contamination.
- Bunging to reduce faecal spillage during processing.
- Carcass trimming and steam vacuuming to remove visible contamination.
- Chilling to prevent growth of *Salmonella*.
- Practices to prevent carcass cross-contamination in the chilling room.

7.3 HAZARD-BASED INTERVENTIONS FOR PORK UP TO STEP 20

7.3.1 Pre-harvest interventions (Step 1)

The experts agreed biosecurity is an important GFP, where external measures can help to reduce the likelihood of introduction of *Salmonella* to *Salmonella*-negative farms and internal measures might reduce the prevalence in finishers in infected herds.

Pre-harvest interventions may have a limited effect on the reduction of *Salmonella* on carcasses unless control measures are also taken post-harvest. If a regulator, risk manager or farmer wishes to lower the within-herd prevalence of *Salmonella*, then the following on-farm interventions could be considered as potential hazard-based interventions for the control of *Salmonella*:

Feed management, such as feeding meal (vs pellets), was recommended for consideration to reduce the *Salmonella* prevalence in finisher pigs.

Acidification of feed and water using organic acids, e.g. lactic or formic acids, was recommended for consideration to reduce *Salmonella* in infected herds.

The effect of organic acids depends on other factors (e.g. *Salmonella* contamination, diseases, feed type and other management factors). Organic acids can reduce feed contamination, although use of acids alone might have a limited effect on reducing *Salmonella* in finisher pigs.

Vaccination could be recommended for consideration as a hazard-based intervention for the control of *Salmonella* on-farm; however, the following should be considered if vaccination is used as a food safety measure:

- Vaccine type, age of pigs at application, and *Salmonella* serovars targeted.
- Potential for interference with herd monitoring based on serology.
- Cost-effectiveness of a vaccination programme, which is situation dependent.
- Effects should ideally be assessed in pigs as close to the time of slaughter as possible.

Vaccination may reduce the prevalence of *Salmonella* from 26.7% to 0.7–22.1% for individual faecal prevalence in finisher pigs, 47.5% to 12.2–43.8% for lymph node prevalence at slaughter, and 46.3% to 13.4–43.4% for caecal prevalence at slaughter. No consistent effect was found on reduction of pen faecal culture prevalence.

7.3.2 Scalding and singeing (Steps 7 and 10)

Scalding and singeing are process steps that were also recommended for consideration as potential hazard-based interventions for the control of *Salmonella* due to the dual benefits provided.

Scalding: The efficacy for inactivation of *Salmonella* depends on the temperature and time (or equivalent combinations), e.g. 61°C during 8 minutes or 70°C for 2–3 minutes. There was extensive evidence of reduction in *Salmonella* prevalence on carcasses during scalding, with *Salmonella* prevalence reductions of from 35% to 1.5%.

Singeing: There was extensive evidence of reduction of *Salmonella* prevalence on carcasses during singeing with *Salmonella* prevalence reductions from 18.1% to 5.4%. A 2 $\log_{10}$ CFU/cm^2 reduction in *Salmonella* concentration on carcasses after singeing was estimated in a single study.

7.3.3 Pre-chill carcass interventions (between steps 16 and 18)

Carcass decontamination treatments with proven efficacy were recommended for consideration as potential hazard-based interventions for the control of *Salmonella* before chilling. The efficacy of carcass decontamination treatments was dependent on a number of factors, as outlined for beef carcasses. Decontamination treatments recommended by the experts were:

Hot water washes and steam pasteurization The experts agreed the carcass surface temperature should reach at least 70°C during treatment to effectively reduce *Salmonella*.

Hot water washes and steam pasteurization (hot water at 74 to 81°C for 5 to 15 sec or steam at 82–85°C for 60 sec) were shown in controlled trials conducted under commercial conditions to reduce the prevalence of *Salmonella* from 4.3% to 1.6%, with unpublished data from commercial settings indicating that reductions could be from 13% to 1%. The experts considered that realistic reductions could be up to 2 $\log_{10}$ CFU/cm^2, which was achieved in challenge trial studies.

Organic acid washes Efficacy of organic acid washes depends on the same process parameters and requirement for uniform carcass surface coverage as for beef.

Organic acids resulted in a reduction of *Salmonella* prevalence on carcasses of from 7% to 2%. In challenge studies conducted under laboratory and pilot plant conditions, organic acid washes resulted in up to 1.8 $\log_{10}$ CFU/cm^2 reductions; however, the experts concluded that a realistic expectation was up to a 0.5 to 1 $\log_{10}$ CFU/cm^2 reduction.

7.4 GHP-BASED CONTROL MEASURES FOR PORK UP TO STEP 20

These are the same as for beef, with the addition of:

- Feed withdrawal to reduce rupturing of the intestines and intestinal spillage during carcass dressing.
- Hygiene during de-hairing (Step 8) and polishing (Step 11) to limit cross- and re-contamination of carcasses that might negate reductions achieved during scalding and singeing.
- Full carcass steam vacuuming as a potential alternative to hot-water washes in small establishments with limited resources.

7.5 HAZARD-BASED INTERVENTIONS FOR BEEF AND PORK FROM PACKAGING TO CONSUMPTION (STEPS 19–26 FOR BEEF; STEPS 21–25 FOR PORK)

Irradiation was recommended for consideration as a potential hazard-based intervention for the control of *Salmonella* in beef and pork products.

Irradiation of pork minced meat resulted in D_{10} values (kGy) of 0.403–0.860 for *S.* Typhimurium. Irradiation of ground beef has resulted in D_{10} values (kGy) of 0.618–0.661 for *Salmonella* spp., with differences possible between serovars.

7.6 GHP-BASED CONTROL MEASURES FOR BEEF AND PORK FROM PACKAGING TO CONSUMPTION (STEPS 19–26 FOR BEEF; STEPS 21–25 FOR PORK)

The systematic review scope did not cover interventions post-packaging; however, the experts recommended the following GHPs for consideration:

Cold chain management Beef and pork should be held below 5°C to prevent growth of *Salmonella*.

HACCP and hygiene prerequisites HACCP-based principles and hygiene prerequisites should be practiced during distribution, retail and in all settings where food is prepared for consumption.

Cooking Beef and pork products should be cooked to a temperature that is sufficient to inactivate *Salmonella*.

8

Limitations and Caveats

There are several limitations associated with the systematic review (presented as Annex 1) that was used as a primary resource by experts in their deliberations about the efficacy of interventions. The review was conducted using pre-tested search algorithms, but it is possible that some important search terms were missed. In addition, only two bibliographic databases were searched for references as part of the rapid approach used. For these reasons, the review could have missed some potentially relevant literature that might have resulted in additional evidence, though a search verification strategy was implemented to attempt to mitigate this potential bias. Another related limitation is that only literature published in English, French, and Spanish was considered for inclusion, which could have resulted in missed articles or underrepresentation of evidence from some geographical regions.

Another limitation of the systematic review is that results of studies on the efficacy of interventions were synthesized across relatively broad intervention categories. This was conducted for pragmatic reasons in order to facilitate summarization and presentation of intervention efficacy trends from a large body of literature. However, as a consequence, details such as intervention application parameters (e.g. dose; treatment duration) and differences in study sampling and laboratory methods were not investigated as possible sources of variation in intervention effects across studies. It is likely that these and other study factors could contribute to the heterogeneity in effects observed for many intervention categories, but it was considered beyond the scope of the review to investigate these factors in detail.

The systematic review only included research that directly measured the impact of interventions on *Salmonella*. Research that measured intervention effects on other pathogens or surrogate organisms (e.g. non-pathogenic *Escherichia coli*) was not included because it was not known to what extent these results might reflect and correspond similarly to the control of *Salmonella*. For example, the behaviour of surrogates such as *E. coli* can differ when compared with different *Salmonella* serovars, intervention processes and meat substrates (Niebuhr *et al.*, 2008). As a result, during the meeting, the experts initially considered the results of the systematic review, and if data specific to *Salmonella* were lacking, considered their experiential knowledge and, where it was agreed to be appropriate, any additional intervention efficacy data on other organisms that might be expected to behave similarly to *Salmonella*.

While data on *Salmonella* were preferred for the purposes of this meeting, the experts recognize that in practice, *Salmonella* is not usually present on beef and pork carcasses in adequate concentrations for enumeration under good hygiene and manufacturing conditions. As a result, surrogate organisms may be necessary for monitoring and validation of interventions. Surrogates for *Salmonella* are ideally strains that retain all of the characteristics of *Salmonella,* or have more robust characteristics, and are non-pathogenic. Where necessary, operators should select an appropriate surrogate for *Salmonella* that meets the requirements of their local plant conditions and situation.

There is a need for caution in interpretation of reported intervention efficacy data due to possible methodological limitations of studies in detecting and isolating injured *Salmonella* cells, or cells that are recovering from injury following exposure to chemical or physical interventions for inactivation. The experts noted that this was of particular concern in consideration of the potential effect of chilling to control *Salmonella*, as evidence for this effect was brought to the meeting based on observations on *E. coli* (Mellefont, Kocharunchitt and Ross, 2015), with similar unpublished results noted for *Salmonella* (Tom Ross, pers. comm.). The experts noted that these findings might also apply to interventions other than chilling, but there was no available evidence to support this.

References cited in the Report of the Expert Meeting

Arthur, T.M., Bosilevac, J.M., Brichta-Harhay, D., Kalchayanand, N., King, D.A., Shackelford, S.D., Wheeler, T.L. & Koohmaraie, M. 2008. Source tracking of *Escherichia coli* O157:H7 and *Salmonella* contamination in the lairage environment at commercial U.S. beef processing plants and identification of an effective intervention. *Journal of Food Protection*, 71(9): 1752–1760.

Baer, A.A., Miller, M.J. & Dilger, A.C. 2013. Pathogens of interest to the pork industry: a review of research on interventions to assure food safety. *Comprehensive Reviews in Food Science and Food Safety*, 12(2): 183–217.

Belluco, S., Cibin, V., Davies, R., Ricci, A. & Wales, A. (eds.). 2015. *A review of the scientific literature on the control of* Salmonella *spp. in food-producing animals other than poultry*. World Organisation for Animal Health (OIE), Paris, France.

Buncic, S. & Sofos, J. 2012. Interventions to control *Salmonella* contamination during poultry, cattle and pig slaughter. *Food Research International*, 45(2): 641–655.

Dey, M., Mayo, J.A., Saville, D., Wolyniak, C. & Klontz, K.C. 2013. Recalls of foods due to microbiological contamination classified by the U.S. Food and Drug Administration, fiscal years 2003 through 2011. *Journal of Food Protection*, 76(6): 932–938.

Duggan, S.J., Mannion, C., Prendergast, D.M., Leonard, N., Fanning, S., Gonzales-Barron, U., Egan, J., Butler, F. & Duffy, G. 2010. Tracking the *Salmonella* status of pigs and pork from lairage through the slaughter process in the Republic of Ireland. *Journal of Food Protection*, 73(12): 2148-2160.

EFSA BIOHAZ Panel [European Food Safety Authority Panel on Biological Hazards). 2014. Scientific opinion on the public health risks related to the maintenance of the cold chain during storage and transport of meat. Part 1 (meat of domestic ungulates). *EFSA Journal*, 12: 3601.

FAO [Food and Agriculture Organization of the United Nations]. 2009. Report of the FAO expert meeting on how to feed the world in 2050. Rome, Italy. Available at: http://www.fao.org/fileadmin/templates/wsfs/docs/EM_report/EM_report.pdf

Farkas, J. 1998. Irradiation as a method for decontaminating food: a review. *International Journal of Food Microbiology*, 44(3): 189–204.

Food Control Consultants Consortium. 2013. Analysis of the costs and benefits of setting a target for the reduction of *Salmonella* in pigs at slaughterhouse level. The European Commission Directorate-General for Health and Consumers. SANCO/E2/

SI2.534057 (available at: http://ec.europa.eu/food/food/biosafety/salmonella/docs/fattening_pigs_slaughthouse.pdf).

Friesema, I.H.M., Schimmer, B., Ros, J.A., Ober, H.J., Heck, M.E.O.C., Swaan, C.M., de Jager, C.M., Peran i Sala, R.M. & van Pelt, W. 2012. A regional *Salmonella enterica* serovar Typhimurium outbreak associated with raw beef products, The Netherlands, 2010. *Foodborne Pathogens and Disease,* 9(2): 102–107.

Guyatt, G.H., Oxman, A.D., Schünemann, H.J., Tugwell, P. & Knottnerus, A. 2011. GRADE guidelines: a new series of articles in the Journal of Clinical Epidemiology. *Journal of Clinical Epidemiology*, 64(4): 380–382.

Higgins, J.P.T. & Green, S. 2011. *Cochrane handbook for systematic reviews of interventions: version 5.1.0.* The Cochrane Collaboration. Available at: http://handbook.cochrane.org/

Keithlin, J., Sargeant, J.M., Thomas, M.K. & Fazil, A. 2015. Systematic review and meta-analysis of the proportion of non-typhoidal *Salmonella* cases that develop chronic sequelae. *Epidemiology and Infection*, 143(7): 1333–1351.

Laufer, A.S., Grass, J., Holt, K., Whichard, J.M., Griffin, P.M. & Gould, L.H. 2015. Outbreaks of *Salmonella* infections attributed to beef – United States, 1973-2011. *Epidemiology and Infection*, 143(9): 2003–2013.

Majowicz, S.E., Musto, J., Scallan, E., Angulo, F.J., Kirk, M., O'Brien, S.J., Jones, T.F., Fazil, A. & Hoekstra, R.M. 2010. The global burden of non-typhoidal *Salmonella* gastroenteritis. *Clinical Infectious Diseases*, 50(6): 882–889.

Mannion, C., Fanning, J., McLernon, J., Lendrum, L., Gutierrez, M., Duggan, S. & Egan, J. 2012. The role of transport, lairage and slaughter processes in the dissemination of *Salmonella* spp. in pigs in Ireland. *Food Research International,* 45(2): 871–879.

Mellefont, L.A., Kocharunchitt, C. & Ross, T. 2015. Combined effect of chilling and desiccation on survival of *Escherichia coli* suggests a transient loss of culturability. *International Journal of Food Microbiology*, 208: 1–10.

Mindlin, M.J., Lang, N., Maguire, H., Walsh, B., Verlander, N.Q., Lane, C., Taylor, C., Bishop, L.A. & Crook, P.D. 2013. Outbreak investigation and case-control study: penta-resistant *Salmonella* Typhimurium DT104 associated with biltong in London in 2008. *Epidemiology and Infection,* 141(9): 1920–1927.

Niebuhr, S.E., Laury, A., Acuff, G.R. & Dickson, J.S. 2008. Evaluation of nonpathogenic surrogate bacteria as process validation indicators for *Salmonella enterica* for selected antimicrobial treatments, cold storage, and fermentation in meat. *Journal of Food Protection,* 71(4): 714–718.

Pires, S.M., Vieira, A.R., Hald, T. & Cole, D. 2014. Source attribution of human salmonellosis: an overview of methods and estimates. *Foodborne Pathogens and Disease,* 11(9): 667–676.

Raguenaud, M.E., Le Hello, S., Salah, S., Weill, F.X., Brisabois, A., Delmas, A. & Germonneau, P. 2012. Epidemiological and microbiological investigation of a large outbreak of monophasic *Salmonella* Typhimurium 4,5,12:i:- in schools associated with imported beef in Poitiers, France, October 2010. *Eurosurveillance*, 17(40): 20289.

Rajić, A. & Young, I. 2013. *Knowledge synthesis, transfer and exchange in agri-food public health: A handbook for science-to-policy professionals.* Guelph, Canada, University of Guelph Available at: https://atrium.lib.uoguelph.ca/xmlui/handle/10214/7293

Rivera-Betancourt, M., Shackelford, S.D., Arthur, T.M., Westmoreland, K.E., Bellinger, G., Rossman, M., Reagan, J.O. & Koohmaraie, M. 2004. Prevalence of *Escherichia coli* O157:H7, *Listeria monocytogenes*, and *Salmonella* in two geographically distant commercial beef processing plants in the United States. *Journal of Food Protection,* 67(2): 295–302.

Sargeant, J.M., Kelton, D.F. & O'Connor, A.M. 2014. Study designs and systematic reviews of interventions: building evidence across study designs. *Zoonoses and Public Health,* 61(S1): 10–17.

WHO. 2006. *Five keys to safer food manual.* Geneva, Switzerland.

WHO. 2014. *WHO handbook for guideline development – 2nd edition.* Geneva, Switzerland.

Annexes

Annex 1

Systematic Review Report

Rapid Systematic Review of the Efficacy of Interventions to Control *Salmonella* in Beef and Pork

Prepared by
Ian Young, FAO, Rome, Italy,
Barbara Wilhelm, University of Guelph, Canada,
Sarah Cahill, FAO, Rome, Italy,
Rei Nakagawa, WHO, Geneva, Switzerland
Patricia Desmarchelier, Food Safety Principles, Australia
Andrijana Rajić, FAO, Rome, Italy,

RSR1. Background

In November 2013, the 45th Session of the Codex Committee on Food Hygiene (CCFH) agreed to develop *Draft Guidelines for the Control of Non-typhoidal* Salmonella *spp. in Beef and Pork Meat* and initiated the process by establishing an electronic working group. The 46th Session, held in November 2014, requested that the Food and Agriculture Organization of the United Nations (FAO) and World Health Organization (WHO) undertake work to provide the Committee with scientific advice on this matter. Specifically, the Committee requested that a systematic review be conducted to identify relevant intervention and mitigation measures for control of *Salmonella* in fresh beef and pork, from the primary production stage to consumption, to provide evidence-informed inputs into this guideline.

RSR2. Methods

RSR2.1 REVIEW APPROACH

Reflecting the rapid timeline, limited resources for review conduct, and complex, policy-driven review scope and mandate, a mixed-method knowledge synthesis approach was adopted for this systematic review (Mays, Pope and Popay, 2005; Rajić and Young, 2013; WHO, 2014). Specifically, the review considered all available evidence on intervention efficacy available in the public domain, including previously published systematic reviews, risk assessments, and stochastic models, as well as primary research. The review was streamlined using a rapid approach to ensure that preliminary findings would be available within approximately two months to inform a CCFH physical working group meeting in May 2015, with updates and finalization over the following four months to inform a follow-up expert meeting in September 2015. Specifically, the search was targeted; detailed analyses were prioritized; and only one reviewer conducted the relevance confirmation and data extraction steps.

RSR2.2 REVIEW QUESTION, SCOPE, AND ELIGIBILITY CRITERIA

The review question was: "What is the efficacy of all possible interventions to control *Salmonella* in fresh beef and pork, from primary production to consumption?" The population of interest included all swine and cattle (beef and dairy) produced for domestic meat consumption, including their carcasses at processing and finished products. Any interventions applied throughout the farm-to-fork chain for fresh beef and pork were considered relevant. Relevant outcome measures for on-farm interventions included faecal shedding, serology, and tissue and organ contamination (e.g. lymph nodes). Any measure of carcass or product contamination was considered relevant during and post-processing (including environmental samples). All diagnostic test methods were considered relevant (e.g. culture, PCR, ELISA), including prevalence (positive/negative) and concentration (CFU) measures. In line with draft CCFH Guidelines, interventions on manufactured (e.g. cured, dried, fermented) and ready-to-eat beef and pork products such as sausages and salamis, and studies measuring milk outcomes in dairy cattle, were excluded from the review scope. All study designs and publication types were considered relevant as long as they were published in English, French or Spanish.

RSR2.3 REVIEW TEAM

The review was conducted by a core team consisting of the JEMRA secretariat (one WHO and two FAO representatives, and two project consultants). The main review activities were implemented by the consultants, with project oversight and coordination from the core team. The team met regularly throughout the review to clarify any questions and areas of discrepancy. Prior to implementing the review, the review protocol and proposed approach, scope, and inclusion criteria were shared with core members of the CCFH working group for feedback.

RSR2.4 SEARCH STRATEGY

Two comprehensive search algorithms were developed to identify relevant literature on beef and pork commodities, respectively. Algorithms were developed by extracting key words from a selection of 10 known relevant articles per commodity, and by reviewing and adapting search strategies and key terms of previously published reviews and risk assessments on this and similar topics (Denagamage *et al.*, 2007, 2010; O'Connor *et al.*, 2008; Friendship *et al.*, 2009; Greig *et al.*, 2012; Wilhelm *et al.*, 2012; Baer, Miller and Dilger, 2013; Wheeler, Kalchayanand and Bosilevac, 2014; Li, Kundu and Holley, 2015). Key terms were combined using the Boolean operator "OR" into categories for Pathogen (*Salmonella* terms), Population (beef and pork terms), and Intervention (intervention terms), and the categories were combined using the "AND" operator (see RSR Annex 1A for the full search algorithms). Algorithms were pre-tested in Scopus to ensure that a known list of 15 relevant articles could be sufficiently identified.

Final searches were implemented in the bibliographic databases Scopus and CAB Abstracts on 11 February 2015. For the beef search, no publication year or other restrictions were imposed. For the pork search, only literature from 2009-2015 was searched, as a citation list of potentially relevant articles covering any intervention against *Salmonella* in the pork chain was already available from a previously conducted scoping review that included a broad search strategy of literature published up to 2009 (Wilhelm *et al.*, 2012). Search verification was conducted by reviewing relevant conference proceedings, through targeted searches in Google to identify potential grey literature (e.g. government and industry reports), and by reviewing the reference lists of a selection of relevant articles (see RSR Annex 1A for details). In addition, any documents and/or unpublished data received from the complementary FAO/WHO "Call for Data" were reviewed for potential inclusion in this review.

RSR2.5 RELEVANCE SCREENING

The relevance of each unique citation was assessed at the title and abstract level using an *a priori* developed form (RSR Annex 1B). The form included one key question to determine the citation's relevance to review question and eligibility criteria. After this stage, the review team decided to alter the eligibility criteria and exclude all research published prior to 1990 from further consideration in the review, as evidence on interventions published prior to this period was not considered reflective of current industry conditions and practices.

RSR2.6 RELEVANCE CONFIRMATION AND PRIORITIZATION

Citations passing the relevance-screening step were procured as full articles and confirmed for relevance using another pre-specified form (RSR Annex 1C). This form was used to characterize articles according to the document type, study design, commodity (beef or pork), point in food chain, and intervention categories investigated. Results from this stage were used by the review team, in consultation with the CCFH working group chairs, to prioritize more detailed data extraction and summarization according to the availability and applicability of evidence for each commodity and point in food chain. For on-farm interventions, only studies measuring outcomes in animals at market-weight or slaughter were considered for detailed data extraction, risk-of-bias assessment, and quantitative meta-analysis; those measuring outcomes in young animals (without direct relevance to food safety) were summarized narratively only, unless this was the only data available for a particular intervention category. Detailed analyses focused only on controlled trials when available, otherwise challenge trials were considered. For interventions at processing, all experimental study designs (controlled trials, challenge trials and quasi-experimental studies) were considered for detailed data extraction and risk-of-bias assessment, with the exception of studies measuring the effect of packaging and final product preservation technique interventions, which were summarized narratively only. Only a limited number of studies were identified evaluating post-processing interventions, as the search strategy was not targeted to properly identify such studies; therefore, these were excluded from further analysis and summarization to avoid possible misrepresentation of interventions at this stage.

RSR2.7 DATA EXTRACTION AND RISK-OF-BIAS ASSESSMENT

Detailed data extractions and risk-of-bias assessments were conducted for prioritized articles using pre-specified tools (RSR Annexes 1D–1G). The data extraction tool included targeted questions about intervention and population descriptions, outcomes measured, diagnostic test methods, and intervention efficacy results. The risk-of-bias tool for primary research studies was adapted from the Cochrane Collaboration's recommended tools for randomized and non-randomized study designs (Higgins and Green, 2011; Cochrane Effective Practice and Organisation of Care Group, 2013). Systematic reviews were assessed for their reporting reliability using the previously validated AMSTAR tool (Shea *et al.*, 2009). Risk assessments and models were evaluated for key reporting reliability criteria using a tool modified from previously suggested quality criteria for risk assessments (Lammerding, 2007).

RSR2.8 REVIEW MANAGEMENT

References identified through the searches were uploaded to RefWorks (Thomson ResearchSoft, Philadelphia, PA), de-duplicated using both the automatic function and manually, then imported into the systematic-review software program DistillerSR (Evidence Partners, Ottawa, ON). DistillerSR was used for relevance screening, relevance confirmation, and risk-of-bias assessment, while detailed data extraction was conducted in Microsoft Excel 2010 (Microsoft Corporation, Redmond, WA). All review forms were pre-tested before use on a selection of relevant citations and articles by two independent reviewers (30 for relevance screening and confirmation, and five for data extraction and risk-of-bias assessment). Reviewing for relevance screening proceeded only when consistent inclusion and exclusion agreement was achieved between reviewers (kappa >0.8). For other steps, pre-tests were used to ensure consistent interpretation of the tools, and, if needed, to improve their clarity. Two independent reviewers conducted relevance screening and risk-of-bias assessment, while only one conducted relevance confirmation and data extraction. Any disagreements between reviewers were resolved through discussion.

RSR2.9 DATA ANALYSIS

Results of previous systematic reviews, risk assessments and stochastic models, and non-prioritized study designs were summarized narratively (Mays, Pope and Popay, 2005). Data from primary research studies of prioritized designs were stratified into comparable subgroups for meta-analysis and descriptive summarization (Borenstein *et al.*, 2009; Higgins and Green, 2011). Data were first stratified by point in the farm-to-fork chain, then study design, then into specific intervention categories that were defined based on insights from previous systematic and narrative literature reviews (Denagamage *et al.*, 2007, 2010; O'Connor *et al.*, 2008; Wilhelm *et al.*, 2012; Baer, Miller and Dilger, 2013; Wheeler, Kalchayanand and Bosilevac, 2014), and finally by different outcome measures (e.g. faecal, serology). Random-effects meta-analysis was conducted in each subgroup when sufficiently reported data were available from ≥2 studies. For processing interventions, challenge trials were only summarized using a median and range of intervention effects due to a lack of reporting of required data, while at farm level, they were considered for meta-analysis only when controlled trials were not available for a particular intervention category. The unit of analysis in all cases was individual trials reported within studies. Prevalence (positive vs negative) data were summarized using the odds ratio (OR) measure, while concentration data were analysed as mean changes in log bacterial counts (e.g. log CFU/cm^2) (Borenstein *et al.*, 2009; Higgins and Green, 2011). All meta-analysis models were conducted using the DerSimonian and Laird method (DerSimonian and Laird, 1986). For interventions at processing, several challenge trials were identified that compared more than one intervention group (e.g. multiple combinations of different chemical washes) to a single control group. In these cases, to avoid counting the same samples more than once in the same summary, multiple treatment groups within the same intervention category in these studies were collapsed together by averaging the intervention effects and summing the sample sizes.

Heterogeneity in all meta-analyses was measured using I^2, which indicates the proportion of variation in effect estimates across studies that is due to heterogeneity rather than sampling error (Higgins *et al.*, 2003). Heterogeneity was considered high and average estimates of effect were not shown when $I^2>60\%$ (Higgins *et al.*, 2003; Higgins and Green, 2011). In these cases, a median and range of effect estimates from individual studies in the meta-analysis subgroup was shown instead, as presenting pooled meta-analysis estimates in the presence of so much variation can be misleading (Higgins and Thompson, 2002). Meta-analysis effect estimates were considered significant if the 95% confidence intervals (CI) excluded the null. Begg's adjusted rank correlation and Egger's regression tests were used to test for possible publication bias on meta-analysis data subsets if there were ≥10 trials and if heterogeneity was not significant (Sterne *et al.*, 2011). For these tests, $P<0.05$ was considered significant. Meta-analysis was conducted using CMA software (Comprehensive Meta-Analysis Version 3, Biostat, Inc., Englewood, NJ) for on-farm studies, and R Version 3.1.3 (R Core Team, 2015) for processing studies.

RSR2.10 GRADE ASSESSMENT

The Cochrane Collaboration's Grades of Recommendation, Assessment, Development and Evaluation (GRADE) approach was used to assess the confidence in the estimated measures of intervention effect for each data subgroup (Guyatt *et al.*, 2011; Wilhelm *et al.*, 2012; WHO, 2014), in terms of how well they might be excepted to represent actual intervention effects in practice or in future studies. The GRADE tool and criteria were modified to suit the needs of the topic and study designs applicable to the research question (Higgins and Green, 2011; Sargeant, Kelton and O'Connor, 2014). The GRADE approach classifies the confidence in findings from each subgroup into one of four levels: one=very low (the true effect is likely to be substantially different from the measured estimate); two=low (the true effect may be substantially different from the measured estimate); three=moderate (the true effect is likely to be close to the measured estimate, but there is a possibility that it is substantially different); four=high (there is strong confidence that the true effect lies close to that of the measured estimate). Due to inherent differences in strength of evidence by study design (Guyatt *et al.*, 2011; Sargeant, Kelton and O'Connor, 2014), controlled trials started at level four (high) and other study designs at level three (moderate). Five downgrading and three upgrading criteria were then assessed for each subgroup, which could lead to reducing or increasing, respectively, the pre-defined GRADE levels (RSR Annex 1H).

RSR2.11 KNOWLEDGE TRANSFER OF RESULTS

Evidence from each data subgroup was compiled and summarized by point-in-chain and commodity into "summary cards" to enhance the relevance and uptake of the results by end-users. The Summary Card approach has been used previously in food safety knowledge synthesis and risk analysis contexts to provide decision-makers and risk managers with complex scientific information in a more practical and interpretable format than traditional technical reports (Ruzante *et al.*, 2010; FAO/WHO, 2014). Summary Cards included the following sections: summary of key findings; intervention description; detailed intervention efficacy summaries; and references.

The first 1 or 2 pages of each Summary Card provides users with an "up-front" synopsis of the key results and findings of the evidence covered in the card, including key take-home messages and practical implications for the CCFH guideline development. The next section provides users with a description and definitions of the intervention categories that are summarized in the card. The detailed evidence section provides a narrative and descriptive summary of the evidence identified for each study design, separated by the main intervention category (farm level) or applicable point in the processing chain (processing level). These summaries are supplemented with "Summary-of-findings" tables that outline the quantitative results and level of confidence in the estimates of intervention effect (GRADE rating) for each evidence subgroup where a quantitative analysis was conducted (Higgins and Green, 2011).

RSR2.12 HOW TO INTERPRET SUMMARY-OF-FINDINGS TABLES AND GRADE RATINGS

The "summary-of-findings" tables include one or more of the following categories: intervention and outcome sample; comparison group; number of trials and studies; quantitative intervention effect; and GRADE rating. The intervention sample indicates the sample type or animal subjects to which the intervention was applied, and the outcome sample indicates the sample type that was subsequently measured for *Salmonella* contamination (in some cases these two samples types were the same). The comparison group refers to the control group to which the intervention is compared, either: (1) no treatment; (2) a reference or standard treatment; or (3) the pre-intervention sample for quasi-experiments. The number of trials indicates the number of unique intervention comparisons included in the specific data subgroup; in some cases studies contained more than one trial.

For studies measuring prevalence outcomes (positive vs negative), results are presented as ORs for the effect of a reduced (or increased) prevalence in the intervention vs control group. For studies that measured concentration outcomes (e.g. log CFU/cm^2), intervention efficacy results are presented as log reductions in the intervention compared with the control group. For interventions with multiple trials, the values presented in these columns refer to the average meta-estimate and 95% confidence interval (CI) from random-effects meta-analysis if heterogeneity was not significant (I^2<60%). If heterogeneity was significant, the median and range of ORs or log reduction values from individual studies is presented instead. If only one trial was available, the values refer to the OR or log reduction value and 95% CI from the individual study.

For studies measuring prevalence outcomes (positive vs negative), an additional column is presented that presents the intervention effect in more intuitive terms (Higgins and Green, 2011), as the change in a given baseline level of *Salmonella* prevalence due to the intervention in the included studies. In this column, the median prevalence of *Salmonella* in the comparison group trials (assumed control risk, ACR) is shown as a reference or baseline prevalence for the given data subgroup. The percentage of samples positive for *Salmonella* in the intervention group was calculated using the following formula (Higgins and Green, 2011): ((OR × ACR) / (1 -ACR + [OR × ACR])). The OR value used in this formula corresponds to the one presented in the intervention effect column, as described above. This value is accompanied by a 95% CI in brackets if heterogeneity was not significant. If heterogeneity was significant, the range of expected reductions from the individual studies is shown in brackets instead to illustrate the variability in intervention effects across studies.

The final column in all tables shows the GRADE rating for the level of confidence in the estimated intervention effect (as very low; low; moderate; or high). The GRADE ratings should not be interpreted as an indication of the overall quality of evidence of the results or confidence in whether or not an intervention is effective, but as a level of confidence that one can place in the estimated, quantitative intervention effect in terms of how closely it would be expected to match the actual intervention effect in practice or future studies. This level of confidence is derived from the factors considered in the GRADE assessment, namely:

- Possible risks of bias in the included research (such as confounding factors).
- How well the effects of the included studies agree with each other (heterogeneity) in both magnitude and direction.
- The power to detect an intervention effect if one exists (sample size of included studies).
- The degree to which the studies are reflective of the topic of interest and target conditions (commercial, real-life application of interventions).
- The possibility that the included evidence is a biased subset of all possible evidence about the efficacy of the intervention (publication bias).

In rare circumstances, when the studies are well-conducted and reported and there is no evidence of risks of bias, other factors might increase the confidence in the estimated intervention effect: (1) presence of a dose-response relationship; (2) all possible confounding factors would likely underestimate rather than overestimate the intervention effect if one exists; or (3) presence of a very large intervention effect.

Taking all of the above factors into account, even when confidence in the estimated intervention effect is high, it is possible that one could expect a different outcome or level of effect in future studies or in practice due to differences (variation) in the specific intervention application circumstances and setting (e.g. different intervention protocols, *Salmonella* strains and their behaviour, environmental factors). Recognizing the importance of these various factors, we have developed a short summary document that outlines some of the possible factors that might contribute to the variation in intervention effects in different settings and that should be considered when making decisions about the possible applicability, transferability, and effectiveness of adopting interventions in a given situation or context (RSR Annex AI).

RSR2.13 REFERENCES

Baer, A.A., Miller, M.J. & Dilger, A.C. 2013. Pathogens of interest to the pork industry: A review of research on interventions to assure food safety. *Comprehensive Reviews in Food Science and Food Safety*. 12(2): 183-217.

Borenstein, M., Hedges, L.V., Higgins, J.P.T. & Rothstein, H.R. 2009. *Introduction to Meta-Analysis*. John Wiley & Sons, Ltd., Chichester, UK.

Cochrane Effective Practice and Organisation of Care Group. 2013. Suggested risk of bias criteria for EPOC reviews. http://epoc.cochrane.org/sites/epoc.cochrane.org/files/uploads/14%20Suggested%20risk%20of%20bias%20criteria%20for%20EPOC%20reviews%202013%2008%2012_0.pdf.

Denagamage, T., O'Connor, A., Sargeant, J. & McKean, J. 2010. The association between sub-therapeutic antibiotics and *Salmonella* Typhimurium in market-weight swine: A systematic review and summation of evidence from 1950 to 2007. *Zoonoses and Public Health,* 57, e14-22.

Denagamage, T.N., O'Connor, A.M., Sargeant, J.M., Rajic, A. & McKean, J.D., 2007. Efficacy of vaccination to reduce *Salmonella* prevalence in live and slaughtered swine: A systematic review of literature from 1979 to 2007. *Foodborne Pathogens and Disease,* 4: 539–549.

DerSimonian, R. & Laird, N. 1986. Meta-analysis in clinical trials. *Controlled Clinical Trials,* 7(3) 177–188.

FAO/WHO, 2014. Ranking of low moisture foods in support of microbiological risk management: Report of an FAO/WHO consultation process. Preliminary report. ftp://ftp.fao.org/codex/meetings/CCFH/CCFH46/LMF%20Part%201-3%2030Oct%202014.pdf.

Friendship, R.M., Mounchili, A., McEwen, S. & Raji, A. 2009. Critical review of on-farm intervention strategies against *Salmonella*. http://development.bpex.org.uk/articles/298614/Critical_review_of_onfarm_intervention_strategies_against_Salmonella.aspx.

Greig, J.D., Waddell, L., Wilhelm, B., Wilkins, W., Bucher, O., Parker, S. & Raji, A. 2012. The efficacy of interventions applied during primary processing on contamination of beef carcasses with *Escherichia coli*: A systematic review-meta-analysis of the published research. *Food Control,* 27: 385–397.

Guyatt, G.H., Oxman, A.D., Schünemann, H.J., Tugwell, P. & Knottnerus, A. 2011. GRADE guidelines: A new series of articles in the Journal of Clinical Epidemiology. *Journal of Clinical Epidemiology,* 64: 380–382.

Higgins, J.P.T. & Thompson, S.G. 2002. Quantifying heterogeneity in a meta-analysis. *Statistics in Medicine*, 21(11): 1539–1558.

Higgins, J.P.T. & Green, S. (Eds.). 2011. Cochrane Handbook for Systematic Reviews of Interventions: Version 5.1.0. The Cochrane Collaboration. http://handbook.cochrane.org/.

Higgins, J.P.T., Thompson, S.G., Deeks, J.J. & Altman, D.G. 2003. Measuring inconsistency in meta-analyses. *British Medical Journal,* 327: 557–560.

Lammerding, A. 2007. Using microbiological risk assessment (MRA) in food safety management. International Life Sciences Institute (ILSI) Europe. Available at: www.ilsi.org/Europe/Publications/R2007Usi_Mic.pdf.

Li, M., Malladi, S., Hurd, H.S., Goldsmith, T.J., Brichta-Harhay, D.M. & Loneragan, G.H. 2015. *Salmonella* spp. in lymph nodes of fed and cull cattle: Relative assessment of risk to ground beef. *Food Control,* 50: 423–434.

Mays, N., Pope, C. & Popay, J. 2005. Systematically reviewing qualitative and quantitative evidence to inform management and policy-making in the health field. *Journal of Health Services Research & Policy,* 10(Suppl. 1): 6–20.

O'Connor, A.M., Denagamage, T., Sargeant, J.M., Rajic, A. & McKean, J. 2008. Feeding management practices and feed characteristics associated with *Salmonella* prevalence in live and slaughtered market-weight finisher swine: A systematic review and summation of evidence from 1950 to 2005. *Preventive Veterinary Medicine,* 87: 213–228.

R Core Team. 2015. R: A language and environment for statistical computing. Vienna, Austria. http://www.R-project.org/.

Raji, A. & Young, I. 2013. Knowledge synthesis, transfer and exchange in agri-food public health: A handbook for science-to-policy professionals. https://atrium.lib.uoguelph.ca/xmlui/handle/10214/7293.

Ruzante, J.M., Davidson, V.J., Caswell, J., Fazil, A., Cranfield, J.A.L., Henson, S.J., Anders, S.M., Schmidt, C. & Farber, J.M. 2010. A multifactorial risk prioritization framework for foodborne pathogens. *Risk Analysis,* 30: 724–742.

Sargeant, J.M., Kelton, D.F., O'Connor, A.M., 2014. Study designs and systematic reviews of interventions: Building evidence across study designs. *Zoonoses and Public Health,* 61(Suppl. 1): 10–17.

Shea, B.J., Hamel, C., Wells, G.A., Bouter, L.M., Kristjansson, E., Grimshaw, J., Henry, D.A. & Boers, M. 2009. AMSTAR is a reliable and valid measurement tool to assess the methodological quality of systematic reviews. *Journal of Clinical Epidemiology,* 62(10): 1013–1020.

Sterne, J.A., Sutton, A.J., Ioannidis, J.P., Terrin, N., Jones, D.R., Lau, J., Carpenter, J., Rücker, G., Harbord, R.M., Schmid, C.H., Tetzlaff, J., Deeks, J.J., Peters, J., Macaskill, P., Schwarzer, G., Duval, S., Altman, D.G., Moher, D. & Higgins, J.P. 2011. Recommendations for examining and interpreting funnel plot asymmetry in meta-analyses of randomised controlled trials. *British Medical Journal [BMJ]*, 343: Article d4002.

Wheeler, T.L., Kalchayanand, N. & Bosilevac, J.M. 2014. Pre- and post-harvest interventions to reduce pathogen contamination in the U.S. beef industry. *Meat Science*, 98(3; Special Issue): 372–382.

Wilhelm, B., Rajic, A., Parker, S., Waddell, L., Sanchez, J., Fazil, A., Wilkins, W. & McEwen, S.A. 2012. Assessment of the efficacy and quality of evidence for five on-farm interventions for *Salmonella* reduction in grow-finish swine: A systematic review and meta-analysis. *Preventive Veterinary Medicine*, 107(1-2): 1–20.

WHO [World Health Organization]. 2014. WHO handbook for guideline development - 2nd ed. http://apps.who.int/iris/bitstream/10665/75146/1/9789241548441_eng.pdf.

RSR3. Results of Review Process and Key Characteristics of Relevant Articles

A flow chart of the knowledge synthesis process for this review is shown below in RSR Figure 1. Key characteristics of all 520 relevant articles are shown in RSR Table 1. More articles were identified investigating interventions for control of *Salmonella* in the pork (n=309) vs beef (n=216) chain. While studies investigating interventions in the pork chain had a more global distribution (RSR Table 1), most studies (76%) investigating interventions in the beef chain were conducted in North America. The most common study design for both commodities was challenge trials (RSR Table 1). Most studies (70%) investigating interventions in the pork chain were conducted at the farm level, while most beef studies (64%) investigated interventions at processing.

Detailed results of the efficacy of interventions are presented as six Summary Cards:

- SC1 Pork – on-farm interventions
- SC2 Pork – transport, lairage, and logistic slaughter interventions
- SC3 Pork – processing interventions
- SC4 Beef – on-farm interventions
- SC5 Beef – transport and lairage interventions
- SC6 Beef – processing interventions

Detailed GRADE assessment results for each Summary Card are presented in Rapid Systematic Review, Appendix J.

RSR FIGURE 1. Review Flow Chart

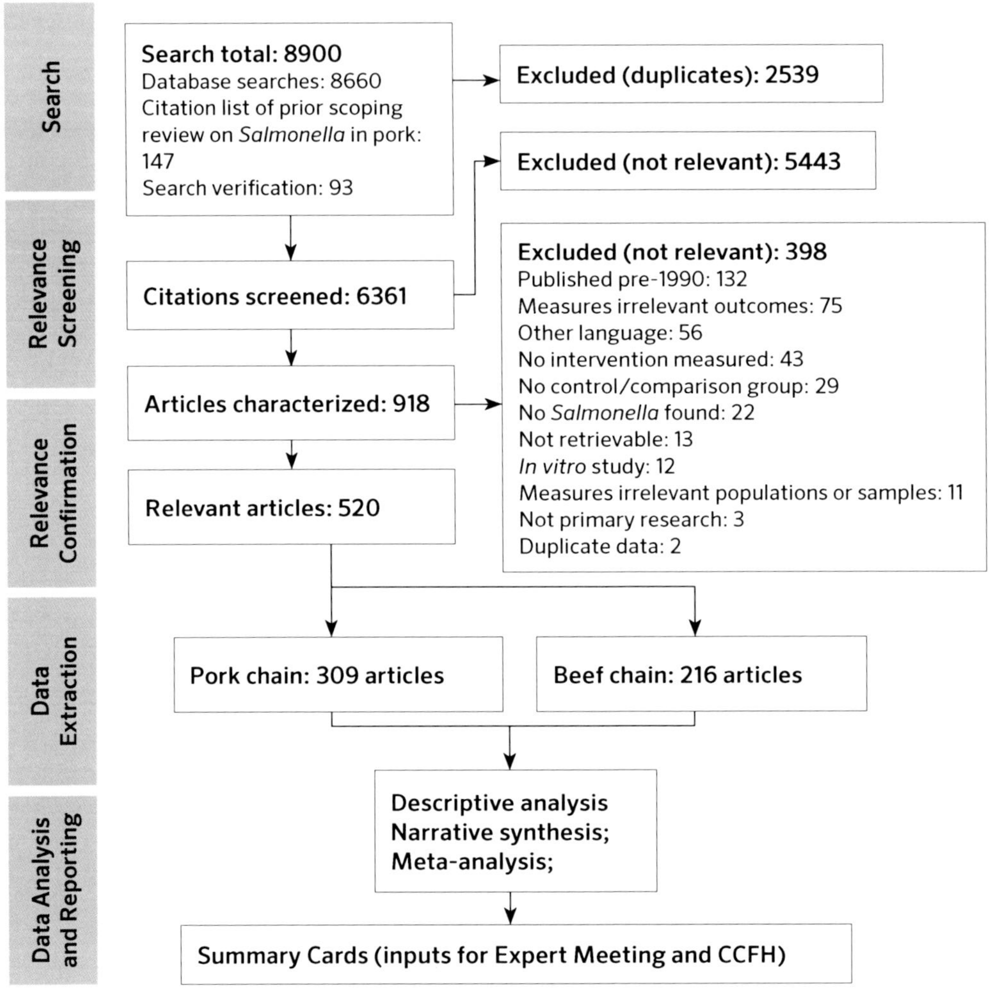

TABLE RSR 1. Key Characteristics of Relevant Articles

Characteristics	Pork (N=309)		Beef (N=216)	
	No.	%	No.	%
Region				
Europe	154	49.8	24	11.1
North America	108	35.0	163	75.5
Asia/Middle East	34	11.0	8	3.7
Central-South America/Caribbean	10	3.2	6	2.8
Australia/South Pacific	2	0.6	7	3.2
Africa	1	0.3	8	3.7
Document type				
Journal article	237	76.7	206	95.4
Conference proceedings paper/abstract	67	21.7	5	2.3
Report	4	1.3	4	1.9
Thesis	1	0.3	1	0.5
Study design				
Challenge trial	105	34.0	131	60.6
Controlled trial	77	24.9	59	27.3
Cross-sectional	54	17.5	19	8.8
Quasi-experiment	30	9.7	21	9.7
Risk assessment/model	30	9.7	10	4.6
Systematic review/meta-analysis	9	2.9	1	0.5
Cohort	8	2.6	5	2.3
Case-control	1	0.3	3	1.4
Study conditions				
Commercial/field conditions	166	53.7	69	31.9
Laboratory conditions	58	18.8	81	37.5
Research farm/pilot plant	33	10.7	51	23.6
Not reported	35	11.3	5	2.3
Intervention type:				
On-Farm:	216	69.9	71	32.9
Biosecurity/management practices	68	22.0	23	10.6
Feed/water acidification	51	16.5	2	0.9
Feed characteristics/management	51	16.5	20	9.3
Antimicrobials	32	10.4	12	5.6

Competitive exclusion/probiotics/ prebiotics	32	10.4	6	2.8
Vaccination	40	12.9	28	13.0
Other:	49	15.9	10	4.6
Transport/lairage/logistic slaughter	34	11.0	4	1.9
Processing:	92	29.8	138	63.9
Standard processing procedures/GHPs	44	14.2	15	6.9
Carcass/product washes, rinses, sprays	36	11.7	91	42.1
Packaging-based interventions	8	2.6	31	14.4
Cleaning/disinfection	7	2.3	2	0.9
Irradiation	3	1.0	4	1.9
Other	7	2.3	8	3.7
Post-processing to consumer	6	1.9	8	3.7

SC0. Summary Card Preface: How to Interpret Summary Cards

The Summary Cards are organized as follows:

1. Summary of key findings:

- 1- to 2-page synopsis of the key results and findings of the evidence covered in the card, including key take-home messages and overall implications.

2. Intervention description:

- A description and definitions of the intervention categories that are summarized in the card.

3. Detailed intervention efficacy summaries

- A narrative and descriptive summary of the evidence identified for each study design, separated by the main intervention category (farm-level cards) or applicable point in the food chain (post-farm-level cards). Includes "summary-of-findings" tables that outline the quantitative results and level of confidence in the applicability of the findings (GRADE rating) for each evidence subgroup where a quantitative analysis or summary was conducted.

4. References

"Summary of findings" tables:

These tables include one or more of the following categories: intervention and outcome sample; comparison group; number of trials and studies; quantitative intervention effect; GRADE rating; and references.

The **intervention sample** indicates the sample type or animal subjects to which the intervention was applied, and the outcome sample indicates the sample type that was subsequently measured for *Salmonella* contamination (in some cases these two samples types were the same).

The **comparison group** refers to the control group to which the intervention is compared, either: (1) no treatment; (2) a reference or standard treatment; or (3) the before or pre-intervention sample for quasi-experiments and some challenge trials.

The **number of trials** indicates the number of unique intervention comparisons included in the specific data subgroup; in some cases studies contained more than one trial.

The **intervention effect** is presented as an odds ratio (OR) for studies measuring *prevalence outcomes* (positive vs negative) or as log reductions for studies that measured *concentration outcomes* (e.g. log CFU/cm^2) in the intervention vs comparison group. These estimates of intervention effect represent one of the following values:

- Average estimate and 95% confidence interval (CI) from meta-analysis if there was no significant variability in the magnitude or direction of effect across studies.
- Median and range of effects from the individual studies when there was significant variability in the effect across studies.
- Result of the individual study with a 95% CI when only one study was available.

The **percentage of samples positive for *Salmonella* in the study population** is shown for studies measuring prevalence outcomes (positive vs negative). These values show how the prevalence of *Salmonella* changes from a given baseline level (median prevalence in the comparison groups of included studies) due to the intervention. A measure of variability is provided for this change from baseline, which represents a 95% CI in cases of no significant variability between studies, or a range of reductions from the included studies in cases when the magnitude of intervention effect varied widely between studies.

GRADE ratings:

A GRADE rating is shown for each intervention subgroup in all tables. This rating represents the level of confidence that one can place in the estimated, quantitative intervention effect in terms of how closely it would be expected to match the actual intervention effect in practice or in future studies. This level of confidence is rated at one of four levels (very low; low; moderate; or high). Controlled trials start at a level of "high" as they provide the strongest evidence for measuring the effect of an intervention, while other study designs start at a level of "moderate".

The level of confidence can be reduced due to one or more of five criteria:

- **Possible risks of bias** in the included research (such as confounding factors).
- How well the effects of the included studies agree with each other (**heterogeneity/variability**) in both magnitude and direction.
- The power to detect an intervention effect if one exists (**sample size of included studies**).
- The degree to which the studies are reflective of the topic of interest and target conditions (**representativeness of commercial, real-life application of interventions**).
- The possibility that the included evidence is a biased subset of all possible evidence about the efficacy of the intervention (**publication bias**).

In rare circumstances, when the studies are well-conducted and reported and there is no evidence of risks of bias, other factors might increase the confidence in the estimated intervention effect: (1) presence of a dose-response relationship; (2) any possible confounding factors would likely underestimate rather than overestimate the intervention effect if one exists; or (3) presence of a very large intervention effect.

Taking all of the above factors into account, a low or very low confidence level does not necessarily mean that an intervention is not effective, as in many cases it may simply mean that one cannot expect the same specific level or magnitude of reduction in *Salmonella* in practice as shown in the studies due to one or more of the factors above, such as variability in the study conditions and how well they might represent real-world settings.

Similarly, even when confidence in the estimated intervention effect is high, it is possible that one could expect a different outcome or level of effect in future studies or in practice due to differences (variation) in the specific intervention application circumstances and setting (e.g. different intervention protocols, *Salmonella* strains and their behaviour, environmental factors). Users should refer to RSR Annex 11, which outlines some of these possible factors that should be considered when making decisions about the possible applicability, transferability and effectiveness of adopting interventions in a given situation or context.

SUMMARY CARD SC1:
Pork Chain – On-Farm Interventions

SC1.1 SUMMARY OF KEY FINDINGS

This Summary Card covers the evidence supporting on-farm interventions for mitigation of *Salmonella* in pigs. Overall, consistent with previous systematic reviews on this topic, we captured a relatively small volume of literature investigating similar outcomes, in the population of greatest interest (i.e. finisher pigs/carcasses), which limited our ability to perform meta-analysis (MA) for estimation of pooled summary estimates of effect. Where possible, we selected data from controlled trials for MA to calculate a summary estimate of effect. However, for major topics which were not underpinned by evidence from controlled trials, we used data from challenge trials for MA.

SC1.1.1 Antimicrobials, including bacteriophage therapy

- Evidence regarding the effects of antimicrobial use in pigs on *Salmonella* prevalence or load was inconsistent. Two systematic reviews investigated antimicrobial use and its effect on *Salmonella*: one reported that no conclusion could be made given the inconsistent findings across literature containing important methodological limitations; and the other one reported a harmful treatment effect for one specific antimicrobial administered orally, with high heterogeneity across studies and low confidence partly reflecting relatively small sample sizes. The four challenge trials captured were the only studies that reported a significantly protective treatment effect. Controlled trials reported inconsistent findings, as did cross-sectional studies.
- Evidence for bacteriophage therapy was also inconsistent. We estimated non-significant MA summary estimates for treatment effect for phage therapy on individual rectal faecal sample, or lymph node prevalence, in challenged weaner and market pigs. In contrast, two challenge trials (Callaway *et al.*, 2011; Saez *et al.*, 2011) reported a significantly reduced ileal load in phage-treated weaner pigs relative to controls.

SC1.1.2 Biosecurity

- External biosecurity measures – including rodent control programmes, use of a hygienic lock, restriction of farm visitors, and restriction of introduction of semen or boars – were significantly associated with reduced odds of *Salmonella* on-farm in observational studies, using both culture and serology as outcome measures.

- Internal biosecurity measures were investigated in a small number of controlled trials, as well as observational studies. Evidence from the one controlled trial investigating the use of cleaning and disinfection reported inconsistent findings across four sampling events. The three controlled trials investigating use of strategic movement, however, yielded a strongly protective summary estimate of effect (odds ratio (OR) = 0.05, 95% confidence Interval (CI) (0.01–0.27)), with low confidence in the results, partly reflecting relatively small sample size. Four stochastic models consistently reported that application of external and/or internal biosecurity measures could reduce on-farm *Salmonella* prevalence, but could not reduce prevalence to negligible levels without the concurrent application of other interventions.

SC1.1.3 Feed/water acidification

- Six controlled trial studies investigated the feeding of organic acids using serology as the outcome measure, with 15/19 trials from the six studies reporting a protective effect (median = 0.46, range (0.01–6.82)), with low confidence. Three controlled trials investigated the effect of organic acids on *Salmonella* prevalence in carcass lymph nodes, with 7/13 trials indicating a protective treatment effect (OR = 0.36, 95% CI (0.19–0.67)), with moderate confidence. Findings when assaying individual finisher faecal culture, or rectal content at slaughter, were inconsistent or non-significant, respectively, both with low confidence.

SC1.1.4 Feed management

- Several potential strategies were captured. Feeding meal as opposed to pellets was investigated in three controlled trials, with a protective treatment effect (median OR = 0.30, range (0.19–1.79)) on *Salmonella* seroprevalence in finishers, and low confidence, partly reflecting inconsistent findings. However, the protective association between feeding meal, relative to pellets, and *Salmonella* prevalence in finishers, was supported by 11 cross-sectional studies. Feeding coarse vs finely-ground feed was associated with a consistently protective effect on *Salmonella* prevalence in three trials from two studies (median OR = 0.17, range (0.12–0.65)), with moderate confidence.

SC1.15 Manipulation of gut microbiota

- Three controlled trials reported a significant reduction in faecal load in finisher pigs fed both prebiotic and probiotic concurrently, and two reported a significant reduction in prevalence of faecal shedding in pigs fed a commercial prebiotic. In contrast, a controlled trial investigating use of a direct-fed microbial reported no significant treatment effect. Challenge trials reported inconsistent findings, and three observational studies reported a non-significant association between use of prebiotics or probiotics and *Salmonella* prevalence in finishers.

SC1.1.6 Vaccinations

- The efficacy of vaccination in reducing *Salmonella* prevalence was investigated in four controlled trials from two studies measuring individual faecal prevalence, with a significant protective, heterogeneous summary estimate of effect (median OR = 0.22, range (0.02–0.78)), with moderate confidence. Treatment effect when measured at the pen level was inconsistent. In contrast, treatment effect was significantly protective although heterogeneous when assaying either individual carcass lymph nodes (median OR = 0.41, range (0.12–0.86)) or caecal content (median OR = 0.53, range (0.18–0.89)) as the outcome measure, with moderate confidence.

SC1.1.7 Multiple interventions concurrently

- A European Union (EU) risk assessment of *Salmonella* in slaughter and breeding pigs suggested that a prioritized list of interventions should be implemented to reduce the risk of salmonellosis from pigs (EFSA, 2010), with high prevalence in breeder pigs to be targeted first, followed by feed, and then environmental contamination and external sources such as rodents and birds. This prioritization of breeding herds is supported by both European and North American observational studies identifying breeding sows as an important source of shedding (Cardinale *et al.*, 2010; Wilkins *et al.*, 2010).

SC1.1.8 Overall implications

- While interventions from each area were reported efficacious in some primary research settings, the effectiveness of biosecurity and vaccination measures was additionally supported by stochastic modelling and risk assessment. Significant reduction of lymph node contamination was supported by summary estimates of treatment effect, for feed/water acidification, and vaccination.
- Many of the datasets underpinning on-farm interventions were supported by datasets with a low level of confidence in the MA summary estimates of effect, which is not an indication of the quality of individual studies, but reflects that the reported estimates are likely to change with further research.
- Several relevant risk assessments and stochastic models were identified (n=24), often describing multiple levels of the pork production chain. Across these studies, the generalizability of data and assumptions from target population to others was frequently unclear (n=14 of 24 studies), making the wider applicability of these studies' findings uncertain.

- Results from stochastic models studying the farm-to-fork continuum were consistent in identifying post-farm levels of the chain as the areas containing points for more effective applications of intervention, regardless of outcome modelled (e.g. carcass prevalence vs human clinical cases of salmonellosis).

SC1.2. INTERVENTION DESCRIPTION

This Summary Card summarizes the evidence for a range of interventions that can be implemented at on-farm level to reduce *Salmonella* contamination of pork. Specific categories of interventions covered in this Summary Card include:

- **Antimicrobials:** any substance of natural, semi-synthetic, or synthetic origin, that kills or inhibits the growth of a microorganism but causes little or no damage to the host (Giguere *et al.*, 2006). Includes those administered parenterally and orally, for treatment or prophylaxis. Therapy employing bacteriophages, i.e. naturally occurring viruses capable of infecting and killing bacteria, is included within this category.
- **Biosecurity:** has been defined as the implementation of measures that reduce the risk of introduction and spread of disease agents (FAO/OIE/World Bank, 2010). Includes, but is not limited to, sanitation, biosafety, disinfection, hygiene and hygiene barriers, all-in-all-out production, depopulation, staff and the environment, litter testing and treatment, and pest control. Biosecurity may consist of external (targeting prevention of introduction of targeted pathogens to the farm or unit) or internal (aimed at reducing spread of pathogens on-farm) procedures.
- **Feed/water acidification:** addition of organic acids such as lactic acid, to feed or water.Includes 'nutraceuticals' such as copper, chromium, zinc, betaine or carnitine.
- **Feed management**: includes various feed strategies such as administration of coarse or finely-ground feed, fermented feed, or liquid feed. Also included in this category is deliberate withholding of feed in the hours immediately prior to transport to slaughter.
- **Manipulation of gut microbiota:** includes use of probiotics, prebiotics, and synbiotics. *Probiotics* are living microorganisms that are fed to animals to colonize the gut environment to encourage a better microbial balance. A *prebiotic* may be defined as a non-digestible food ingredient that beneficially affects the host by selectively stimulating the favourable growth and activity of one or a limited number of bacteria in the colon. The term *synbiotic* describes a combination of probiotic and prebiotic approaches. It includes application of protective bacterial species or cultures to out-compete and prevent *Salmonella* colonization in pigs. It can also include specific bacterial species or cultures,

or caecal contents or other materials from animals or the environment that contain many different (known or unknown) bacterial species.

- **Vaccination:** refers to immunization of the subject using either autogenous or commercial *Salmonella* vaccines.

Evidence was identified for each of these six categories of interventions, which are presented and discussed separately below.

SC1.3 ANTIMICROBIALS

Twenty-eight primary research studies investigated the use of antimicrobials for reduction of *Salmonella* in pork, including eight controlled trials, nine challenge trials, and 11 cross-sectional studies, as well as two systematic reviews.

SC1.3.1 Antimicrobials

Denagamage *et al.* (2010), using systematic review methodology, reported that no conclusion could be made regarding the association of sub-therapeutic use of antimicrobials on *Salmonella* Typhimurium in market pigs, given the inconsistent findings across literature containing important methodological limitations. Wilhelm *et al.* (2012) reported a significantly harmful treatment effect (OR range: 14 (95% CI 1.9–108); 1.0 (0.43–2.5)) for one specific antimicrobial (oral tetracycline), using faecal *Salmonella* spp. shedding as the outcome measure. High heterogeneity across studies was found, with low confidence in the results. The systematic reviews received scores of 9/10 relevant fields for Denagamage *et al.* (2010), and 9/11 for Wilhelm *et al.* (2012)), using the AMSTAR quality assessment tool.

Eight controlled trials were captured and investigated a variety of antimicrobials (aureomycin, bacitracin, chlortetracycline, flavomycin, tylosin). One reported a significantly ($P < 0.05$) harmful treatment effect (Funk *et al.*, 2007), while six other studies reported no significant treatment effect (Funk *et al.*, 2006; Kim *et al.*, 2014; LeJeune *et al.*, 2006; Roesler *et al.*, 2005; Wagner *et al.*, 2008; Wells *et al.*, 2013) and one reported a protective treatment effect on small intestinal *Salmonella* load (Wang *et al.*, 2007).

In contrast, four challenge studies (Ebner and Mathew, 2000; Letellier *et al.*, 2000; Mathew *et al.*, 2009; Robbins *et al.*, 2013) all reported a significantly protective treatment effect of various antimicrobials on faecal shedding.

Observational studies reported inconsistent findings, with four cross-sectional studies (Correge, Hemonic and Gouvars, 2009; Hotes *et al.*, 2010; Leontides,

Summary-of-findings table for phage therapy

TABLE SC1.1 Challenge trials investigating phage therapy, measuring prevalence

Intervention	Population/ sample	No. trials/ studies	Odds ratio for intervention effect[1]	% Samples *Salmonella* positive in study population[3]		GRADE rating[2]	References
				No treatment	With treatment		
Phage therapy	Weaners (Lymph node culture)	3/3	MA = 0.39 (95% CI: 0.12-1.22)	45.8%	24.8% (95% CI: 9.2-50.8)	Very low	Callaway *et al.*, 2011; Saez *et al.*, 2011; Wall *et al.*, 2010
	Weaners (Faecal culture)	2/2	1.00	100%	100%	Very low	Saez *et al.*, 2011; Wall*et al.*, 2010

NOTES: MA = meta-analysis average estimate from random-effects model; CI = confidence interval.

(1) For interventions with multiple trials, refers to the average meta-analysis estimate and 95% CI from random-effects meta-analysis if heterogeneity was not significant (I^2<60%). If heterogeneity was significant, the median and range of odds ratios or log reduction values from individual studies are presented instead. If only one trial was available, refers to the odds ratio or log reduction value and 95% CI from the individual study.

(2) GRADE rating descriptions: very low = the true effect is likely to be substantially different from the measured estimate; low = the true effect may be substantially different from the measured estimate; moderate = the true effect is likely to be close to the measured estimate, but there is a possibility that it is substantially different; high = there is strong confidence that the true effect lies close to that of the measured estimate.

(3) The percentage of samples positive for *Salmonella* in the comparison group (assumed control risk, ACR) refers to the median prevalence in the comparison group trials. The percentage in the intervention/treatment group was calculated using the following formula for the odds ratio (OR) measure of effect: (OR × ACR / 1 -ACR + [OR × ACR]), and provides an indication of the effect of the intervention on changing a given baseline level of *Salmonella* prevalence as reported in the included studies. The odds ratio used in this formula corresponds to the odds ratio presented in the intervention effect column as described under (1) above. If heterogeneity was not significant (I^2<60%), the estimate is presented with a 95% CI in brackets. If heterogeneity was significant, the median estimate is presented with the range of individual study effects in brackets to illustrate the variability in results across studies.

Grafanakis and Genigeorgis, 2003; Van der Wolf *et al.*, 2001a) reporting a significantly harmful treatment effect on prevalence of *Salmonella* in finishers, four (Farzan *et al.*, 2006; Garcia-Feliz *et al.*, 2009; Hautekiet *et al.*, 2008; Lo Fo Wong *et al.*, 2004) a non-significant treatment effect ($P > 0.05$) on *Salmonella* prevalence in finishers, with two (Correia-Gomes *et al.*, 2012, 2013) reporting a non-significant treatment effect in breeder herds, using both serology and faecal culture as the outcome measures. Tenhagen *et al.* (2009) reported a non-significant association between antimicrobial use (tetracycline, amoxicillin, and colistin) and the culture of *Salmonella* spp. from lymph nodes of German slaughter pigs, but a significant association between antimicrobial use and *Salmonella* Typhimurium. Farzan *et al.* (2006) reported a significant positive relationship between daily antimicrobial usage and crude optical density of the individual pig, using an enzyme linked immunosorbent assay (ELISA) and multivariable mixed linear regression with farm as a random variable.

SC1.3.2 Phage therapy

One controlled trial and five challenge studies investigated the use of bacteriophage therapy for reduction of *Salmonella* in pigs. Three challenge studies measured *Salmonella* prevalence or load in weaner pigs (Callaway *et al.*, 2011; Saez *et al.*, 2011; Wall *et al.*, 2010); one controlled trial (Yan *et al.*, 2012) and one challenge study (Gebru *et al.*, 2010) sampled grower pigs; and two challenge studies sampled market pigs (Albino *et al.*, 2014; Wall *et al.*, 2010). Yan *et al.* (2012) reported significantly reduced faecal *Salmonella* spp. load in grower pigs fed a basal diet plus 0.25 g/kg, or 0.50 g/kg bacteriophage (1×10^8 PFU/g), relative to controls.

Albino *et al.* (2014) reported a non-significant reduction in mean *Salmonella* concentration in ileal and caecal content of 90–100 kg pigs treated with various concentrations of a phage pool. Wall *et al.* (2010) administered 15 ml of 109 PFU/ml of a phage cocktail orally every two hours for six hours, to market pigs, and reported no significant treatment effect on prevalence, assaying either ileo-caecal lymph nodes or faecal samples. However, caecal *Salmonella* concentrations were significantly reduced in the treatment group. Callaway *et al.* (2011) administered 3×10^9 PFU of phage cocktail orally to weaner pigs, at 24 and 48 hours after *Salmonella* Typhimurium challenge. Upon sacrifice at 96 hours post-challenge, there was a non-significant difference in lymph node and caecal content prevalence between treatment and control groups, but a significant reduction in rectal shedding in the treatment group. Using a semi-quantitative shedding score, Gebru *et al.* (2010) reported significantly reduced faecal shedding of *Salmonella* Typhimurium in grower pigs (initially 38.7 kg ±6.7 kg) administered 3×10^9 PFU of bacteriophage/kg feed, sampled 14 days after challenge. Saez *et al.* (2011) reported a non-significant reduction in lymph node prevalence, relative to controls, but a significant reduction in ileal content load, between pigs either fed 5×10^{11} PFU, microencapsulated, in-feed, per day for 5 days; or administered 5×10^{11} PFU orally by gavage every 2 hr for 6 hr immediately post-challenge.

Three studies investigated the effect of phage therapy on prevalence of *Salmonella* lymph node contamination in weaner pigs, with a non-significant summary estimate of treatment effect (SC1 Table 1). Similarly, two studies investigated the effect of phage therapy on faecal *Salmonella* shedding in weaner pigs, with a non-significant summary estimate of effect. Both summary estimated received a GRADE rating of 'very low confidence' reflecting relatively small sample sizes and also sampling of animals younger than this review's preferred age group (i.e. finisher/market animals).

SC1.4 BIOSECURITY

Sixty-eight relevant studies were found investigating biosecurity interventions for mitigation of *Salmonella* in pigs: five controlled trials, three quasi-experiments, two cohort studies, one case-control study, and 40 cross-sectional studies, as well as 16 stochastic models and one systematic review. More experimental studies tended to examine internal biosecurity measures (i.e. those measures to potentially mitigate within-farm spread), relative to external biosecurity measures, (i.e. those attempting to prevent pathogen entry to the farm).

SC1.4.1 External biosecurity practices

Cross-sectional studies in this dataset investigated several external biosecurity measures. Three cross-sectional studies (Cardinale *et al.*, 2010; Choe, Hassan and Loh, 2011; Vico and Mainar-Jaime, 2012) investigated the association between feral bird access to the pigs and farm *Salmonella* status, and all reported that not allowing bird access to pigs had a significantly protective effect on *Salmonella* culture or serology in finishers. In two cross-sectional studies of Portuguese breeding farms (Correia-Gomes *et al.*, 2012; 2013), a significant association was reported between presence of rodents in the pig barn and *Salmonella* culture in breeding animals' faecal samples. Similarly, Vico, Mainr-Jaime and Aussere (2011) reported that Spanish pig farms not employing rodent control programmes had significantly increased odds (OR = 4.3, 95% CI (2.2–8.4)) of *Salmonella* contamination of lymph nodes in slaughter pigs, and Kich *et al.* (2001) reported an association between implementation of rodent control programmes and reduced *Salmonella* seroprevalence in Brazilian finishing herds. However, Garcia-Feliz *et al.* (2009) reported a non-significant association between presence of a rodent control programme on-farm, and *Salmonella* faecal shedding in Spanish finisher pigs. Wales *et al.* (2009) reported that *Salmonella* serovars isolated in local wildlife were often also isolated from the pigs on-farm in the area.

Use of a hygienic lock facility was reported to be significantly protective of *Salmonella* seropositivity in finishers in three studies (Hautekiet *et al.*, 2008; Lo Fo Wong *et al.*, 2004; van der Wolf *et al.*, 2001a). The protective effect of footwear disinfection on odds of *Salmonella* culture or seropositives in finishers was reported by two studies (Choe, Hassan and Loh, 2011; Hautekiet *et al.*, 2008). In contrast, Vico and Mainar-Jaime (2012) reported a significant harmful association between use of footwear exclusive to the farm and *Salmonella* seropositives; Garcia-Feliz *et al.* (2009) reported a non-significant association. Failure to disinfect farm vehicle wheels was associated with increased odds of *Salmonella* seropositives in two studies (Choe, Hassan and Loh, 2011; Twomey *et al.*, 2010a). Simons *et al.* (2009) reported a cost analysis examining several potential farm-level *Salmonella* interventions and concluded that rodent control was less cost effective than vaccination or administration of organic acids in reducing human salmonellosis in the UK.

Policy regarding introduction of animals and/or semen was reported to be a significant predictor of *Salmonella* on-farm. Correia-Gomes *et al.* (2012; 2013) report a significant association between introducing semen or boars from off-farm, with increased odds of *Salmonella* faecal culture in breeding production holdings. Similarly, two studies reported significantly increased odds of *Salmonella* seropositives in finishers from farms obtaining pigs from multiple sources (Correge, Hemonic and Gouvars, 2009; Lo Fo Wong *et al.*, 2004). Dahl (2009) reported significantly increased odds of *Salmonella* seropositives in Danish slaughter pigs in conventional, relative to specific-pathogen-free farms (OR = 2.65, 95% CI (1.55–4.50)). Burns *et al.* (2013) report a cross-sectional study of 10 Irish pig farms with historically high *Salmonella* seroprevalence, in which the *Salmonella* Typhimurium monophasic variant predominant from faecal samples on-farm was recovered from feed samples from four farms, suggesting that contaminated feed could have been an infection source.

Bahnson *et al.* (2007), sampling 115 herds, reported significantly increased odds of *Salmonella* in ileo-caecal lymph nodes of slaughter pigs from farms allowing visitors with same-day contact with other pig herds (OR = 2.2–95% CI (1.15–4.3)). Smith, Clough and Cook (2010), sampling 566 UK pig holdings, reported that farms with more than five pig deliveries per year, or more than six deadstock visits per year, were associated with significantly increased mean *Salmonella* meat juice ELISA S/P ratios.

Hill *et al.* (2011a) reported a stochastic model for farm-level *Salmonella* transmission. Authors report that the breeding sow was potentially an important infection source, and that reduction of potential external sources of contamination should only be targeted in situations of relatively low breeding herd prevalence. Soumpasis, Alban and Butler (2012) described a stochastic model of *Salmonella* Typhimurium infection on pig farms, with parameters for employing internal and biosecurity procedures, and/or a hypothetical 100% protective vaccine. The model predicts that perfect external biosecurity measures will not eliminate *Salmonella* prevalence on a positive farm without internal measures, since the *Salmonella* bacteria spread rapidly room-to-room. Similarly Lurette, Belloc and Keeling (2011) described a network model based upon recorded pig movements within France. The model predicts that once infection is endemic within a production system, animal movement restrictions need to be augmented by within-herd controls to reduce *Salmonella* prevalence.

SC1.4.2 Internal biosecurity practices

Strategic movement of pigs at weaning was studied in two controlled trials (Table SC1.2) with each trial reporting a significant protective treatment effect (Dahl *et al.*, 1997; Nietfeld *et al.*, 1998), although Nietfeld *et al.*, note that the 'treatment' group, as with the control group, received an antimicrobial (Carbadox™) in the ration and may have received benefit from this.

Cleaning and disinfection, which may be considered an external or internal measure depending on the specific setting, was investigated in one systematic review (Wilhelm *et al.*, 2012) that reported a non-significant summary estimate of effect for pen disinfection, with significant heterogeneity and low confidence. Cook (2004) reported a controlled trial with four sampling events, using pooled faecal culture as the outcome measure, with inconsistent findings (median OR = 0.90, Range (0.27–2.65)). In contrast, van der Heijden *et al.* (2006) investigated the improvement of hygiene and management by using a HACCP control programme, reporting a decline in mean herd-level ELISA S/P ratio in a year-long study of five 'high risk' herds. Bode, Baier and Blaha (2007) described a quasi-experiment in which a suite of cleaning and disinfection measures were introduced into one German production system consisting of a breeding sow herd, nursery and three finisher barns, having persistently high *Salmonella* meat juice seroprevalence in finishers despite apparent good hygiene. Authors reported that stringent application of the disinfection measures would be required for sustained reduction in *Salmonella* prevalence. Mannion *et al.* (2007) reported a quasi-experiment investigating the effectiveness of cleaning and disinfection practices on both low- and high-risk herds. The investigators reported that in low-risk herds, the cleaning and disinfection procedures reduced the levels of *Salmonella* on pen floors. In contrast, in high-risk herds, levels of *Salmonella* in the pens were increased after cleaning; this was also true of feeders.

Observational studies reported inconsistent findings regarding the effectiveness of pen disinfection in reduction of *Salmonella* in finishers. Funk *et al.* (1999), in a study of cohorts of United States of America finisher pigs, assigned a score of 0 (clean) to 3 (greatest faecal accumulation) weekly for 14 weeks of finishing period, for 200 pens. The investigators reported an inverse relationship between faecal accumulation score in the finisher pen, and the relative risk of a finisher pig shedding *Salmonella*. Choe, Hassan and Loh (2011) reported significantly increased odds of *Salmonella* faecal culture in finishers in farms employing less frequent pen cleaning; in contrast, Correge and Hemonic (2012) reported that the persistence of *Salmonella* in the pens themselves was influenced not by cleaning and disinfection protocols, but by the bacteriological status of the sows, and serological status of finishers, in farrow-to-finish farms. A French cohort study of 119 farrow-to-finish farms reported a significant association between hygiene score in both farrowing and post-weaning areas of the production unit, and *Salmonella* culture on-farm (Fravalo *et al.*, 2003a). Leontides, Grafanakis and Genigeorgis (2003) reported a non-significant association between pen disinfection and ELISA seropositivity in finishers from 59 Greek herds. Ikwap *et al.* (2014), in a survey of piglets and weaners from 93 Ugandan herds, reported significantly reduced odds of *Salmonella* on-farms regularly cleaning feeders every two days, relative to daily (OR=0.18; 95% CI 0.05–0.72)). Tessier *et al.* (2013), in a survey of 50 farrow-to-finish and multiplier herds in Reunion Island, reported a significant ($P < 0.05$) protective association between cleaning and disinfection of the ceiling and natural service area, and positive *Salmonella* culture from finishers.

Berriman *et al.* (2013a) described a stochastic compartmental model of *Salmonella* transmission within UK pig herds, in both slatted and non-slatted barns. The authors concluded that hygiene measures (i.e. cleaning and disinfection) alone were insufficient to eliminate *Salmonella* on infected farms. Gautam *et al.* (2014) described a modified Susceptible-Infected-Recovered-Susceptible (SIRS) model, investigating the effects of enhanced cleaning on *Salmonella* prevalence in grow-finishers. The authors concluded that increased frequency and effectiveness of cleaning may reduce *Salmonella* prevalence in pigs at time of slaughter, but will not eliminate it. Hill *et al.* (2011b) described a stochastic model studying the effect of on-farm and abattoir interventions in reducing human salmonellosis. Running scenarios for two hypothetical high- and low-prevalence EU member states, the model predicted no significant change in incidence of human salmonellosis would be achieved by increased cleaning and disinfection on-farm and during transport, in the absence of other measures. Hotes, Traulsen and Kriete (2011) described a stochastic model of non-clinical *Salmonella* transmission within a pork supply chain. Their findings suggest hygiene control strategies in farrowing units were associated with a significant decrease in *Salmonella* prevalence after lairage at the abattoir; similar measures applied only to finishers were less effective.

Erdman *et al.* (2005) performed a pig depopulation quasi-experiment and reported a significantly reduced odds of *Salmonella* culture from finisher pooled faecal pen cultures and slaughter pig ELISA 'positives' post- vs pre-treatment.

Funk, Davies and Gebreyes (2001) investigated risk factors for individual faecal *Salmonella* in finishers in a cohort study of three-site systems. Presence of a toilet on-site was used as a proxy measure for owners' attitudes towards hygiene; finisher pigs on sites with a toilet had significantly reduced odds of faecal samples culturing 'positive' for *Salmonella*.

In contrast, several observational studies reported a beneficial effect associated with management procedures which potentially enhanced pig *Salmonella* exposure. A German farm-level case-control study (Gotter *et al.*, 2012) reported that farm staff wearing dirty boots was significantly associated with reduced *Sal-*

Summary-of-findings table for biosecurity

TABLE SC1.2. Controlled trials studying strategic movement

Intervention	Population (Sample)	No. trials/ studies	Odds ratio for intervention effect(1)	% Samples *Salmonella* positive in study population(2)		GRADE rating(3)	References
				Without treatment	With treatment		
Strategic movement	Finishers (Caecal content/ Faecal swab)	3/2	MA = 0.05 (95% CI: 0.01–0.27)	34.8%	2.5% (95% CI: 0.5-12.6%)	Low	Dahl *et al.*, 1997; Nietfeld *et al.*, 1998

NOTES: MA = meta-analysis average estimate from random-effects model; CI = confidence interval.

(1) Refers to the average meta-analysis estimate and 95% CI from random-effects meta-analysis, as heterogeneity was not significant (I^2<60%).

(2) The percentage of samples positive for *Salmonella* in the comparison group (assumed control risk, ACR) refers to the median prevalence in the comparison group trials. The percentage in the intervention/treatment group was calculated using the following formula for the odds ratio (OR) measure of effect: (OR × ACR / 1 -ACR + [OR × ACR]), and provides an indication of the effect of the intervention on changing a given baseline level of *Salmonella* prevalence as reported in the included studies. The odds ratio used in this formula corresponds to the odds ratio presented in the intervention effect column as described under (1) above. If heterogeneity was not significant (I^2<60%), the estimate is presented with a 95% CI in brackets. If heterogeneity was significant, the median estimate is presented with the range of individual study effects in brackets to illustrate the variability in results across studies.

(3) GRADE rating descriptions: very low = the true effect is likely to be substantially different from the measured estimate; low = the true effect may be substantially different from the measured estimate; moderate = the true effect is likely to be close to the measured estimate, but there is a possibility that it is substantially different; high = there is strong confidence that the true effect lies close to that of the measured estimate.

monella prevalence assaying carcass meat juice ELISA, although pig contact with other animal species significantly increased the odds of a farm being categorized as 'positive'. Similar findings were reported by Davies *et al.* (1997), who reported a lower farm-level prevalence of *Salmonella* in farrow-to-finish farms reporting a continuous flow of pigs, relative to participating farms using all-in-all-out flow. As well, van der Wolf *et al.* (2001a) reported that omission of disinfection after pressure washing a compartment as part of an all-in-all-out procedure was significantly associated with reduced *Salmonella* seroprevalence.

Cross-sectional studies were also employed to investigate several aspects of the farm site. Four studies, including the EU baseline *Salmonella* survey conducted in 2008, reported that presence of fully slatted flooring was associated with significantly reduced odds of *Salmonella* seropositives in finishers, relative to other flooring types (EFSA, 2011; Hotes *et al.*, 2010; Smith *et al.*, 2010b; Twomey *et al.*, 2010a); Leontides, Grafanakis and Genigeorgis (2003) reported a non-significant association between flooring type and ELISA seropositivity. Davies (1998), in a study of six *Salmonella* positive farms using open flush gutters, hypothesized that the increased risk of *Salmonella* shedding by pigs in barns with open gutters may not be due to the use of recycled lagoon water for flushing, but to inefficient removal of faecal matter. In contrast, in a survey of 65 herds, Kich *et al.* (2001)

reported increased odds of *Salmonella* seroprevalence in Brazilian finishers associated with presence of flush gutter systems. Potential snout-to-snout contact between pens was reported to be associated with increased odds of *Salmonella* in finishers, measured by both serology (Lo Fo Wong *et al.*, 2004) and faecal culture (Wilkins *et al.*, 2010).

Baptista *et al.* (2010) described a survey of 50 Portuguese pig herds, assaying pig faeces, lymph nodes or serum ELISA, in which farms were categorized for overall biosecurity based on multiple criteria, including provision of boots for visitors; requiring shower-in; cleaning and disinfection of loading bay after use; access of other animals such as cats to livestock; and rodent control programmes in place. The authors reported that using logistic regression analysis, 'factor 1' (or overall poor biosecurity) was significantly ($P = 0.04$) associated with *Salmonella* in pigs on-farm.

Continuous pig flow was reported to be associated with increased odds of both faecal *Salmonella* spp. shedding (Farzan *et al.*, 2010; 2006) and increased mean herd S/P value (Hautekiet *et al.*, 2008), while Twomey *et al.* (2010b) reported reduced odds of *Salmonella* on-farms practicing all-in-all-out flow (OR = 0.34, 95% CI (0.18–0.65)). In contrast, Davies *et al.* (1997) reported more North Carolina farms practising all-in-all-out flow were *Salmonella* culture positive at the herd level, relative to farrow-to-finish farms. Rostagno, Hurd and McKean, 2009 reported that those pigs marketed in later 'pulls' from a finisher pen had significantly greater *Salmonella* faecal shedding, and also meat juice 'positives', relative to the first group to be pulled. Hill *et al.* (2007) reported a stochastic model for farm *Salmonella* infection which predicted that average prevalence of meat-juice positive pigs was five percent greater on continuous farms relative to units practising all-in-all-out pig flow, although over time the variation between individual farms was much greater than that between the two management groups. Lurette *et al.* (2011b) describe a stochastic model of a farrow to finish herd with two components: animal dynamics, and *Salmonella* transmission. The authors conclude that all-in all-out pig movement had little impact on reducing *Salmonella* prevalence compared with reducing the mean slaughter age and improving the efficacy of room decontamination. Berriman *et al.* (2013b) described stochastic models investigating *Salmonella* spread on-farm for both slotted and non-slotted floors. The authors report that rate of shedding was an important parameter in predicting control of pathogen spread.

SC1.5 FEED AND WATER ACIDIFICATION

O'Connor *et al.* (2008), using systematic review methodology, reported no strong evidence for an association between presence of *Salmonella* and feed acidification. In contrast, Wilhelm *et al.* (2012) used systematic review-cum-meta-analysis methodology, and reported a protective, but heterogeneous, treatment effect for the inclusion of organic acids in finisher rations for reduction of *Salmonella,* with low confidence. The reviews met 7/10 (O'Connor *et al.*, 2008) and 9/11 (Wilhelm *et al.*, 2012) AMSTAR reporting criteria.

Forty primary research studies captured in this review investigated inclusion of organic acids in feed: 14 controlled trials, one quasi-experiment, 16 challenge trials, and nine cross-sectional studies, as well as five risk assessments or stochastic models.

Three controlled trials (Arguello *et al.*, 2013a; Cook *et al.*, 2006; Willamil *et al.*, 2011) assayed individual finisher faecal prevalence, with inconsistent treatment effect reported (median OR = 0.08–range (0.01–26)) (Table SC1.3). Inconsistent results may have reflected the varied effects of post-weaning multi-systemic wasting syndrome (PWMS) across several of the treatment groups. When the trials potentially influenced by PWMS were removed from the dataset, the treatment effect showed a significantly greater magnitude, with reduced heterogeneity (OR = 0.30, 95% CI (0.18–0.49)).

Dos Santos *et al.* (2007) report a trial investigating feeding of non-protected or lipid micro-encapsulated lactic and formic acid to finishing pigs. The authors reported a significant reduction in *Salmonella* seroprevalence in the non-protected acid treatment group, and also a non-significant change in prevalence of faecal shedding during transport and lairage (in comparison with control and protected acid treatment groups, which showed significantly increased faecal shedding during this period). Jorgensen *et al.* (2001) reported a non-significant ($P = 0.07$) association between feeding 2.8% lactic acid and *Salmonella* pen faecal prevalence in weaners on high-risk farms. Similarly, Kristensen, Jorgensen and Boes (2005) reported a non-significant association between feeding 1.0% benzoic acid or 0.5% formic acid plus 0.5% lactic acid to weaner pigs, and faecal *Salmonella* prevalence.

Six controlled trials investigated the feeding of organic acids and measured finisher serum ELISA (Arguello *et al.*, 2013a; Creus *et al.*, 2007; Cook *et al.*, 2006, van der Wolfe *et al.*, 2001b; Wingstrand *et al.*, 1997; Willamil *et al.*, 2011), with 15 of 19 trials reporting a protective effect, with low confidence in the results, partly reflecting the inconsistent findings across trials (Table SC1.3). Again, inconsistent results may have in part reflected the varied effects of PWMS across some treatment groups. However, when the trials potentially influenced by PWMS were removed from this dataset, the treatment effect showed a non-significant change in magnitude, and the dataset still had significant heterogeneity of treatment effect across trials. Van der Heijden *et al.* (2006) reported a field trial investigating feeding of either 0.85%

Ramf™ in-feed, or 0.2% Selko-pH™ in water, each to 10 infected herds over a one year study period. The authors report a significant decline in mean herd ELISA S/P ratio during the study period.

Three controlled trials (Arguello *et al.*, 2013a; Creus *et al.*, 2007; De Busser *et al.*, 2009) measured the effectiveness of organic acids to reduce *Salmonella* prevalence in carcass lymph nodes, with seven of 13 trials indicating a significant protective treatment effect for a pooled summary OR of 0.36 (95% CI: 0.19–0.67)

Summary-of-findings tables for feed and water acidification

TABLE SC1.3. Controlled trials investigating feed and water acidification, sampling finisher pigs

Intervention	Population (Sample)	No. trials/ studies	Odds ratio for intervention effect(1)	% Samples *Salmonella* positive in study population(2)		GRADE rating(3)	References
				No treatment	With treatment		
Acidification	Finishers (Faecal culture)	9/3	Median = 0.08 (Range: 0.01-26.0)	20.1%	2.0% (Range: 0.3-86.7)	Low	Arguello *et al.*, 2013a; Cook *et al.*, 2006; Willamil *et al.*, 2011
Acidification	Finishers (Serum ELISA)	19/6	Median = 0.46 (Range: 0.01-6.82)	39.0%	22.7% (95% CI: 0.6-81.3)	Low	Arguello *et al.*, 2013a; Creus *et al.*, 2007; Cook *et al.*, 2006; Van der Wolf *et al.*, 2001b; Willamil *et al.*, 2011; Wingstrand *et al*, 1997;

NOTES: MA = meta-analysis average estimate from random-effects model; CI = confidence interval.

(1) For interventions with multiple trials, refers to the average meta-analysis estimate and 95% CI from random-effects meta-analysis if heterogeneity was not significant (I^2<60%). If heterogeneity was significant, the median and range of odds ratios from individual studies are presented instead. If only one trial was available, refers to the odds ratio and 95% confidence interval from the individual study.

(2) The percentage of samples positive for *Salmonella* in the comparison group (assumed control risk, ACR) refers to the median prevalence in the comparison group trials. The percentage in the intervention/treatment group was calculated using the following formula for the odds ratio (OR) measure of effect: (OR × ACR / 1 - ACR + [OR × ACR]), and provides an indication of the effect of the intervention on changing a given baseline level of *Salmonella* prevalence as reported in the included studies. The odds ratio used in this formula corresponds to the odds ratio presented in the intervention effect column as described under (1) above. If heterogeneity was not significant (I^2<60%), the estimate is presented with a 95% CI in brackets. If heterogeneity was significant, the median estimate is presented with the range of individual study effects in brackets to illustrate the variability in results across studies.

(3) GRADE rating descriptions: very low = the true effect is likely to be substantially different from the measured estimate; low = the true effect may be substantially different from the measured estimate; moderate = the true effect is likely to be close to the measured estimate, but there is a possibility that it is substantially different; high = there is strong confidence that the true effect lies close to that of the measured estimate.

(Table SC1.4). Three studies (Creus *et al.*, 2007; De Busser *et al.*, 2009; Visscher *et al.*, 2009) reported a non-significant summary estimate of effect on rectal content (Table SC1.4). Correge *et al.* (2012) reported a significant protective treatment effect on *Salmonella* excretion after administering a mixture of organic acids, sodium salts, and bioflavonoids to sows for four weeks in water, using a quasi-experimental design.

Van der Heijden *et al.* (2006) reported a field trial investigating feeding of either 0.85% Ramf™ in-feed, or 0.2% Selko-pH™ in water, to 10 infected herds, each, over a one-year study period. Authors reported a significant decline in mean herd ELISA S/P ratio. Visscher *et al.* (2009) described a trial in which pigs received either fine (control) or coarse (treatment) textured feed, as well as organic acids;

TABLE SC1.4. Controlled trials investigating feed and water acidification, sampling carcasses

Intervention	Population (Sample)	No. trials/ studies	Odds ratio for intervention effect(1)	% Samples *Salmonella* positive in study population(2)		GRADE rating(3)	References
				No treatment	With treatment		
Acidification	Carcasses (Lymph nodes)	13/3	MA = 0.36 (95% CI: 0.19–0.67)	22.6%	9.5% (95% CI: 5.3–16.3)	Moderate	Arguello *et al.*, 2013a; Creus *et al.*, 2007, De Busser *et al.*, 2009
Acidification	Carcasses (Rectal content)	12/3	MA = 0.56 (95% CI: 0.20–1.53)	10.0%	5.9% (95% CI: 1.8–14.5)	Low	Creus *et al.*, 2007; De Busser *et al.*, 2009; Visscher *et al.*, 2009

NOTES: MA = meta-analysis average estimate from random-effects model; CI = confidence interval.

(1) For interventions with multiple trials, refers to the average meta-analysis estimate and 95% CI from random-effects meta-analysis if heterogeneity was not significant (I^2<60%). If heterogeneity was significant, the median and range of odds ratios from individual studies are presented instead. If only one trial was available, refers to the odds ratio and 95% confidence interval from the individual study.

(2) The percentage of samples positive for *Salmonella* in the comparison group (assumed control risk, ACR) refers to the median prevalence in the comparison group trials. The percentage in the intervention/treatment group was calculated using the following formula for the odds ratio (OR) measure of effect: (OR × ACR / 1-ACR + [OR × ACR]), and provides an indication of the effect of the intervention on changing a given baseline level of *Salmonella* prevalence as reported in the included studies. The odds ratio used in this formula corresponds to the odds ratio presented in the intervention effect column as described under (1) above. If heterogeneity was not significant (I^2<60%), the estimate is presented with a 95% CI in brackets. If heterogeneity was significant, the median estimate is presented with the range of individual study effects in brackets to illustrate the variability in results across studies.

(3) GRADE rating descriptions: very low = the true effect is likely to be substantially different from the measured estimate; low = the true effect may be substantially different from the measured estimate; moderate = the true effect is likely to be close to the measured estimate, but there is a possibility that it is substantially different; high = there is strong confidence that the true effect lies close to that of the measured estimate.

this investigation is further described in the 'feed management' section. Trawinska *et al.* (2012) described a trial in which pre-partum gilts received 2.5 g/100 kg body weight/day ascorbic acid; the authors report a non-significant treatment effect on faecal shedding of *Salmonella*, relative to controls.

Five challenge trials reported a significant protective treatment effect on *Salmonella* shedding in weaner pigs when administered organic acids (Gebru *et al.*, 2010; Michiels *et al.*, 2012; Sweeney *et al.*, 2011; Tanaka *et al.*, 2010; Taube *et al.*, 2009). In contrast, nine other challenge trials reported a non-significant treatment effect (Ahmed *et al.*, 2014a; Boyen *et al.*, 2008; Calveyra *et al.*, 2012; De Ridder *et al.*, 2013a, b; Howard, Hurd and Gailey, 2003; Letellier *et al.*, 2000; Martin-Pelaez *et al.*, 2010; Walsh *et al.*, 2012). Rajtak *et al.* (2012) studied the survival of *Salmonella* strains spiked in faeces from pigs fed organic acids, and concluded that rates of decline were significantly greater for *Salmonella* isolates in faeces from pigs supplemented with potassium diformate, relative to controls. Van Winsen *et al.* (2002) described a challenge study investigating particularly the lactic acid component of fermented feed; this trial is included with other studies of fermented feed in the 'feed management' section of this report.

Eight cross-sectional studies reported a non-significant association between feeding organic acids and *Salmonella* prevalence, measured by faecal culture or serology, in finishers (Bruhn, Rushton and Reist, 2011; Correia-Gomes *et al.*, 2011; Correge, Hemonic and Gouvars, 2009; EFSA, 2011; Garcia-Feliz *et al.*, 2009; Hotes *et al.*, 2010; Leontides, Grafanakis and Genigeorgis, 2003; Lo Fo Wong *et al.*, 2004), while one (Hotes *et al.*, 2010) reported a significant positive association between feed or water acidification and seropositivity, measuring meat juice ELISA (OR = 1.8, 95% CI (1.3–2.49)). The pH of liquid feed was a variable captured by the cross-sectional survey of van der Wolfe *et al.* (2001a) but was non-significant in the final multivariable model.

EFSA, in outlining a scientific opinion on a quantitative microbial risk assessment (QMRA) of *Salmonella* in slaughter and breeder pigs (EFSA, 2010), quoted the QMRA itself, describing the potential for on-farm interventions such as feed or water acidification to change the pig's *Salmonella* dose-response relationship by half a log (thus achieving significant reduction in human cases):

"This type of effect has rarely been described in the literature and it is debatable whether such an effect could be achieved consistently at a national-herd level."

Similarly, Hill *et al.* (2011b) described risk model outputs from two scenarios for fictional high- and low-prevalence EU member states. Modifying pig resistance by increasing required infectious dose by one log was predicted to significantly reduce incidence of human salmonellosis; however the authors reported no evidence that on-farm interventions such as administration of organic acids were capable of achieving a change in dose-response of this magnitude. Hill *et al.* (2011c) described a QMRA model which predicted that maximum reduction in cases of human salmonellosis could be achieved by implementing on-farm interventions such as feeding organic acids concurrently with processing-level interventions.

Goldbach and Alban (2006) described a stochastic cost-benefit model of the Danish pork production industry, comparing the effects of feeding home-prepared feed, feeding organic acid, hot water decontamination at processing, and sanitary slaughter. The authors concluded that hot water decontamination was the only economically viable option. Similarly, Simons et al. (2009) described a model comparing the cost effectiveness of implementation of selected interventions to reduce *Salmonella* in the UK pork production chain. Use of organic acids on-farm was predicted to be significantly less cost-effective relative to the processing interventions (bunging, carcass washing, carcass washing and drying) investigated.

SC1.6 FEED MANAGEMENT

O'Connor *et al.* (2008) investigated several potential feeding management interventions (feed withdrawal from swine prior to slaughter; heat treatment of feed; pelletized feed vs mash; coarse vs fine grind; and wet vs dry feeds) using systematic review methodology. Strongest evidence was found for a protective association between feeding meal relative to pelleted feed, but a low level of confidence was expressed, given some of the underpinning studies' relatively low evidentiary value or sample size. Wilhelm *et al.* (2012) also used systematic review and meta-analysis methodology to investigate five selected on-farm interventions, including feed management, for reduction of *Salmonella* in pigs. Strongest evidence was again reported for feeding meal relative to pellets (OR = 0.21, 95% CI (0.14–0.31)) with non-significant heterogeneity. Using a modification of the GRADE approach for assessing quality of evidence, the dataset underpinning the intervention 'meal vs pellets' was ranked 'low' confidence, meaning that further research is likely to revise this estimate.

Forty-five primary research studies identified in this review investigated feed management interventions for *Salmonella* reduction in pigs: eight controlled trials, one quasi-experiment, five challenge trials, two cohort studies, one case-control study, and 28 cross-sectional studies, as well as four stochastic models. Additionally, five controlled trials and three challenge trials were found investigating various feed additives.

The effect of feeding meal relative to pellets was investigated in four controlled trials sampling finishers and measuring serology (Hansen *et al.*, 2001; Jorgensen, Dahl and Wingstrand, 1999; Jorgensen *et al.*, 2003; Kjeldsen and Dahl, 1999), in one controlled trial measuring rectal and faecal culture of slaughter pigs (Lebel *et al.*, 2013), and in one controlled trial sampling sows and measuring faecal culture (Kjaersgaard *et al.*, 2002). The median OR in finishers, measuring serology, was 0.30 (range 0.19–1.79), with a protective treatment effect reported in four of five trials (Table SC1.5). The farm that was the subject of the fourth trial became classified as 'high risk' during the study period. Removing this from the dataset yielded a pooled summary OR estimate of 0.52 (95% CI: 0.35–0.77) with low heterogeneity. This observation highlights the potential for variation in treatment effect both within our dataset, and in future application of the interventions, resulting from the presence or absence of known and unknown covariates. Additionally, all four finisher investigations noted reduced average daily gain (ADG) in the treated groups.

Kjaersgaard *et al.* (2002), sampling sows in a controlled trial, reported a significantly greater individual-level faecal *Salmonella* prevalence in the treatment group receiving meal (authors state the meal was not heat-treated and may have been a source of infection). In contrast, Lebel *et al.* (2013) reported a significant reduction in *Salmonella* prevalence, measured via culture of rectal and caecal content of slaughter pigs naturally challenged with *Salmonella* post-weaning, associated with feeding meal as opposed to pelleted feed. Rajtak *et al.* (2012), describing a survival study in which *Salmonella* strains were spiked in faeces of pigs fed meal or pellets, reported a reduction ($P < 0.10$) in the rate of decline in the faeces of pigs fed pellets, relative to meal. A protective effect of feeding meal, relative to pellets, was also reported in eleven cross-sectional studies (Dahl, 2009; Davies *et al.*, 1997; EFSA, 2011; Farzan *et al.*, 2010; Garcia-Feliz *et al.*, 2009; Hautekiet *et al.*, 2008; Hotes *et al.*, 2010; Leontides, Grafanakis and Genigeorgis, 2003; Vico and Mainar-Jaime, 2012; Wilkins *et al.*, 2010; Lo Fo Wong *et al.*, 2004) measuring both faecal and serological *Salmonella* prevalence in finishers. In contrast, Gotter *et al.* (2012) reported a non-significant association between *Salmonella* feeding pellets and presence of *Salmonella* on German pig farms in a case-control study.

The effect of feed texture (fine vs coarse) was investigated sampling finishers (Kjeldsen and Dahl, 1999) and slaughter pigs (Visscher *et al.*, 2009) assaying serum and meat juice ELISA, respectively. Feeding coarsely ground feed was associated with a significantly protective treatment effect across three trials (Table SC1.5), with moderate confidence. In contrast, Taube *et al.* (2009) reported a challenge trial conducted in weaned pigs with a non-significant treatment effect reported for coarse diet relative to the control diet. However, authors note that the study diets did not differ as much as intended in particle size.

In a controlled trial investigating the effect of withholding feed for 0, 12 or 24 hours prior to slaughter, Morrow *et al.* (2002) report a non-significant treatment effect, with *Salmonella* prevalence in caecal contents as the outcome measure. Rostagno, Eicher and Lay (2009) reported a challenge trial investigating the effects of 12-hour feed withdrawal, two hours of transportation, or both, on *Salmonella* prevalence in slaughter pigs measuring ileal, caecal, and rectal contents and mesenteric lymph nodes. No significant treatment effect was reported using prevalence as the outcome measure. However, feed withhold, with or without transport, was significantly associated with higher *Salmonella* concentrations in ileal contents.

Feeding of wet and/or fermented feed was investigated in eight cross-sectional studies (Bahnson *et al.*, 2006; Correge, Hemonic and Gouvars, 2009; Dahl, 2009; Farzan *et al.*, 2006; Hotes *et al.*, 2010; Stojanac, Stevancevic and Stancic, 2013.; Twomey *et al.*, 2010a; van der Wolf *et al.*, 2001a), which reported a significant protective treatment effect. In contrast, Van Winsen *et al.* (2002) described a seeder challenge trial reporting no significant treatment effect for feeding of fermented ration on the prevalence of *Salmonella* shedding. Fravalo *et al.* (2003b) reported a cohort study of 119 French farrow-to-finish herds, with a non-significant association between dry feeding and *Salmonella* prevalence, relative to wet feeding, measuring faecal samples in finisher pigs. A non-significant association between wet feeding and *Salmonella* prevalence was reported by three cross-sectional studies (Cardinale *et al.*, 2010; Smith *et al.*, 2010b; Twomey *et al.*, 2010b) measuring meat juice ELISA or faecal samples.

The use of home-mixed as opposed to purchased feed was reported associated with significantly reduced odds of *Salmonella* in cross-sectional studies measuring faecal culture (Correia-Gomes *et al.*, 2012; 2013), meat juice ELISA (Smith *et al.*, 2010a), or lymph node culture (Tenhagen *et al.*, 2009). Pieper *et al.* (2012) described a study investigating the effect of feeding hull-less, relative to common barley, to weaner pigs and reported significantly reduced rectal shedding six days after *Salmonella* Typhimurium challenge, but non-significant differences were noted when measuring mesenteric lymph nodes, or ileal, caecal, or colonic content, in treatment relative to control groups. Similarly, Smith, Clough, and Cook (2010) reported that the percentage of barley in the finisher diet was significantly associated with reduced *Salmonella* meat juice ELISA S/P ratio in slaughter pigs. Funk, Davies and Gebreyes (2001), in a cohort study, reported that finishers in cohorts with relatively greater *Salmonella* prevalence had significantly less efficient feed conversion (FC) relative to the lower prevalence (<18.75%) cohorts (OR = 13.48, 95% CI (1.22–149.46)).

SC1.6.1 Other feed additives

Four studies investigated addition of chlorate to feed or water for reduction of *Salmonella* prevalence or load: one controlled trial (Patchanee, Crenshaw and Bahnson, 2007) and three challenge trials (Anderson *et al.*, 2004; 2001; Callaway *et al.*, 2003), of which two (Anderson *et al.*, 2004; Callaway *et al.*, 2003) sampled market pigs.

Patchanee, Crenshaw and Bahnson (2007) administered oral chlorate daily for 5 days, to weaner pigs originating from sows known to be shedding *Salmonella*. The authors reported a significant reduction in load of faecal *Salmonella* shedding at five days post-weaning, and reduced caecal content load, as well as prevalence of lymph

Summary-of-findings tables for feed management interventions

TABLE SC1.5. Controlled trials investigating feed texture

Intervention	Population (Sample)	No. trials/ studies	Odds ratio for intervention effect[(1)]	% Samples *Salmonella* positive in study population[(2)]		GRADE rating[(3)]	References
				No treatment	With treatment		
Meal vs pellets	Finishers (Serum ELISA)	5/4	Median = 0.30 (Range: 0.19-1.79)	32.8%	22.7% (Range: 12.4-46.6)	Low	Hansen *et al.*, 2005; Jorgensen, Dahl and Wingstrand, 1999; Jorgensen *et al.*, 2003; Kjeldsen and Dahl,1999
Coarse vs fine grind	Finishers Carcasses Serum / meat juice (ELISA)	3/2	Median = 0.17 (Range: 0.12-0.65)	6.5%	1.2% (Range: 0.8-4.3)	Moderate	*Kjeldsen and Dahl, 1999; Visscher et al., 2009*

NOTES: CI = confidence interval.

(1) For interventions with multiple trials, refers to the average meta-analysis estimate and 95% CI from random-effects meta-analysis if heterogeneity was not significant (I^2<60%). If heterogeneity was significant, the median and range of odds ratios from individual studies are presented instead. If only one trial was available, refers to the odds ratio and 95% confidence interval from the individual study.

(2) The percentage of samples positive for *Salmonella* in the comparison group (assumed control risk, ACR) refers to the median prevalence in the comparison group trials. The percentage in the intervention/treatment group was calculated using the following formula for the odds ratio (OR) measure of effect: (OR × ACR / 1 -ACR + [OR × ACR]), and provides an indication of the effect of the intervention on changing a given baseline level of *Salmonella* prevalence as reported in the included studies. The odds ratio used in this formula corresponds to the odds ratio presented in the intervention effect column as described under (1) above. If heterogeneity was not significant (I^2<60%), the estimate is presented with a 95% CI in brackets. If heterogeneity was significant, the median estimate is presented with the range of individual study effects in brackets to illustrate the variability in results across studies.

(3) GRADE rating descriptions: very low = the true effect is likely to be substantially different from the measured estimate; low = the true effect may be substantially different from the measured estimate; moderate = the true effect is likely to be close to the measured estimate, but there is a possibility that it is substantially different; high = there is strong confidence that the true effect lies close to that of the measured estimate.

node contamination, at 14 days post-weaning. Anderson *et al.* (2001) reported a significant ($P < 0.05$) treatment effect on caecal *Salmonella* load in weaner pigs after administering 100 millimoles of sodium chlorate by gavage, at 8 and 16 hrs after *Salmonella* Typhimurium challenge, and also reported a significant treatment by time-after-treatment interaction. Anderson *et al.* (2004) reported significantly reduced caecal *Salmonella* load in weaner pigs after administration of drinking water containing either 30–40, or 60–80 mg/kg body weight chlorate, using experimental chlorate preparation (ECP) in drinking water, for 24 or 36 hours, starting 24 hours after challenge. The authors also reported significantly reduced prevalence of *Salmonella* in caecal and rectal contents relative to control animals, with a non-significant effect on prevalence of lymph node contamination.

In the same study, the authors reported a significant treatment effect of chlorate on caecal *Salmonella* load in market pigs, with a non-significant treatment effect on rectal faecal load, or prevalence measuring caecal or rectal content, or lymph node contamination. In contrast, Callaway *et al.* (2003) reported a non-significant reduction, after four hours of exposure to feed containing ECP, on *Salmonella* prevalence in the tonsils of *Salmonella* Typhimurium challenged market pigs. The authors also reported no difference in lymph node or intestinal content *Salmonella* status between treatment and control groups. However, in a quasi-experiment using 10 naturally *Salmonell*a infected pigs, the same authors reported that ECP significantly reduced caecal *Salmonella* colonization.

The use of several other feed additives was also reported. Two controlled trials (Chu *et al.*, 2013a, b) reported a significantly protective treatment effect when 0.3% bamboo charcoal was added to the ration of finisher pigs for reduction of faecal *Salmonella* load; ADG and FC were significantly improved in all treatment groups. Jung *et al.* (2011) reported feeding a mixture of *Coptidis rhizoma*, *Lonicerae flos* and *Paeonia japonica* (1:1:1, v/v/v) methanol extracts and dioctahedral smectite (0.5–1.0%) mixed with feed, to five-week-old weaned pigs. The authors reported a significantly reduced faecal *Salmonella* load from days two to seven (termination) of the trial. Tallarico *et al.* (2013) investigated feeding encapsulated formic and citric acid plus essential oils (FORMA XOLTM) added to diet at 4 kg/ton for the first 4 weeks, then 1 kg/ton for the remainder of the trial, to finisher pigs. The authors reported a significantly protective treatment effect on carcass lymph node *Salmonella* prevalence, with a non-significant effect on seroprevalence. Edrington *et al.* (2006), describing a challenge trial, reported a significant treatment effect of feeding ractopamine (Paylean 9™ at 18 g/ton) on faecal *Salmonella* shedding and prevalence of liver contamination with the challenge strain, in grower pigs.

Five risk analyses investigated potential feeding management strategies. EFSA's scientific opinion on a QMRA of *Salmonella* in slaughter and breeder pigs concluded that contaminated feed was an important potential source of *Salmonella* infection, particularly in situations of relatively lower prevalence in the breeding herd; however, its complete elimination from feed might not be practical. Hill *et al.* (2011a) reported a stochastic model for *Salmonella* transmission on pig farms. The model predicted that in situations of lower prevalence in the breeding herd, introduction of *Salmonella* on-farm via contaminated feed became a relatively more important transmission pathway. In such low prevalence situations, strategies to mitigate transmission via contaminated feed could be more effective than in the baseline scenario of higher prevalence in the breeding herd. Similar to the observations described for feeding of organic acids, Hill *et al.* (2011b) described a *Salmonella* transmission model and concluded that modifying the pig *Salmonella* dose-response curve by the 1 to 2 logs required to significantly reduce human salmonellosis would be very difficult to achieve nationally via feed management strategies such as wet feeding.

Goldbach and Alban (2006) described a stochastic cost-benefit model of the Danish pork production industry, concluding that of the options investigated, including use of home-prepared feeds, only hot-water carcass decontamination was an economically viable option for *Salmonella* reduction.

SC1.7 MANIPULATION OF GUT MICROBIOTA

Use of prebiotics, probiotics, or synbiotics for reduction of *Salmonella* in pigs was investigated in 31 primary research studies: nine controlled trials, 21 challenge trials, and three cross-sectional studies.

Suryanarayana, Sreedhar and Babu (2013) sampled pigs into the finisher stage and reported a significant reduction in mean faecal *Salmonella* load in finishers receiving both the prebiotic (fructo-oligosaccharides) and probiotic (*Saccharomyces cerevisiae*) treatments relative to un-supplemented controls. Improved FC and ADG were also reported for all three (prebiotic/probiotic/both) treatment groups. Vico, Mainr-Jaime and Aussere (2011) investigated feeding a galactomannan (Salmosan™, fed at 0.5 kg or 3.0 kg/ton of feed) during the entire fattening period. The investigators reported no significant treatment effect on seroprevalence, faecal shedding, or mesenteric lymph node prevalence at the lower dosage, but a significant treatment effect on all three measures at the higher (3.0 kg/ton) dosage. Similarly, Mainar-Jaime *et al.* (2013) reported a controlled trial in which weaner pigs were fed a galactomannan (Salmosan™, fed at 0.5 kg, 2.0 kg, or 3.0 kg/ton of feed). The investigators reported a significant treatment effect for the two groups receiving the higher dosages, using both mesenteric lymph node and faecal culture prevalence at slaughter as the outcome measure. In contrast, LeJeune *et al.* (2006)

investigated pigs receiving direct-fed microbials (*Lactobacillus acidophilus, Streptococcus faecalis, Bifidobacterium thermophilum, Bifidobacterium pseudolongium*) for six weeks post-weaning, compared with pigs the same age fed sub-therapeutic antimicrobials, or basal diet, and reported no significant treatment effect on *Salmonella* prevalence, with improved ADG in the antimicrobial-fed group relative to the probiotic group.

Application of the treatment in five controlled trials was either during the suckling stage, or at weaning, with sampling shortly thereafter. Zhang *et al.* (2014) described an investigation of feeding multi-enzymes derived from *Bacillus subtilis, Bacillus amyloliquefaciens* and *Trichoderma* spp. to pigs 35-65 days of age, reporting significantly reduced *Salmonella* concentration in faecal shedding, with a significant dose-response relationship demonstrated to increasing doses of multi-enzymes (ranging from 100 to 350 mg/kg of diet).

Maneewan *et al.* (2011) reported a significantly reduced *Salmonella* load in 11-day-old piglets fed *B. subtilis* MP9 (10^{11} CFU/ml.) and *B. subtilis* MP10 (10^{11} CFU/ml) relative to controls. The effect of feeding dietary yeast to young pigs was investigated in two controlled trials. Upadrasta *et al.* (2013) reported a 'significantly reduced *Salmonella* level' in pigs treated from four weeks to seven weeks of age with spent cider yeast, relative to controls. In contrast, Weedman *et al.* (2011) reported a non-significant treatment effect on *Salmonella* when 3-day-old pigs were fed either milk or milk supplemented with yeast culture (0.1 g of yeast culture product per kg of body weight) from days 4 to 21. Wells *et al.* (2005) investigated feeding non-fat dried skim milk powder, which was roughly 50% lactose, as 10% of grower/grower-finisher/finisher diets in a controlled trial, reporting a non-significant treatment effect with faecal culture as the outcome measure.

Of the 19 challenge trials identified, 15 sampled animals 65 days of age or younger. However, Calveyra *et al.* (2012) investigated pigs treated with a probiotic starting at 43 days old, challenged with *Salmonella* Typhimurium, and followed for an additional 28 days. The authors reported no significant treatment effect on faecal prevalence, but noted a trend towards reduced load of shedding. Gebru *et al.* (2010) investigated grower pigs (mean body weight 38 ±6.7 kg) fed six different treatments, including a probiotic (*Lactobacillus plantarum* CJLP56) for two weeks prior and two weeks after oral challenge with *Salmonella* Typhimurium. At day 14 post-challenge, shedding scores in the treatment groups were significantly reduced relative to the control group, with probiotic-treated pigs having significantly better ADG relative to control pigs over the four-week trial. Spiehs, Shurson and Johnston (2008) investigated the use of a probiotic (either *Bacillus* licheniformis and B. subtilis or *Enterococcus faecium*) in finishers (110 kg at the start of the trial). The authors reported no significant difference in faecal shedding across the treatment and control groups.

Four challenge trials investigated *E. faecium*: three (Kreuzer *et al.*, 2012; Spiehs, Shurson and Johnston, 2008; Walsh *et al.*, 2012) reported a non-significant treatment effect on *Salmonella* prevalence or load, while Szabo *et al.* (2009) reported significantly greater faecal shedding and organ colonization in pigs treated with *E. faecium* relative to controls. Studies investigating the feeding of *Lactobacillus* spp. also reported a range of findings: three reported a significant protective treatment effect (Ahmed *et al.*, 2014b; Casey *et al.*, 2007; Naqid *et al.*, 2015) and three a non-significant treatment effect (Afonso *et al.*, 2013; LeJeune *et al.*, 2006; Letellier *et al.*, 2000).

Nisbet *et al.* (1999) described feeding a competitive exclusion culture to pigs at birth and weaning, reducing *Salmonella* Typhimurium shedding post-challenge. Fedorka-Cray *et al.* (1999) reported significantly reduced *Salmonella* load in caecal contents of treated pigs. Genovese *et al.* (2003) similarly reported significant treatment effects of competitive exclusion cultures on caecal colonization and faecal shedding in suckling and weaned pigs. In contrast, Anderson *et al.* (1999) reported widespread *Salmonella* shedding in both treatment and control groups receiving porcine-derived competitive exclusion culture; no statistical analysis was presented due to potential confounding of treatment effects by litter effects. Bergeron *et al.* (2013) described a trial in which weaned pigs were fed *Pediococcus acidilactici* as well as other 'functional foods' including cranberry and challenged at 49 days of age with *Salmonella* Typhimurium; the authors did not report a significant treatment effect of diet on *Salmonella* faecal shedding. Mroz and Toride (2002) reported a non-significant treatment effect on *Salmonella* Typhimurium shedding post-challenge when feeding dried bacterial cells to weaned piglets. Sweeney *et al.* (2011) described feeding a commercial mixture of herbs and organic acids to weaner pigs, reporting a significant treatment effect on concentration of *Salmonella* in tonsils at slaughter on day 32 post-challenge.

Three observational studies reported a non-significant association between use of prebiotic or probiotics on-farm and *Salmonella* prevalence (EFSA, 2011; Leontides, Grafanakis and Genigeorgis, 2003; Lo Fo Wong *et al.*, 2004).

SC1.8 VACCINATION

Vaccination of pigs for reduction of *Salmonella* was investigated in three systematic reviews (Denagamage *et al.*, 2007; Wilhelm *et al.*, 2012; Wisener *et al.*, 2014). Denagamage *et al.* (2007) concluded that there was evidence that some vaccines were effective; however, because of the methodological quality of the underpinning evidence, this conclusion might be incorrect. Wilhelm *et al.* (2012) concluded that evidence regarding vaccine effectiveness in finisher pigs was inconsistent. Wisener *et al.* (2014) investigated the use of challenge trials to assess *Salmonella* vaccine effectiveness in pigs, concluding that challenge trials tend to report a more favourable outcome relative to controlled trials. The systematic reviews received reporting reliability scores of 7/10 relevant fields for Denagamage *et al.* (2007), and 9/11 for Wilhelm *et al.* (2012) and Wisener *et al.* (2014) using the AMSTAR quality assessment tool.

Vaccination for mitigation of *Salmonella* was investigated in 29 primary research studies found in this review: 10 controlled trials and 19 challenge trials.

Faecal samples from individual market pigs were assayed in two studies with four trials (Arguello *et al.*, 2013b; De Ridder *et al.*, 2014) investigating a live attenuated

Summary-of-findings tables for vaccination

TABLE SC1.6 Controlled trials investigating vaccination, measuring outcomes in finisher pigs

Intervention	Population/ sample	No. trials/ studies	Odds ratio for intervention effect(1)	% Samples *Salmonella* positive in study population(2)		GRADE rating(3)	References
				No treatment	With treatment		
Vaccine	Individual faecal culture	4/2	Median = 0.22 (Range: 0.02-0.78)	26.7%	7.4% (Range: 0.7-22.1)	Moderate	Arguello *et al.*, 2013b; De Ridder *et al.*, 2014
Vaccine	Pen faecal culture	4/2	Median = 1.41 (Range: 0.17-4.42)	22.5%	29.0% (Range: 5.3-56.2)	Very Low	Arguello *et al.*, 2013b; Farzan and Friendship, 2010

NOTES: MA = meta-analysis average estimate from random-effects model; CI = confidence interval.

(1) For interventions with multiple trials, refers to the average meta-analysis estimate and 95% CI from random-effects meta-analysis if heterogeneity was not significant (I^2<60%). If heterogeneity was significant, the median and range of odds ratios from individual studies are presented instead. If only one trial was available, refers to the odds ratio and 95% confidence interval from the individual study.

(2) The percentage of samples positive for *Salmonella* in the comparison group (assumed control risk, ACR) refers to the median prevalence in the comparison group trials. The percentage in the intervention/treatment group was calculated using the following formula for the odds ratio (OR) measure of effect: (OR × ACR / 1 -ACR + [OR × ACR]), and provides an indication of the effect of the intervention on changing a given baseline level of *Salmonella* prevalence as reported in the included studies. The odds ratio used in this formula corresponds to the odds ratio presented in the intervention effect column as described under (1) above. If heterogeneity was not significant (I^2<60%), the estimate is presented with a 95% CI in brackets. If heterogeneity was significant, the median estimate is presented with the range of individual study effects in brackets to illustrate the variability in results across studies.

(3) GRADE rating descriptions: very low = the true effect is likely to be substantially different from the measured estimate; low = the true effect may be substantially different from the measured estimate; moderate = the true effect is likely to be close to the measured estimate, but there is a possibility that it is substantially different; high = there is strong confidence that the true effect lies close to that of the measured estimate.

and an inactivated *Salmonella* Typhimurium vaccine, respectively, yielding a significant median protective OR of 0.22 (Range 0.02–0.78), with moderate confidence (Table SC1.6). Two controlled trials assayed pooled finisher pen faecal samples, with opposing findings: one reported a significant protective treatment effect for an autogenous *Salmonella* Typhimurium vaccine (Arguello *et al.*, 2013b) and one a non-significant effect for both an autogenous *Salmonella* Typhimurium and a commercial *Salmonella* Choleraesuis vaccine (Farzan and Friendship, 2010), with very low confidence, partly reflecting the inconsistency of findings (Table SC1.6). Farzan and Friendship (2010) noted several potential problems with their study: the control group had the lowest prevalence of *Salmonella* of any group prior to initiation of the study; pigs had probably been exposed to the *Salmonella* on-farm prior to vaccination; and given that three different *Salmonella* serovars had been identified on-farm, perhaps a multivalent autogenous vaccine might have been more appropriate.

TABLE SC1.7 Controlled trials measuring outcomes in pork carcasses

Intervention	Population/ sample	No. trials/ studies	Odds ratio for intervention effect[1]	% Samples *Salmonella* positive in study population[2]		GRADE rating[3]	References
				No treatment	With treatment		
Vaccine	Lymph nodes	6/4	Median = 0.41 (Range: 0.12-0.86)	47.5%	27.1% (Range: 12.2-43.8)	Moderate	Arguello *et al.*, 2013b; De Ridder *et al.*, 2014; Maes *et al.*, 2001; Schwarz *et al.*, 2011
Vaccine	Caecal content	3/2	Median = 0.53 (Range: 0.18-0.89)	46.3%	31.3% (Range: 13.4-43.4)	Moderate	Arguello *et al.*, 2013b; De Ridder *et al.*, 2014

NOTES:

(1) For interventions with multiple trials, refers to the average meta-analysis estimate and 95% CI from random-effects meta-analysis if heterogeneity was not significant (I^2<60%). If heterogeneity was significant, the median and range of odds ratios from individual studies are presented instead. If only one trial was available, refers to the odds ratio and 95% confidence interval from the individual study.

(2) The percentage of samples positive for *Salmonella* in the comparison group (assumed control risk, ACR) refers to the median prevalence in the comparison group trials. The percentage in the intervention/treatment group was calculated using the following formula for the odds ratio (OR) measure of effect: (OR × ACR / 1 - ACR + [OR × ACR]), and provides an indication of the effect of the intervention on changing a given baseline level of *Salmonella* prevalence as reported in the included studies. The odds ratio used in this formula corresponds to the odds ratio presented in the intervention effect column as described under (1) above. If heterogeneity was not significant (I^2<60%), the estimate is presented with a 95% CI in brackets. If heterogeneity was significant, the median estimate is presented with the range of individual study effects in brackets to illustrate the variability in results across studies.

(3) GRADE rating descriptions: very low = the true effect is likely to be substantially different from the measured estimate; low = the true effect may be substantially different from the measured estimate; moderate = the true effect is likely to be close to the measured estimate, but there is a possibility that it is substantially different; high = there is strong confidence that the true effect lies close to that of the measured estimate.

Four studies measured *Salmonella* prevalence in carcass lymph nodes (Arguello *et al.*, 2013b; De Ridder *et al.*, 2014; Maes *et al.*, 2001; Schwarz *et al.*, 2011) and two (Arguello *et al.*, 2013b; De Ridder *et al.*, 2014) measured prevalence in caecal contents at slaughter to assess vaccine efficacy. All ten individual trials reported a significant protective treatment effect, although meta-analysis indicated a high degree of heterogeneity in the treatment effect across studies in both datasets (Table SC1.7), with moderate confidence.

Liu *et al.* (2011) reported a trial in which 2-month-old pigs were immunized with an autogenous *Salmonella* Typhimurium vaccine. Vaccinated pigs had significantly greater mean optical densities of *Salmonella* immunoglobulins relative to control pigs, and consistent with other investigators (Arguello *et al.*, 2013b; Farzan and Friendship, 2010), the authors also reported improved ADG in the vaccinated group. Kolb *et al.* (2011) reported that vaccinating suckling pigs orally with a modified live *Salmonella* vaccine significantly reduced peritoneal swab and ileo-caecal lymph node prevalence. Marier *et al.* (2009) administered an attenuated live *Salmonella* Typhimurium vaccine to sows on a farm with a history of *Salmonella* in finishers. The authors reported that there was no significant treatment effect on faecal shedding in sows or piglets, or *Salmonella* spp. prevalence in ileo-caecal lymph nodes or meat juice ELISA of slaughter pigs. However, there was significantly reduced *Salmonella* spp. prevalence in caecal content of slaughter pigs. Additionally, the authors reported that the dominant serovar on-farm was *Salmonella* Derby, and speculated that the findings reflected the limited cross-protection of the *S.* Typhimurium vaccine.

Roesler *et al.* (2006) reported a controlled trial in which 25 sows were immunized with an autogenous *Salmonella* Typhimurium vaccine and the performance of the pigs in their next litters was compared with that of pigs derived from 37 sows receiving oral enrofloxacin from 14 days pre-farrowing to weaning. *Salmonella* Typhimurium was not detected in the faeces of any of the vaccinated sows' offspring, from birth to slaughter; in contrast, 47.4% of the pigs in the unvaccinated group shed the bacterium.

The challenge trials identified investigated both modified live and killed vaccines, measuring both *Salmonella* culture and serology outcomes in a range of ages (1–20 weeks). All of the studies measuring culture of faeces or organs reported a significant protective effect of vaccination on *Salmonella* (Bearson and Bearson, 2011; De Ridder *et al.*, 2013a, b; Foss *et al.*, 2013; Gradassi *et al.*, 2013; Groninga *et al.*, 2000; Haneda *et al.*, 2011; Hur *et al.*, 2011; Jordan, Kaiser and Cline, 2013; Kolb *et al.*, 2003; Letellier *et al.*, 2000; Leyman *et al.*, 2012; Matiasovic *et al.*, 2013; Neubauer, Roof and Kolb, 2005; Roesler *et al.*, 2004, 2011; Ruggeri *et al.*, 2014; Springer *et al.*, 2001). The difference in findings across challenge trials relative to controlled trials in this dataset highlights the potential importance of some contextual predictors of effect (such as baseline prevalence, co-infection with other pathogens, and the effect of herd immunity) on controlled trial results, and supports the observations of Wisener *et al.* (2014).

As mentioned previously, Hill *et al.* (2011b) described a *Salmonella* transmission model and concluded that modifying the pig dose-response curve by the 1 to 2 logs required to significantly reduce human salmonellosis would be very difficult to achieve nationally via on-farm strategies such as vaccination.

Hotes, Traulsen and Kriete (2011) described a stochastic model of non-clinical *Salmonella* transmission within a pork supply chain, reporting that vaccination of sows and piglets was an effective control measure to decrease slaughter pig *Salmonella* prevalence, if a large enough proportion of farms adopted vaccination. Soumpasis, Alban and Butler (2012) used stochastic modelling to investigate potential effects of vaccination on *Salmonella* prevalence in a pig farm. The authors concluded that vaccination, even with a vaccine of hypothetically perfect effectiveness, was not sufficient to eliminate *Salmonella* infection on a high-risk farm.

SC1.9 MULTIPLE STRATEGIES CONCURRENTLY OR COMPARATIVELY

Bode, Baier and Blaha (2007) applied a suite of practices (intensified cleaning and disinfection practices; improved rodent control; switched to chlorinated drinking water; added Formi [0.6–1.2% potassium formate] to finisher ration; switching to roughest possible feed grind; increased barley to 35% of ration) to three German category three (high risk) farms using a quasi-experimental (i.e. 'before-and-after') study design. Individual finisher faecal *Salmonella* prevalence was reduced in two of three farms.

Alban and Stärk (2005) described a stochastic model of the Danish pig production system, with the individual pig as the unit of interest. The variables with maximum effect on the *Salmonella* prevalence on the final carcass were: (1) number of herds with a high prevalence of *Salmonella*; (2) singeing efficiency; (3) contamination and cross-contamination at degutting; and (4) cross-contamination during handling. The largest reduction was observed when several factors were improved concurrently.

Bollaerts *et al.* (2010) described a stochastic risk model for *Salmonella* prevalence in Belgian minced pork, from primary production to preparation and consumption. Reducing seroprevalence of pigs at primary production was less effective in reducing human risk relative to strategies applied at processing.

The risk assessment reported by EFSA (2010) on *Salmonella* in slaughter or breeder pigs considered the potential for application of multiple interventions pre-slaughter and concluded that:

> *'farm and transport interventions are likely to vary in their ability to change slaughter pig prevalence by a sufficient amount to change numbers of (human) salmonellosis cases. However, a combination of farm interventions applied across a large proportion of farms is likely to have a cumulative effect in reducing slaughter pig prevalence'.*

SC1.10 MISCELLANEOUS PREDICTORS

Observational studies have identified several factors that may not be directly amenable to an intervention but are important to understand as they influence other intervention strategies' effectiveness. For example, the importance of sow shedding on nursery and finisher pig prevalence has been identified by a large baseline study of *Salmonella* prevalence in European herds (EFSA, 2011) and also in cross-sectional studies both in Europe (Correge and Hemonic, 2012) and North America (Wilkins *et al.*, 2010). Similarly, Hill *et al.* (2011a) described a farm-level stochastic model of *Salmonella* transmission in pigs for individual EU member states, in which the most influential predictor was sow *Salmonella* prevalence. Larger farms have been reported to have greater odds of *Salmonella* in finishers, measuring serology or culture, in five cross-sectional studies (Bruhn, Rushton and Reist, 2011; Correia-Gomes *et al.*, 2012, 2013; Garcia-Feliz *et al.*, 2009; Van der Wolf *et al.*, 2001a). Ikwap *et al.* (2014), in a survey of Ugandan pig farms, reported that an 'intensive' method of pig farming was associated with a significantly reduced odds of positive *Salmonella* culture in weaner pigs, relative to 'tethering and roaming' (OR=0.11; 95% CI 0.02–0.64), and 'intensive' methods also were associated with significantly reduced odds relative to 'semi-intensive' methods (OR=0.12; 95% CI (0.01–0.96)).

SC1.11 References cited in Summary Card 1

Afonso, E.R., Parazzi, L.J., Marino, C.T., Martins, S.M.M.K., Araujo, L.F., Araujo, C.S.D., Vilela, F.G. & Moretti, A.D. 2013. Probiotics association in the suckling and nursery in piglets challenged with *Salmonella* Typhimurium. *Brazilian Archives of Biology and Technology,* 56(2): 249–258.

Ahmed, S.T., Hwang, J.A., Hoon, J., Mun, H.S. & Yang, C.J. 2014a. Comparison of single and blend acidifiers as alternative to antibiotics on growth performance, faecal microflora, and humoral immunity in weaned piglets. *Asian-Australasian Journal of Animal Sciences*, 27(1): 93–100.

Ahmed, S.T., Hoon, J., Mun, H.S. & Yang, C.J. 2014b. Evaluation of *Lactobacillus* and Bacillus-based probiotics as alternatives to antibiotics in enteric microbial challenged weaned piglets. *African Journal of Microbiology Research,* 8(1): 96–104.

Alban, L. & Stärk, K.D.C. 2005. Where should the effort be put to reduce the *Salmonella* prevalence in the slaughtered swine carcass effectively? *Preventive Veterinary Medicine,* 68(1): 63–79.

Albino, L.A.A., Rostagno, M.H., Hungaro, H.M. & Mendonca, R.C.S. 2014. Isolation, characterization, and application of bacteriophages for *Salmonella* spp. biocontrol in pigs. *Foodborne Pathogens and Disease,* 11(8): 602–609.

Anderson, R.C., Stanker, L.H., Young, C.R., Buckley, S.A., Genovese, K.J., Harvey, R.B., DeLoach, J.R., Kelth, N.K. & Nisbet, D.J. 1999. Effect of competitive exclusion treatment on colonization of early-weaned pigs by *Salmonella* serovar Choleraesuis. *Swine Health and Production,* 7(4): 155–160.

Anderson, R.C., Buckley, S.A., Callaway, T.R., Genovese, K.J., Kubena, L.F., Harvey, R.B. & Nisbet, D.J. 2001. Effect of sodium chlorate on *Salmonella* Typhimurium concentrations in the weaned pig gut. *Journal of Food Protection,* 64(2): 255–258.

Anderson, R.C., Hume, M.E., Genovese, K.J., Callaway,T.R., Jung, Y.S., Edrington,T.S., Poole, T.L., Harvey,R.B., Bischoff, K.M., Nisbet, D.J. 2004. Effect of drinking-water administration of experimental chlorate ion preparations on *Salmonella enterica* serovar Typhimurium colonization in weaned and finished pigs. *Veterinary Research Communications,* 28(3): 179–189

Arguello, H., Carvajal, A., Costillas, S. & Rubio, P. 2013a. Effect of the addition of organic acids in drinking water or feed during part of the finishing period on the prevalence of *Salmonella* in finishing pigs. *Foodborne Pathogens and Disease,* 10(10): 842–849.

Arguello, H., Carvajal, A., Naharro, G. & Rubio, P. 2013b. Evaluation of protection conferred by a *Salmonella* Typhimurium inactivated vaccine in *Salmonella*-infected finishing pig farms. *Comparative Immunology Microbiology and Infectious Diseases,* 36(5): 489–498.

Bahnson, P.B., Troutt, H.F., Weigel, R.M., Miller, G.Y. & Isaacson, R.E. 2007. Risk factors for the detection of *Salmonella* in ileo-colic lymph nodes in United States of America slaughtered pigs. pp. 73–75, *in:* Proceedings of the 7th International Symposium on the Epidemiology and Control of Foodborne Pathogens in Pork. 9–11 May 2007. Verona, Italy.

Bahnson, P.B., Fedorka-Cray, P., Ladely, S.R. & Mateus-Pinilla, N. 2006. Herd-level risk factors for *Salmonella enterica* subsp. *enterica* in U.S. market pigs. *Preventive Veterinary Medicine,* 76(3–4): 249–262.

Baptista, F.M., Alban, L., Nielsen, L.R., Domingos, I., Pomba, C. & Almeida, V. 2010. Use of herd information for predicting *Salmonella* status in pig herds. *Zoonoses and Public Health,* 57(Suppl.1): 49–59.

Bearson, B.L. & Bearson, S.M.D. 2011. Serological response of swine to a *Salmonella enterica* serovar Typhimurium strain that reduces gastrointestinal colonization, faecal shedding and disease due to virulent *Salmonella* Typhimurium. pp. 228–230, *in:* Proceedings of 9th International Conference on the Epidemiology and Control of Biological, Chemical and Physical Hazards in Pigs and Pork. Maastricht, The Netherlands.

Bergeron, N., Lessard, M., Quessy, S. & Letellier, A. 2013. Evaluation of the impact of functional foods on the course of *Salmonella* infection in piglets. pp. 151–153, *in:* Proceedings of the 10th International Conference on the Epidemiology and Control of Biological, Chemical and Physical Hazards in Pigs and Pork. Portland, ME, USA.

Berriman, A.D.C., Clancy, D., Clough, H.E., Armstrong, D. & Christley, R.M. 2013a. Effectiveness of simulated interventions in reducing the estimated prevalence of *Salmonella* in UK pig herds. *PLoS ONE,* 8: Article e66054.

Berriman, A.D.C., Clancy, D., Clough, H.E. & Christley, R.M. 2013b. Semi-stochastic models for *Salmonella* infection within finishing pig units in the UK. *Mathematical Biosciences,* 245(2): 148–156.

Bode, K., Baier, S. & Blaha, T. 2007. Specific cleaning and disinfection procedures for *Salmonella* infected pig herds. pp. 500–506, *in: Animal health, animal welfare and biosecurity.* Proceedings of 13th International Congress in Animal Hygiene. Tartu, Estonia.

Bollaerts, K., Messens, W., Aerts, M., Dewulf, J., Maes, D., Grijspeerdt, K. & Stede, Y. 2010. Evaluation of scenarios for reducing human salmonellosis through household consumption of fresh minced pork meat. *Risk Analysis,* 30: 853–865.

Boyen, F., Haesebrouck, F., Vanparys, A., Volf, J., Mahu, M., Van Immerseel, F., Rychlik, I., Dewulf, J., Ducatelle, R. & Pasmans, F. 2008. Coated fatty acids alter virulence properties of *Salmonella* Typhimurium and decrease intestinal colonization of pigs. *Veterinary Microbiology,* 132: 319–327.

Bruhn, S., Rushton, J. & Reist, M. 2011. Risk factor analysis and possible control measures for *Salmonella* in Swiss breeding pigs within the framework of an imminent need for a national control programme. Paper presented at International Conference on Animal Health Surveillance (ICAHS), Lyon, France, 17–20 May 2011. Published in *Épidémiologie et Santé Animale,* 59/60: 222–224.

Burns, A.M., Lawlor, P.G., Tiwari, U. & Duffy, G. 2013. Assessing the role of feed as a risk factor for *Salmonella* in pig production. pp. 125–127, *in:* Proceedings of the 10th

International Conference on the Epidemiology and Control of Biological, Chemical and Physical Hazards in Pigs and Pork. Portland, ME, USA.

Callaway, T.R., Anderson, R.C., Edrington, T.S., Genovese, K.J., Stahl, C.H., Jung, Y.S., Bischoff, K.M., Poole, T.L., Harvey, R.B. & Nisbet, D.J. 2003. Experimental chlorate product treatment reduces Salmonella populations in swine during lairage. *Journal of Animal Science*, 81(Suppl.1): 529.

Callaway, T.R., Edrington, T.S., Brabban, A., Kutter, B., Karriker, L., Stahl, C., Wagstrom, E., Anderson, R., Poole, T.L., Genovese, K., Krueger, N., Harvey, R.B. & Nisbet, D.J. 2011. Evaluation of phage treatment as a strategy to reduce *Salmonella* populations in growing swine. *Foodborne Pathogens and Disease,* 8: 261–266.

Calveyra, J.C., Nogueira, M.G., Kich, J.D., Biesus, L.L, Vizzotto, R., Berno, L., Coldebella, A., Lopes, L., Mores, N., Lima, G.J.M.M. & Cardoso, M. 2012. Effect of organic acids and mannanoligosaccharide on excretion of *Salmonella* Typhimurium in experimentally infected growing pigs. *Research in Veterinary Science,* 93: 46–47.

Cardinale, E., Maeder, S., Porphyre, V. & Debin, M. 2010. *Salmonella* in fattening pigs in Reunion Island: herd prevalence and risk factors for infection. *Preventive Veterinary Medicine,* 96: 281–285.

Casey, P.G., Gardiner, G.E., Casey, G., Bradshaw, B., Lawlor, P.G., Lynch, P.B., Leonard, F.C., Stanton, C., Ross, R.P., Fitzgerald, G.F. & Hill, C. 2007. A five-strain probiotic combination reduces pathogen shedding and alleviates disease signs in pigs challenged with *Salmonella enterica* serovar Typhimurium. *Applied and Environmental Microbiology*, 73(6): 1858–1863.

Choe, D.W., Hassan, L. & Loh, T.C. 2011. The prevalence of antimicrobial resistant *Salmonella* spp. and the risk factors associated with their occurrence in finisher pigs in Seberang Perai, Malaysia. *Pertanika Journal of Tropical Agricultural Science,* 34: 303–310.

Chu, G.M., Jung, C.K., Kim, H.Y., Yun, H.J., Kim, J.H., Jung, M.S., Lee, S.J., Song, Y.N., Ibrahim, R.I.H., Cho, J.H., Lee, S.S. & Song, Y.M. 2013a. Effects of bamboo charcoal and bamboo vinegar as antibiotic alternatives on growth performance, immune responses and faecal microflora population in fattening pigs. Animal Science Journal, 84:113–120.

Chu, G.M., Kim, J.H., Kim, H.Y., Ha, J.H., Jung, M.S., Song, Y., Cho, J.H., Lee, S.J., Ibrahim, R.I.H., Lee, S.S. & Song, Y.M. 2013b. Effects of bamboo charcoal on the growth performance, blood characteristics and noxious gas emission in fattening pigs. *Journal of Applied Animal Research,* 41: 48–55.

Cocora, Z.M. & Tibru, I. 2014. The role of feeding type in *Salmonella* spp. transmission. *Bulletin of the University of Agricultural Sciences and Veterinary Medicine Cluj-Napoca,* 71: 499–500.

Cook, A.J.C. 2004. Measuring the impact of *Salmonella* control in finishing pigs - lessons from a pilot study. *Pig Journal,* 53: 157–163.

Cook, A.J.C., Miller, A.J., O'Connor, J., Marier, E.A., Williamson, S.M., Hayden, J., Featherstone, C.A., Twomey, D.F. & Davies, R.H. 2006. Investigating *Salmonella* infection on problem pig farms (ZAP Level 3) - case studies in England. *Pig Journal,* 58: 190–203.

Correge, I. & Hemonic, A. 2012. Influence of the cleaning and disinfection process for pig farms on the efficiency of decontamination and the persistence of *Salmonella. Journées Recherche Porcine*, 44: 97–98.

Correge, I., Hemonic, A. & Gouvars, B., 2009. Conditions d'elevage associees a la seroprevalence salmonelles des porcs en fin d'engraissement. [Farming conditions and practices linked to Salmonella prevalence in slaughter pigs.] Paper presented at Journée Epidemiologie AEEMA-AESA, France, 4–5 June 2009. Published in *Épidémiologie et Santé Animale,* 55: 17–26.

Correge, I., Pupin, P., Beaujean, F., Pinsard, J.L., Gaultier, T. & Remigereau, O. 2012. Impact of administering organic acids, salts of sodium and bioflavonoids on *Salmonella* excretion in sows. *Journées Recherche Porcine,* 44: 193–194.

Correia-Gomes, Mendonça, D., Vieira-Pinto, M. & Niza-Ribeiro, J. 2011. Risk associations for Salmonella spp in pen samples of breeding pigs in Portugal using binomial modeling. pp. 232–235, *in:* Proceedings of 9th International Conference on the Epidemiology and Control of Biological, Chemical and Physical Hazards in Pigs and Pork. Maastricht, The Netherlands.

Correia-Gomes, C., Economou, T., Mendonca, D., Vieira-Pinto, M. & Niza-Ribeiro, J. 2012. Assessing risk profiles for *Salmonella* serotypes in breeding pig operations in Portugal using a Bayesian hierarchical model. *BMC Veterinary Research,* 8: Article Number 226

Correia-Gomes, C., Mendonca, D., Vieira-Pinto, M. & Niza-Ribeiro, J. 2013. Risk factors for *Salmonella* spp. in Portuguese breeding pigs using a multilevel analysis. *Preventive Veterinary Medicine,* 108: 159–166.

Creus, E., Pérez, J.F., Peralta, B., Baucells, F. & Mateu, E. 2007. Effect of acidified feed on the prevalence of *Salmonella* in market-age pigs. *Zoonoses and Public Health,* 54: 314–319.

Dahl, J., Wingstrand, A., Nielsen, B. & Baggesen, D.L. 1997. Elimination of *Salmonella* Typhimurium infection by the strategic movement of pigs. Veterinary Record, 140: 679–681.

Dahl, J. 2009. Herd and pig level risk factors for *Salmonella*-seropositivity in pigs. pp. 50–52, *in:* Proceedings of the 8th International Symposium on the Epidemiology and Control of Foodborne Pathogens in Pork. Quebec City, Canada.

Davies, P.R., Morrow, W.E.M., Jones, F.T., Deen, J., Fedorka-Cray, J.P. & Harris, I.T. 1997. Prevalence of *Salmonella* in finishing swine raised in different production systems in North Carolina, USA. *Epidemiology and Infection,* 119: 237–244.

Davies, P. 1998. Faecal shedding of *Salmonella* by pigs housed in buildings with open-flush gutters. *Journal of Swine Health and Production,* 6: 101–106.

De Busser, E.V., Dewulf, J., Nollet, N., Houf, K., Schwarzer, K., De Sadeleer, L., ,De Zutter, L. & Maes, D. 2009. Effect of organic acids in drinking water during the last 2 weeks prior to slaughter on *Salmonella* shedding by slaughter pigs and contamination of carcasses. *Zoonoses and Public Health,* 56: 129–136.

Denagamage, T.N., O'Connor, A.M., Sargeant, J.M., Rajic, A. & McKean, J.D. 2007. Efficacy of vaccination to reduce *Salmonella* prevalence in live and slaughtered swine: A systematic review of literature from 1979 to 2007. *Foodborne Pathogens and Disease,* 4: 539–549.

Denagamage, T., O'Connor, A., Sargeant, J. & McKean, J. 2010. The association between sub-therapeutic antibiotics and *Salmonella* Typhimurium in market-weight swine: a systematic review and summation of evidence from 1950 to 2007. *Zoonoses and Public Health,* 57(7-8): E14–E22.

De Ridder, L., Maes, D., Dewulf, J., Pasmans, F., Boyen, F., Haesebrouck, F., Meroc, E., Butaye, P. & Van der Stede, Y. 2013a. Evaluation of three intervention strategies to reduce the transmission of *Salmonella* Typhimurium in pigs. *Veterinary Journal,* 197: 613–618.

De Ridder, L., Maes, D., Dewulf, J., Pasmans, F., Boyen, F., Haesebrouck, F., Meroc, E., Roels, S., Leyman, B., Butaye, P. & Van der Stede, Y. 2013b. Effect of a DIVA vaccine with and without in-feed use of coated calcium-butyrate on transmission of *Salmonella* Typhimurium in pigs. *BMC Veterinary Research,* 9: 243–251.

De Ridder, L., Maes, D., Dewulf, J., Butaye, P., Pasmans, F., Boyen, F., Haesebrouck, F. & Van der Stede, Y. 2014. Use of a live attenuated *Salmonella enterica* serovar Typhimurium vaccine on farrow-to-finish pig farms. *Veterinary Journal,* 202: 303–308.

Dos Santos, J., Creus, E., Perez, J.F., Mateu, E. & Martin-Orue, S.M. 2007. Efficacy of a micro-encapsulated or non-encapsulated blend of lactic and formic acid to reduce the prevalence of *Salmonella* in finishing pigs. *Journal of Animal Science,* 85: 32

Ebner, P.D. & Mathew, A.G. 2000. Effects of antibiotic regimens on the faecal shedding patterns of pigs infected with *Salmonella* Typhimurium. *Journal of Food Protection,* 63: 709–714.

Edrington, T.S., Callaway, T.R., Smith, D.K., Genovese, K.J., Anderson, R.C. & Nisbet, D.J. 2006. Effects of ractopamine HCl on *Escherichia coli* O157:H7 and *Salmonella in vitro* and on intestinal populations and faecal shedding in experimentally infected sheep and pigs. *Current Microbiology,* 53: 82–88.

Erdman, M.M., Harris, I.T., Torremorell, M., Wilt, V.M. & Harris, D.L. 2005. Occurrence of *Salmonella* serotype Typhimurium DT104 on a commercial swine farm before, during, and after depopulation and repopulation. *Journal of the American Veterinary Medicine Association*, 227: 460–466.

EFSA [European Food Safety Authority]. 2010. Scientific opinion on a quantitative microbiological risk assessment of *Salmonella* in slaughter and breeder pigs. *EFSA Journal*, 8(4): 1547 [90 p.].

EFSA. 2011. Analysis of the baseline survey on the prevalence of *Salmonella* in holdings with breeding pigs, in the EU, 2008 – Part B: Factors associated with *Salmonella* pen positivity. *EFSA Journal*, 9(7): 2329 [159 p.].

Farzan, A., Friendship, R.M., Dewey, C.E., Warriner, K., Poppe, C. & Klotins, K. 2006. Prevalence of *Salmonella* spp. on Canadian pig farms using liquid or dry feeding. *Preventive Veterinary Medicine*, 73: 241–254.

Farzan, A., Friendship, R.M., Dewey, C.E., Poppe, C. & Funk, J. 2010a. Evaluation of the risk factors for shedding *Salmonella* with or without antimicrobial resistance in swine using multinomial regression method. *Zoonoses and Public Health*, 57: 85–93.

Farzan, A. & Friendship, R.M. 2010b. A clinical field trial to evaluate the efficacy of vaccination in controlling *Salmonella* infection and the association of *Salmonella*-shedding and weight gain in pigs. *Canadian Journal of Veterinary Research*, 74: 258–263.

Fedorka-Cray, P., Bailey, J.S., Stern, N.J., Cox, N.A., Ladely, S.R. & Musgrove, M. 1999. Mucosal competitive exclusion to reduce *Salmonella* in swine. *Journal of Food Protection*, 62: 1376–1380.

FAO [Food and Agriculture Organization of the United Nations]/OIE [World Organisation for Animal Health]/World Bank. 2010. Good practices for biosecurity in the pig sector: issues and options in developing and transition countries. *[FAO] Animal Production and Health Paper*, No. 169. Available at: http://www.fao.org/docrep/012/i1435e/i1435e00.pdf Accessed 2015-11-10

Foss, D.L., Agin, T.S., Bade, D., Dearwester, D.A., Jolie, R., Keich, R.L., Lohse, R.M., Reed, M., Rosey, E.L., Schneider, P.A., Taylor, L.P. & Willy, M.S. 2013. Protective immunity to *Salmonella enterica* is partially serogroup specific. *Veterinary Immunology and Immunopathology*, 155: 76–86.

Fravalo, P., Cariolet, R., Quegiuner, M. & Salvat, G. 2003a. Individual effect of the steps preceding slaughtering on *Salmonella* contamination of pigs. pp. 61–64, *in:* Proceedings of the 5th International Symposium on the Epidemiology and Control of Foodborne Pathogens in Pork. Heraklion-Crete, Greece.

Fravalo, P., Hascoet, Y., Fablet, C., Beloeil, P.A., Madec, F. & Salvat, G. 2003b. Is there a way to act simultaneously against both *Salmonella* and *Listeria monocytogenes* in pig herds? *Sciences des Aliments*, 23(1): 107–111.

Funk, J.A., Davies, P.R., Nichols, M.A. & Morrow, W.E.M. 1999. Evaluation of the association between pen faecal accumulation and prevalence of *Salmonella enterica* shedding in swine. pp. 126–128, *in:* Proceedings of the 3rd International Symposium on the Epidemiology and Control of *Salmonella* in Pork. Washington, DC. USA

Funk, J.A., Davies, P.R. & Gebreyes, W. 2001. Risk factors associated with *Salmonella enterica* prevalence in three-site swine production systems in North Carolina, USA. *Berliner und Munchener Tierarztliche Wochenschrift*, 114(9-10): 335–338.

Funk, J.A., LeJeune, J.T., Wittum, T.E. & Rajala-Schultz, P. 2006. The effect of sub-therapeutic chlortetracycline on antimicrobial resistance in the faecal flora of swine. *Microbial Drug Resistance - Mechanisms Epidemiology and Disease,* 12(3): 210–218.

Funk, J.A., Wittum, T.E., LeJeune, J.T., Rajala-Schultz, P., Bowman, A. & Mack, A. 2007. Evaluation of stocking density and sub-therapeutic chlortetracycline on *Salmonella enterica* subsp. *enterica* shedding in growing swine. *Veterinary Microbiology,* 124(3-4): 202–208.

Garcia-Feliz, C., Carvajal, A., Collazos, J.A. & Rubio, P. 2009. Herd-level risk factors for faecal shedding of *Salmonella enterica* in Spanish fattening pigs. *Preventive Veterinary Medicine,* 91(2-4): 130–136.

Gautam, R., Lahodny Jr., G., Bani-Yaghoub, M., Morley, P.S. & Ivanek, R. 2014. Understanding the role of cleaning in the control of *Salmonella* Typhimurium in grower-finisher pigs: a modelling approach. *Epidemiology and Infection,* 142: 1034–1049.

Gebru, E., Lee, J.S., Son, J.C., Yang, S.Y., Shin, S.A., Kim, B., Kim, M.K. & Park, S.C. 2010. Effect of probiotic-, bacteriophage- or organic acid-supplemented feeds or fermented soybean meal on the growth performance, acute-phase response, and bacterial shedding of grower pigs challenged with *Salmonella enterica* serotype Typhimurium. *Journal of Animal Science,* 88(12): 3880–3886.

Genovese, K.J., Anderson, R.C., Harvey, R.B., Callaway, T.R., Poole, T.L., Edrington, T.S., Fedorka-Cray, P.J. & Nisbet, D.J. 2003. Competitive exclusion of *Salmonella* from the gut of neonatal and weaned pigs. *Journal of Food Protection,* 66(8): 1353–1359.

Giguere, S., Prescott, J.F., Baggot, J.D., Walker, R.D. & Dowling, P.M. 2006. *Antimicrobial therapy in veterinary medicine.* Blackwell Publishing, Ames, Iowa, USA.

Goldbach, S.G. & Alban, L. 2006. A cost-benefit analysis of *Salmonella*-control strategies in Danish pork production. *Preventive Veterinary Medicine,* 77(1-2): 1–14.

Gotter, V., Klein, G., Koesters, S., Kreienbrock, L., Blaha, T. & Campe, A., 2012. Main risk factors for *Salmonella*-infections in pigs in north-western Germany. *Preventive Veterinary Medicine,* 106(3-4): 301–307.

Gradassi, M., Pesciaroli, M., Martinelli, N., Ruggeri, J., Petrucci, P., Hassan, W.H., Raffatellu, M., Scaglione, F.E., Ammendola, S., Battistoni, A., Alborali, G.L. & Pasquali, P. 2013. Attenuated *Salmonella enterica* serovar Typhimurium lacking the ZnuABC transporter: an efficacious orally-administered mucosal vaccine against salmonellosis in pigs. *Vaccine,* 31(36): 3695–3701.

Groninga, J., Springer, E., Braunschmidt, M. & Pankratz, D. 2000. *Salmonella* derby cross-protection study. *Veterinary therapeutics: research in applied veterinary medicine,* 1(1): 59–62

Haneda, T., Okada, N., Kikuchi, Y., Takagi, M., Kurotaki, T., Miki, T., Arai, S. & Danbara, H. 2011. Evaluation of *Salmonella enterica* serovar Typhimurium and Choleraesuis slyA mutant strains for use in live attenuated oral vaccines. *Comparative Immunology Microbiology and Infectious Diseases,* 34(5): 399–409.

Hansen, C.F., Knudsen, K.E.B., Jensen, B.B. & Kjaersgaard, H.D. 2005. Effect of meal feed, potato protein concentrate, barley, beet pellets, and zinc gluconate on *Salmonella* prevalence, gastro-intestinal health and productivity in finishers. *in:* Proceedings of the 4th International Symposium on the Epidemiology and Control of *Salmonella* and Other Food Borne Pathogens in Pork. Sept. 2001. Leipzig, Germany. *Berliner und Munchener Tierarztliche Wochenschrift,* 114(9-10): 321–408.

Hautekiet, V., Geert, V., Marc, V. & Rony, G. 2008. Development of a sanitary risk index for *Salmonella* seroprevalence in Belgian pig farms. *Preventive Veterinary Medicine,* 86: 75–92.

Hill, A.A., Snary, E.L., Alban, L. & Cook, A.J.C. 2007. An updated transmission model for *Salmonella* in grower-finisher pigs. pp. 10–13, *in:* Proceedings of the 7th International Symposium on the Epidemiology and Control of Foodborne Pathogens in Pork. Verona, Italy.

Hill, A., Snary, E., Simons, R., Swart, A., Vigre, H., Domingues, A.R., Hald, T., Evers, E., Tennant, J., Denman, S. & Kelly, L. 2011a. A farm transmission model for *Salmonella* in pigs for individual EU member states. pp. 34–37, *in:* Proceedings of 9th International Conference on the Epidemiology and Control of Biological, Chemical and Physical Hazards in Pigs and Pork. Maastricht, The Netherlands.

Hill, A., Swart, A., Simons, R., Vigre, H., Domingues, A., Hald, T., Evers, E. & Snary, E. 2011b. Assessing the effect of on-farm and abattoir interventions in reducing human salmonellosis from pig meat consumption in the EU. pp. 381–384, *in:* Proceedings of 9th International Conference on the Epidemiology and Control of Biological, Chemical and Physical Hazards in Pigs and Pork. Maastricht, The Netherlands.

Hill, A.A., Snary, E.L., Simons, R.R.L., Swart, A.N., Vigre, H., Domingues, A.R., Hald, T., Evers, E., Tennant, J., Denman, S. & Kelly, L. 2011c. A quantitative microbiological risk assessment for *Salmonella* transmission in individual EU member states. pp. 161–164, *in:* Proceedings of 9th International Conference on the Epidemiology and Control of Biological, Chemical and Physical Hazards in Pigs and Pork. Maastricht, The Netherlands.

Hotes, S., Kemper, N., Traulsen, I., Rave, G. & Krieter, J. 2010. Risk factors for Salmonella infection in fattening pigs – an evaluation of blood and meat juice samples. *Zoonoses and Public Health,* 57(1): 30–38.

Hotes, S., Traulsen, S.I. & Krieter, J. 2011. *Salmonella* control measures with special focus on vaccination and logistic slaughter procedures. *Transboundary and Emerging Diseases,* 58(5): 434–444.

Howard, M.D., Hurd, H.S. & Gailey, J.K. 2003. Efficacy of lactic acid to prevent rapid *Salmonella* infection in market weight swine [Abstract]. *Journal of Animal Science,* 81(Suppl. 1): 369, W301.

Hur, J., Song, S.O., Lim, J.S., Chung, I.K. & Lee, J.H. 2011. Efficacy of a novel virulence gene-deleted *Salmonella* Typhimurium vaccine for protection against *Salmonella* infections in growing piglets. *Veterinary Immunology and Immunopathology,* 139: 250–256.

Ikwap, K., Erume, J., Owiny, J.D.O., Nasinyama, G.W., Melin, L., Bengtsson, B., Lundeheim, N., Fellstrom, C. & Jacobson, M. 2014. *Salmonella* species in piglets and weaners from Uganda: Prevalence, antimicrobial resistance and herd-level risk factors. *Preventive Veterinary Medicine,* 115: 39–47.

Jordan, D., Kaiser, T. & Cline, G. 2013. Efficacy of a *Salmonella* Typhimurium and *Salmonella* Choleraesuis experimental combination vaccine against *S.* Typhimurium challenge in growing pigs. pp. 100–102, *in:* Proceedings of the 10th International Conference on the Epidemiology and Control of Biological, Chemical and Physical Hazards in Pigs and Pork. Portland, ME, USA.

Jorgensen, J., Dahl, J., Wingstrand, A., 1999. The effect of feeding pellets, meal and heat treatment on the *Salmonella*-prevalence in finishing pigs. pp. 308–312, *in:* Proceedings of the 3rd International Symposium on the Epidemiology and Control of *Salmonella* in Pork. Washington, DC., USA.

Jorgensen, L., Boes, J., Kranker, S., Kjaergaard, S. & Wachmann, H. 2001. Effect of an optimised pelleted diet on *Salmonella* prevalence and pig productivity. p. 136, *in:* Proceedings of the 4th International Symposium on the Epidemiology and Control of *Salmonella* and Other Food Borne Pathogens in Pork. Leipzig, Germany.

Jorgensen, L., Boes, J., Kranker, S., Kjaersgaard, H. & Wachmann, H., 2003. Effect of an optimised pelleted diet on *Salmonella* prevalence and pig productivity. pp. 136-

138, *in:* Proceedings of the 5th International Symposium on the Epidemiology and Control of Foodborne Pathogens in Pork. Heraklion, Crete, Greece.

Jung, W.-C., Cha, C.-N., Lee, Y.-E., Yoo, C.-Y., Park, E.-K., Kim, S. & Lee, H.-J. 2011. Anti-diarrheal effects of a combination of Korean traditional herbal extracts and dioctahedral smectite on piglet diarrhea caused by *Escherichia coli* and *Salmonella* Typhimurium. *Pakistan Veterinary Journal*, 31(4): 336–340.

Kich, J.D., Mores, N., Vidal, C.E., Piffer, I., Barioni Jr., W., Amaral, A., Ramminger, L. & Cardosa, M. 2001. Risk factors associated with serological *Salmonella* prevalence in finishing pig herds in southern Brazil. pp. 253–255, *in:* Proceedings of the 4th International Symposium on the Epidemiology and Control of *Salmonella* and Other Food Borne Pathogens in Pork. Leipzig, Germany.

Kim, H.B., Singer, R.S., Borewicz, K., White, B.A., Sreevatsan, S., Johnson, T.J., Espejo, L.A. & Isaacson, R.E. 2014. Effects of tylosin administration on C-reactive protein concentration and carriage of *Salmonella enterica* in pigs. *American Journal of Veterinary Research*, 75(5): 460–467.

Kjaersgaard, H.D., Jorgensen, L., Wachmann, H. & Dahl, J. 2002. Effect of meal feed or pelleted feed on *Salmonella* prevalence in sows. p. 318, in: Proceedings of the 17th International Pig Veterinary Society Congress. Ames, Iowa, USA.

Kjeldsen, N.J. & Dahl, J. 1999. The effect of feeding non-heat treated, non-pelleted feed compared to feeding pelleted, heat-treated feed on the *Salmonella*-prevalence of finishing pigs. pp. 313–316, in: Proceedings of the 3rd International Symposium on the Epidemiology and Control of Salmonella in Pork. Washington, DC, USA.

Kolb, J., Roof, M. & Burkhart, K. 2003. Reduction of *Salmonella* Choleraesuis contamination in pork carcasses by vaccination. pp. 210–212, in: Proceedings of the 5th International Symposium on the Epidemiology and Control of Foodborne Pathogens in Pork. Heraklion-Crete, Greece.

Kolb, J., Oropeza, A., Johnson, D., Frana, T., Edler, R. & Husa, J. 2011. Post-harvest reduction of *Salmonella* by use of vaccination in growing pigs. p. 236, in: Proceedings of 9th International Conference on the Epidemiology and Control of Biological, Chemical and Physical Hazards in Pigs and Pork. Maastricht, The Netherlands.

Kreuzer, S., Janczyk, P., Assmus, J., Schmidt, M.F.G., Brockmann, G.A., Nockler, K., 2012. No beneficial effects evident for Enterococcus faecium NCIMB 10415 in weaned pigs infected with *Salmonella enterica* serovar Typhimurium DT104. *Applied and Environmental Microbiology*, 78, 4816-4825.

Kristensen, C.S., Jorgensen, L. & Boes, J. 2005. The effect of benzoic acid in the feed on *Salmonella* Typhimurium in weaned pigs. pp. 278–279, in: Proceedings of the 6th International Symposium on the Epidemiology and Control of Foodborne Pathogens in Pork. Rohnert Park, CA, USA.

Lebel, P., Fraval, P., Longpre, J., Yergeau, E., Laplante, B. & Letellier, A. 2013. Digestive microbiota changes during application of an effective feed presentation-based mitigation option against *Salmonella* shedding in pigs. pp. 147–150, in: Proceedings of the 10th International Conference on the Epidemiology and Control of Biological, Chemical and Physical Hazards in Pigs and Pork. Portland, ME, USA.

LeJeune, J.T., Kauffman, M.D., Amstutz, M.D. & Ward, L.A. 2006. Limited effects of a commercial direct-fed microbial on weaning pig performance and gastrointestinal microbiology. *Journal of Swine Health and Production*, 14(5): 247–252.

Leontides, L.S., Grafanakis, E. & Genigeorgis, C. 2003. Factors associated with the serological prevalence of *Salmonella enterica* in Greek finishing swineherds. Epidemiology and Infection, 131(1): 599–606.

Letellier, A., Messier, S., Lessard, L. & Quessy, S. 2000. Assessment of various treatments to reduce carriage of *Salmonella* in swine. *Canadian Journal of Veterinary Research - Revue Canadienne de Recherche Veterinaire*, 64(1): 27–31.

Leyman, B., Boyen, F., Verbrugghe, E., Parys, A., Haesebrouck, F. & Pasmans, F. 2012. Vaccination of pigs reduces *Salmonella* Typhimurium numbers in a model mimicking pre-slaughter stress. *Veterinary Journal*, 194(2): 250–252.

Liu, X.B., Liang, A.X., Feng, X.G., Guo, A.Z., Ke, C.Y., Zhang, S.J. & Yang, L.G. 2011. Oral and intranasal administration of somatostatin DNA vaccine mediated by attenuated *Salmonella Enterica* Serovar Typhimurium to promote growth of piglets. *Animal*, 5(8): 1231–1236.

Lo Fo Wong, D.M.A., Dahl, J., Stege, H., Van Der Wolf, P.J., Leontides, L., Von Altrock, A. & Thorberg, B.M. 2004. Herd-level risk factors for subclinical *Salmonella* infection in European finishing-pig herds. *Preventive Veterinary Medicine*, 62(4): 253–266.

Lurette, A., Belloc, C. & Keeling, M. 2011a. Contact structure and Salmonella control in the network of pig movements in France. *Preventive Veterinary Medicine*, 102(1): 30–40.

Lurette, A., Touzeau, S., Ezanno, P., Hoch, T., Seegers, H., Fourichon, C. & Belloc, C. 2011b. Within-herd biosecurity and *Salmonella* seroprevalence in slaughter pigs: A simulation study. *Journal of Animal Science*, 89(7): 2210–2219.

Maes, D., Gibson, K., Trigo, E., Saszak, A., Grass, J., Carlson, A. & Blaha, T. 2001. Evaluation of cross-protection afforded by a *Salmonella* Choleraesuis vaccine against Salmonella infections in pigs under field conditions. *Berliner und Munchener Tierarztliche Wochenschrift*, 114: 339–341.

Mainar-Jaime, R.C., Andres, S., Ausere, M. & Vico, J.P. 2013. Galactomannans for the control of *Salmonella* infections in fattening pigs. pp. 12–14, in: Proceedings of the 10th International Conference on the Epidemiology and Control of Biological, Chemical and Physical Hazards in Pigs and Pork. Portland, ME, USA.

Mannion, C., Lynch, P.B., Egan, J. & Leonard, F.C. 2007. Efficacy of cleaning and disinfection on pig farms in Ireland. *Veterinary Record*, 161(11): 371–375.

Marier, E.A., Sheehan, R., McLaren, I.M., Bianchini, J. Snow, L.C. & Cook, A.J.C. 2009. Prevalence of *Salmonella* in the offspring of sows vaccinated with a live-attenuated *Salmonella* Typhimurium vaccine. pp. 397–400, in: Proceedings of the 8th International Symposium on the Epidemiology and Control of Foodborne Pathogens in Pork. Quebec City, Canada.

Martin-Pelaez, S., Costabile, A., Hoyles, L., Rastall, R.A., Gibson, G.R., Ragione, R.M., Woodward, M.J., Mateu, E. & Martin-Orue, S. 2010. Evaluation of the inclusion of a mixture of organic acids or lactulose into the feed of pigs experimentally challenged with *Salmonella* Typhimurium. *Veterinary Microbiology*, 142(3-4): 337–345.

Matiasovic, J., Kudlackova, H., Babickova, K., Stepanova, H., Volf, J., Rychlik, I., Babak, V. & Faldyna, M. 2013. Impact of maternally-derived antibodies against *Salmonella enterica* serovar Typhimurium on the bacterial load in suckling piglets. *Veterinary Journal*, 196: 114–115.

Maneewan, C., Yamauchi, K., Thirabunyanon, M., Siri, S., Mekbungwan, A. & Thongwittaya, N. 2011. Development of *Bacillus subtilis* MP and effective utilization on productivity and microorganisms in faeces of suckling piglets. *International Journal of Applied Research in Veterinary Medicine*, 9(4): 382–387.

Mathew, A.G., Rattanatabtimtong, S., Nyachoti, C.M. & Fang, L. 2009. Effects of in-feed egg yolk antibodies on Salmonella shedding, bacterial antibiotic resistance, and health of pigs. Journal of Food Protection, 72: 267–273.

Michiels, J., Missotten, J., Rasschaert, G., Dierick, N., Heyndrickx, M. & Smet, S. 2012. Effect of organic acids on *Salmonella* colonization and shedding in weaned piglets in a seeder model. *Journal of Food Protection*, 75(11): 1974–1983.

Morrow, W.E.M., See, M.T., Eisemann, J.H., Davies, P.R. & Zering, K. 2002. Effect of withdrawing feed from swine on meat quality and prevalence of *Salmonella* colonization at slaughter. *Journal of the American Veterinary Medical Association*, 220(4): 497–502.

Mroz, Z. & Toride, Y. 2002. Protection of piglets against *Salmonella* infection with dried bacterial cells. *Journal of Dairy Science*, 85(Suppl 1): 39–40.

Naqid, I.A., Owen, J.P., Maddison, B.C., Gardner, D.S., Foster, N., Tchorzewska, M.A., La Ragione, R.M. & Gough, K.C. 2015. Prebiotic and probiotic agents enhance antibody-based immune responses to *Salmonella* Typhimurium infection in pigs. Animal Feed Science and Technology, 201: 57–65.

Neubauer, A., Roof, M. & Kolb, J. 2005. Vaccine efficacy in swine challenged with a highly virulent S. Typhimurium. pp. 124–125, *in*: Proceedings of the 6th International Symposium on the Epidemiology and Control of Foodborne Pathogens in Pork. Rohnert Park, CA, USA.

Nietfeld, J.C., Feder, I., Kramer, T.T., Schoneweis, D. & Chengappa, M.M. 1998. Preventing *Salmonella* infection in pigs with offsite weaning. *Journal of Swine Health and Production*, 6(1): 27–32.

Nisbet, D.J., Anderson, R.C., Harvey, R.B., Genovese, K.J., DeLoach, J.R. & Stanker, L.H. 1999. Competitive exclusion of *Salmonella* serovar Typhimurium from the gut of early weaned pigs. pp. 80–82, *in*: Proceedings of the 3rd International Symposium on the Epidemiology and Control of Salmonella in Pork. Washington, DC, USA.

O'Connor, A.M., Denagamage, T., Sargeant, J.M., Rajić, A. & McKean, J. 2008. Feeding management practices and feed characteristics associated with *Salmonella* prevalence in live and slaughtered market-weight finisher swine: a systematic review and summation of evidence from 1950 to 2005. *Preventive Veterinary Medicine*, 87(3-4): 213–228.

Patchanee, P., Crenshaw, T.D. & Bahnson, P.B. 2007. Oral sodium chlorate, topical disinfection, and younger weaning age reduce *Salmonella enterica* shedding in pigs. *Journal of Food Protection*, 70(8): 1798–1803.

Pieper, R., Bindelle, J., Malik, G., Marshall, J., Rossnagel, B.G., Leterme, P. & Van Kessel, A.G. 2012. Influence of different carbohydrate composition in barley varieties on *Salmonella* Typhimurium var. Copenhagen colonisation in a "Trojan" challenge model in pigs. *Archives of Animal Nutrition*, 66(3): 163–179.

Rajtak, U., Boland, F., Leonard, N., Bolton, D. & Fanning, S. 2012. Roles of diet and the acid tolerance response in survival of common *Salmonella* serotypes in faeces of finishing pigs. *Applied and Environmental Microbiology*, 78(1): 110–119.

Robbins, R.C., Artuso-Ponte, V., Moeser, A.J., Morrow, W.E.M., Spears, J.W. & Gebreyes, W.A. 2013. Effects of quaternary benzo(c) phenanthridine alkaloids on growth performance, shedding of organisms, and gastrointestinal tract integrity in pigs inoculated with multidrug-resistant *Salmonella* spp. *American Journal of Veterinary Research*, 74(12): 1530–1535.

Roesler, U., Marg, H., Schroder, I., Mauer, S., Arnold, T., Lehmann, J., Truyen, U. & Hensel, A. 2004. Oral vaccination of pigs with an invasive gyrA-cpxA-rpoB *Salmonella* Typhimurium mutant. *Vaccine*, 23: 595–603.

Roesler, U., Von Altrock, A., Heller, P., Bremerich, S., Arnold, T., Lehmann, J., Waldmann, K.-H., Truyen, U. & Hensel, A. 2005. Effects of fluorequinolone treatment acidified feed, and improved hygiene measures on the occurrence of *Salmonella* Typhimurium DT104 in an integrated pig breeding herd. *Journal of Veterinary Medicine Series B – Infectious Diseases and Veterinary Public Health,* 52(2): 69–74.

Roesler, U., Heller, P., Waldmann, K.-H., Truyen, U. & Hensel, A. 2006. Immunization of sows in an integrated pig-breeding herd using a homologous inactivated *Salmonella* vaccine decreases the prevalence of *Salmonella* Typhimurium infection in the offspring. *Journal of Veterinary Medicine Series B – Infectious Diseases and Veterinary Public Health,* 53(5): 224–228.

Roesler, U., Stief, M., Lindner, T., Truyen, U. & Selbitz, H.-J. 2011. Microbial and serological effects of vaccination of sows and suckling piglets with an attenuated live *Salmonella* vaccine. P. 130, *in:* Proceedings of 9th International Conference on the Epidemiology and Control of Biological, Chemical and Physical Hazards in Pigs and Pork. Maastricht, The Netherlands.

Rostagno, M.H., Hurd, H.S. & McKean, J.D. 2009. Split marketing as a risk factor for *Salmonella enterica* infection in swine. *Foodborne Pathogens and Disease,* 6(7): 865–869.

Rostagno, M.H., Eicher, S.D. & Lay, S.D. Jr, 2009. Effect of feed withdrawal and transportation on *Salmonella enterica* infection in market weight pigs. P. 92, *in:* Proceedings of the 8th International Symposium on the Epidemiology and Control of Foodborne Pathogens in Pork. Quebec City, Canada.

Ruggeri, J., Pesciaroli, M., Gaetarelli, B., Scaglione, F.E., Pregel, P., Ammendola, S., Battistoni, A., Bollo, E., Alborali, G.L. & Pasquali, P. 2014. Parenteral administration of attenuated *Salmonella* Typhimurium delta znuABC is protective against salmonellosis in piglets. *Vaccine,* 32: 4032–4038.

Saez, A.C., Zhang, J.Y., Rostagno, M.H. & Ebner, P.D. 2011. Direct feeding of microencapsulated bacteriophages to reduce *Salmonella* colonization in pigs. *Foodborne Pathogens and Disease,* 8(12): 1269–1274.

Schwarz, P., Kich, J.D., Kolb, J. & Cardoso, M. 2011. Use of an avirulent live *Salmonella* Choleraesuis vaccine to reduce the prevalence of *Salmonella* carrier pigs at slaughter. *Veterinary Record,* 169(21): 553–&.

Simons, R., Hill, A., de Souza Monteiro, D., Fraser, R. & Cook, A.J.C. 2009. A cost effectiveness analysis for *Salmonella* interventions in the British pig meat production chain. pp. 377–379, *in:* Proceedings of the 8th International Symposium on the Epidemiology and Control of Foodborne Pathogens in Pork. Quebec City, Canada.

Springer, S., Lindner, T., Steinbach, G. & Selbitz, H.-J. 2001. Investigation of the efficacy of a genetically stable live *Salmonella* Typhimurium vaccine for use in swine. pp. 85–87, *in:* Proceedings of the 4th International Symposium on the Epidemiology and Control of *Salmonella* and Other Food Borne Pathogens in Pork. Leipzig, Germany.

Smith, R.P., Clough, H.E. & Cook, A.J.C. 2010. Analysis of meat juice ELISA results and questionnaire data to investigate farm-level risk factors for Salmonella infection in UK pigs. *Zoonoses and Public Health,* 57(Suppl. 1): 39–48.

Smith, R.P., Sanchez-Vazquez, M., Cook, A.J.C., Clough, H.E. & Edwards, S.A. 2010b. An analysis of quality assurance and zoonoses action plan data from pig herds in the United Kingdom. *Pig Journal,* 64: 43–51.

Soumpasis, I., Alban, L. & Butler, F. 2012. Controlling *Salmonella* infections in pig farms: a framework modelling approach. *Food Research International,* 45: 1139–1148.

Spiehs, M.J., Shurson, G.C. & Johnston, L.J. 2008. Effects of two direct-fed microbials on the ability of pigs to resist an infection with *Salmonella enterica* serovar Typhimurium. *Journal of Swine Health and Production,* 16(1): 27–36.

Springer, S., Lindner, T., Steinbach, G. & Selbitz, H.-J. 2001. Investigation of the efficacy of a genetically stable live *Salmonella* Typhimurium vaccine for use in swine. pp. 85–88, *in:* Proceedings of the 4th International Symposium on the Epidemiology and Control of *Salmonella* and Other Food Borne Pathogens in Pork. Leipzig, Germany.

Stojanac, N., Stevancevic, O. & Stancic, I. 2013. Estimation of the *Salmonella* spp. prevalence in pig farms with dry and wet feeding. *African Journal of Microbiological Research,* 7: 3272–3274.

Suryanarayana, M.V.A.N., Sreedhar, S. & Babu, B.J. 2013. Interactive effect of prebiotic (oligofructose) and probiotic (saccharomyces) feed additives on nutrient utilization, growth, feed conversion, and faecal microbiota population in pigs. *Animal Science Reports,* 7: 107–113.

Sweeney, T., Dillon, S., Fanning, J., Egan, J., O'Shea, C.J., Figat, S., Gutierrez, J.J.M., Mannion, C., Leonard, F. & O'Doherty, J.V. 2011. Evaluation of seaweed-derived polysaccharides on indices of gastrointestinal fermentation and selected populations of microbiota in newly weaned pigs challenged with *Salmonella* Typhimurium. *Animal Feed Science and Technology,* 165(1-2): 85–94.

Szabo, I., Wieler, L.H., Tedin, K., Scharek-Tedin, L., Taras, D., Hensel, A., Appel, B. & Nockler, K. 2009. Influence of a probiotic strain of *Enterococcus faecium* on *Salmonella enterica* serovar Typhimurium DT104 infection in a porcine animal infection model. *Applied Environmental Microbiology,* 75(9): 2621–2628.

Tallarico, N., Franchi, L., Rugna, G., Ramini, M., Luppi, A., Conchello, L. & Merialdi, G. 2013. Efficacy of a blend of encapsulated acids and essential oils to reduce *Salmonella* carriage in slaughtering pigs. p. 179, *in:* Proceedings of the Joint Meeting of the 5th European Symposium of Porcine Health Management and the 50th Anniversary Meeting of the Pig Veterinary Society of Great Britain. Edinburgh, UK.

Tanaka, T., Imai, Y., Kumagae, N. & Sato, S. 2010. The effect of feeding lactic acid to *Salmonella* Typhimurium experimentally infected swine. *Journal of Veterinary Medical Science,* 72(7): 827–831.

Taube, V.A., Neu, M.E., Hassan, Y., Verspohl, J., Beyerbach, M. & Kamphues, J. 2009. Effects of dietary additives (potassium diformate/organic acids) as well as influences of grinding intensity (coarse/fine) of diets for weaned piglets experimentally infected with *Salmonella* Derby or *Escherichia coli. Journal of Animal Physiology and Animal Nutrition,* 93(3): 350–358.

Tenhagen, B.-A., Wegeler, C., Schroeter, A., Dorn, C., Helmuth, R. & Kasbohrer, A. 2009. Association of *Salmonella* spp. in slaughter pigs with farm management factors. pp. 232–234, *in:* Proceedings of the 8th International Symposium on the Epidemiology and Control of Foodborne Pathogens in Pork. Quebec City, Canada.

Tessier, C., Atiana, L., Cardinale, E. & Denis, M. 2013. Longitudinal investigation of *Salmonella* spp. from farm to fork in the pig industry in Reunion Island. pp. 110–112, *in:* Proceedings of the 10th International Conference on the Epidemiology and Control of Biological, Chemical and Physical Hazards in Pigs and Pork. Portland, ME, USA.

Trawinska, B., Lechowski, J., Polonis, A. &. Kowaleczko, M. 2012. Effect of feed supplemented with vitamin C on microbial flora of swine faeces and blood morphology. *Bulletin of The Veterinary Institute in Pulawy,* 56(2): 171–175.

Twomey, D.F., Miller, A.J., Snow, L.C., Armstrong, J.D., Barnes, R.H., Williamson, S.M., Featherstone, C.A., Reichel, R. & Cook, A.J.C. 2010a. Association between biosecurity and *Salmonella* species prevalence on English pig farms. *Veterinary Record,* 166(23): 722–724.

Twomey, D.F., Miller, A.J., Snow, L.C., Cook, A.J.C., Davies, R.H., Williamson, S.M., Reichel, R., Featherstone, C.A. & Armstrong, J.D. 2010b. Biosecurity observations on pig farms with high *Salmonella* seroprevalence in England and Wales. *Pig Journal,* 64: 69.

Upadrasta, A., O'Sullivan, L., O'Sullivan, O., Sexton, N., Lawlor, P.G., Hill, C., Fitzgerald, G.F., Stanton, C. & Ross., R.P. 2013. The effect of dietary supplementation with spent cider yeast on the swine distal gut microbiome. *PLoS ONE,* 8: Article e75714

Van der Heijden, M., Ederveen, W., Niewerth, D. & Frankena, K. 2006. Effectiveness and profitability of different *Salmonella* control programmes in a large field trial. *Pig Journal,* 58: 182.

Van der Wolf, P.J., Wolbers, W.B., Elbers, A.R., Van der Heijden, H.M., Koppen, J.M., Hunneman, W.A., Van Schie, F.W. & Tielen, M.J. 2001a. Herd-level husbandry factors associated with the serological *Salmonella* prevalence in finishing pig herds in the Netherlands. *Veterinary Microbiology,* 78(3): 205–219.

Van Der Wolf, P.J., Van Schie, F.W., Elbers, A.R., Engel, B., Van Der Heijden, H.M., Hunneman, W.A. & Tielen, M.J. 2001b. Administration of acidified drinking water to finishing pigs in order to prevent *Salmonella* infections. *Veterinary Quarterly,* 23(3): 121–125.

Van Winsen, R.L., Keuzenkamp, D., Urlings, B.A.P., Lipman, L.J.A., Snijders, J.A.M., Verheijden, J.H.M. & Van Knapen, F. 2002. Effect of fermented feed on shedding of Enterobacteriaceae by fattening pigs. *Veterinary Microbiology*, 87(3): 267–276.

Vico, J.P., Mainr-Jaime, R.C. & Aussere, M. 2011. The addition of galacto-ologosaccharides on the feed for the control of salmonellosis in fattening pigs. p. 268, *in:* Proceedings of 9th International Conference on the Epidemiology and Control of Biological, Chemical and Physical Hazards in Pigs and Pork. Maastricht, The Netherlands.

Vico, J.P. & Mainar-Jaime, R. 2012. Serological survey of *Salmonella* spp. infection in finishing pigs from northeastern Spain and associated risk factors. *Spanish Journal of Agricultural Research,* 10: 372–382.

Visscher, C.F., Winter, P., Verspohl, J., Stratmann-Selke, J., Upmann, M., Beyerbach, M. & Kamphues, M.J. 2009. Effects of feed particle size at dietary presence of added organic acids on caecal parameters and the prevalence of *Salmonella* in fattening pigs on-farm and at slaughter. *Journal of Animal Physiology and Animal Nutrition,* 93(4): 423–430.

Wagner, B.A., Straw, B.E., Fedorka-Cray, P. & Dargatz, D.A. 2008. Effect of antimicrobial dosage regimen on *Salmonella* and *Escherichia coli* isolates from feeder swine. *Applied and Environmental Microbiology,* 74: 1731–1739.

Wales, A.D., McLaren, I.M., Bedford, S., Carrique-Mas, J., Cook, A.J.C. & Davies, R.H. 2009. Longitudinal survey of the occurrence of *Salmonella* in pigs and the environment of nucleus breeder and multiplier pig herds in England. *Veterinary Record,* 165: 648–657.

Wall, S.K., Zhang, J.Y., Rostagno, M.H. & Ebner, P.D. 2010. Phage therapy to reduce preprocessing *Salmonella* infections in market-weight swine. *Applied and Environmental Microbiology,* 76(1): 48–53.

Walsh, M.C., Rostagno, M.H., Gardiner, G.E., Sutton, A.L., Richert, B.T. & Radcliffe, J.S. 2012. Controlling *Salmonella* infection in weanling pigs through water delivery of direct-fed microbials or organic acids. Part I: Effects on growth performance, microbial populations, and immune status. *Journal of Animal Science,* 90(1): 261–271.

Wang, Y.-Z., Shan, T.-Z., Xu, Z.-R., Feng, J. & Wang, Z.-Q. 2007. Effects of the lactoferrin (LF) on the growth performance, intestinal microflora and morphology of weanling pigs. *Animal Feed Science and Technology,* 135(3-4): 263–272.

Weedman, S.M., Rostagno, M.H., Patterson, J.A., Yoon, I., Fitzner, G. & Eicher, S.D. 2011. Yeast culture supplement during nursing and transport affects immunity and intestinal microbial ecology of weanling pigs. *Journal of Animal Science,* 89(6): 1908–1921.

Wells, J.E., Yen, J.T. & Miller, D.N. 2005. Impact of dried skim milk in production diets on *Lactobacillus* and pathogenic bacterial shedding in growing-finishing swine. *Journal of Applied Microbiology,* 99: 400–407.

Wells, J.E., Kalchayanand, N., Berry, E.D. & Oliver, W.T. 2013. Effects of antimicrobials fed as dietary growth promoters on faecal shedding of *Campylobacter, Salmonella* and Shiga-toxin producing *Escherichia coli* in swine. *Journal of Applied Microbiology,* 114(2): 318–328.

Wilhelm, B., Rajic, A., Parker, S., Waddell, L., Sanchez, J., Fazil, A., Wilkins, W. & McEwen, S.A. 2012. Assessment of the efficacy and quality of evidence for five on-farm interventions for *Salmonella* reduction in grow-finish swine: A systematic review and meta-analysis. *Preventive Veterinary Medicine,* 107: 1–20.

Wilkins, W., Rajic, A., Waldner, C., McFall, M., Chow, E., Muckle, A. & Rosengren, L. 2010. Distribution of *Salmonella* serovars in breeding, nursery, and grow-to-finish pigs, and risk factors for shedding in ten farrow-to-finish swine farms in Alberta and Saskatchewan. *Canadian Journal of Veterinary Research – Revue Canadienne de Recherche Veterinaire,* 74(2): 81–90.

Willamil, J., Creus, E., Perez, J.F., Mateu, E. & Martin-Orue, S.M. 2011. Effect of a microencapsulated feed additive of lactic and formic acid on the prevalence of *Salmonella* in pigs arriving at the abattoir. *Archives of Animal Nutrition,* 65(6): 431–444.

Wingstrand, A., Dahl, J., Thomsen, L.K., Jorgensen, L. & Jensen, B.B. 1997. Influence of dietary administration of organic acids and increased feed structure on *Salmonella.* pp. 170–172, *in:* Proceedings of the 2nd International Symposium on Epidemiology and Control of *Salmonella* in Pork. Copenhagen, Denmark.

Wisener, L.V., Sargeant, J.M., O'Connor, A.M., Faires, M.C. & Glass-Kaastra, S. 2014. The evidentiary value of challenge trials for three pre-harvest food safety topics: a systematic assessment. *Zoonoses and Public Health,* 61(7): 449–476.

Yan, L., Hong, S.M. & Kim, I.H., 2012. Effect of bacteriophage supplementation on the growth performance, nutrient digestibility, blood characteristics, and faecal microbial shedding in growing pigs. *Asian-Australasian Journal of Animal Science,* 25(10): 1451–1456.

Zhang, G.G., Yang, Z.B., Wang, Y., Yang, W.R. & Zhou, H.J. 2014. Effects of dietary supplementation of multi-enzyme on growth performance, nutrient digestibility, small intestinal digestive enzyme activities, and large intestinal selected microbiota in weanling pigs. *Journal of Animal Science,* 92(5): 2063–2069.

SUMMARY CARD SC2:
Pork Chain – Transport, Lairage and Logistic Slaughter Interventions

SC2.1 SUMMARY OF KEY FINDINGS

This Summary Card covers the evidence for interventions implemented variously at transport to slaughter, at lairage or prior to slaughter, for control of *Salmonella* in pork. Overall, we found a relatively small volume of literature investigating similar outcomes in the population of interest (i.e. finisher pigs or carcasses), which limited our ability to perform meta-analysis for estimation of pooled summary estimates of effect.

SC2.1.1 Transport to slaughter

- Pigs transported to slaughter may yield a significantly increased carcass prevalence of *Salmonella* relative to those not experiencing transport, reflecting exposure to increased shedding in transit, and also to off-farm sources of infection. The trailer itself, even after washing and disinfection, may potentially be a source of infection. However, stochastic modelling suggests that even though hygiene during transportation to slaughter could be improved, this is likely to have less impact on carcass prevalence or human salmonellosis than other post-farm interventions.

SC2.1.2 Lairage

- While seven of eight trials investigating avoidance of lairage as an intervention, and measuring caecal content of slaughter pigs, reported a protective treatment effect on *Salmonella* prevalence; four trials measuring lymph node prevalence yielded opposing findings. Similarly, investigations comparing carcass *Salmonella* prevalence in pigs exposed to contaminated or uncontaminated lairage reported opposing findings.

SC2.1.3 Logistic slaughter

- Three primary research studies (controlled trial, quasi-experiment and a cohort study) report a non-significant treatment effect for logistic slaughter on prevalence of *Salmonella* in carcasses. Similarly, five stochastic models investigated logistic slaughter, reporting treatment effects from 'negligible' to 'insufficient to reach target prevalence, in the absence of other measures applied concurrently'.

SC2.1.4 Overall implications

- Although the potential for animals that are not infected with *Salmonella* on-farm to become exposed and/or infected *en route* to slaughter and processing is well-recognized, the evidence supporting interventions for reduction of *Salmonella* at transport, lairage, or in the timing or organization of slaughter (i.e. logistic slaughter), remains inconsistent, and with a 'very low' to 'moderate' level of confidence.
- The level of evidence and inconsistent treatment effects reported reflect the complexity of potential transmission routes in this process. Also, there are a large number of known and unknown predictors and confounders for risk of infection from farm to slaughter, including: animal susceptibility; cut-points for categorization of herd prevalence; extent and intensity of staff training; efficacy of abattoir standard hygienic practice implementation; and time of slaughter relative to elapsed time since start of shift.
- The varied observations from primary research as well as stochastic models measuring the effects of interventions at multiple points in the pork production chain suggest that while transport, lairage, and logistic slaughter interventions may reduce *Salmonella* contamination of carcasses, other areas in the chain (particularly processing) offer more effective intervention points to attempt to ultimately reduce incidence of human salmonellosis. However, more effective reduction in human cases may be achieved by intervening concurrently on-farm; during transport and lairage; and slaughter and processing.

SC2.2 INTERVENTION DESCRIPTION

This Summary Card summarizes the evidence for interventions that can be implemented during transport and at lairage, and policies to segregate slaughter of herds with different contamination levels to reduce *Salmonella* contamination of pork. Specific categories of interventions covered in this Summary Card are:

- **Transport to slaughter:** refers to the transportation of market animals from the final farm production unit to the abattoir.
- **Lairage:** refers to holding facilities at the abattoir for animals awaiting slaughter.
- **Logistic slaughter:** refers to the strategic scheduling of slaughter of animals from herds categorized as 'low risk' for the pathogen prior to animals from herds categorized as 'higher risk', to minimize the potential for cross-contamination.

SC2.3 TRANSPORT TO SLAUGHTER

Primary research investigating transport to slaughter included one controlled trial, two quasi-experiments, and one case-control study, as well as six stochastic models.

Hurd *et al.* (2002) investigated bacterial flora in pigs experiencing or not experiencing transport and lairage prior to slaughter. *Salmonella* prevalence was significantly less in those animals not experiencing transport and lairage, regardless of sample matrix assayed (faecal material, caecal contents, or carcass lymph nodes). Serotyping of isolates demonstrated a greater abundance of strains at the abattoir relative to on-farm, suggesting that some animals were infected from sources off-farm.

Rajkowski, Eblen and Laubauch (1998), using a quasi-experimental design, sampled stock trailer floors prior to and after cleaning and disinfection, across all four seasons and after both short (<500 miles) and long (>500 miles) hauls. The authors report no significant difference in *Salmonella* prevalence by season or length of haul, with contamination in 80% of bedding sampled. Similarly, Dorr, Lowman and Gebreyes (2005) sampled ten stock trailers from each of four truck wash stations in North Carolina, United States of America, all using different methods of disinfection. While the post-wash prevalence in trucks from one station (using recycled water and Virkon™ disinfectant) was 0%, two stations had trucks with detectable *Salmonella* post-wash (20% and 45%, using recycled water with phenol, and fresh water with soap and phenol, respectively) and one station, using and recycled water and phenol, had 100% *Salmonella* prevalence in trucks post-wash (an actual increase in prevalence after washing-disinfection).

Mannion *et al.* (2007) similarly used quasi-experimental design to study effectiveness of truck washing and disinfection in removing *Salmonella*. The authors reported that of 108 samples taken post-wash after carrying Irish high-risk herds and 54 taken post-wash after carrying low-risk herds, 18% and 6%, respectively, were still *Salmonella* positive. The authors concluded that current washing protocols were inadequate.

Rostagno and Lay (2011) described a study in which market pigs were mixed for six hours with another pig, or transported for an hour, or both, prior to being challenged with a low dose (10^4 CFUs) of *Salmonella* Typhimurium, with slaughter six hours post-challenge. The authors reported that all treatment groups had significantly greater concentrations of *Salmonella* in the ileum, with pigs receiving both treatments (i.e. mixing and transport) also having significantly greater concentrations of *Salmonella* in caecal contents, and a non-significant treatment effect on lymph node *Salmonella* concentrations. The authors concluded that pre-slaughter stress increased animal susceptibility to low-dose *Salmonella* challenge.

Rostagno, Eicher and Lay (2009) described a study in which market pigs were randomly assigned to 12 hours of feed withdrawal, or two hours of transportation, or both, three days after intra-nasal challenge with *Salmonella* Typhimurium. The authors reported that the treatment group receiving both feed withdrawal and transportation had significantly greater *Salmonella* concentration in ileal contents.

Gotter *et al.* (2012) reported a German farm-level case-control study assaying slaughter pigs' meat juice ELISA, investigating risk factors for farm-level *Salmonella* infection. The authors reported that movement of individual animals during the fattening period (although transportation is not described as part of 'movement') increased the odds of a farm being positive (odds ratio (OR) = 5.3, 95% confidence intervals (CI) (1.35–20.35)). Failure to use a separate transporter for each age group of animals was significantly associated with increased odds of *Salmonella* on-farm (OR = 11.4, 95% CI (1.94–66.18)). Unexpectedly, failure to clean the transporter was significantly associated with reduced odds of *Salmonella* on-farm (OR = 0.2, 95% CI (0.05–0.72)).

Alban and Stärk (2005) describe a stochastic model of *Salmonella* prevalence from farm to carcass. The authors reported no predicted significant effect on *Salmonella* prevalence for separate transportation of herds of different risk levels.

The European Food Safety Authority (EFSA) reported a quantitative risk assessment of *Salmonella* in slaughter and breeder pigs (EFSA, 2010). Transport and lairage did not have a major effect on the modelled outcome (human cases) in any of the scenarios, even when modelling improved hygiene practices with 100% uptake and effectiveness. However, the authors noted that 'neglecting the pre-harvest stage inevitably leads to maintaining the flow of *Salmonella* into the slaughterhouse' (EFSA, 2010).

Hill *et al.* (2011) described stochastic modelling of a transport and lairage model for *Salmonella* transmission in pigs for individual European Union member states. For both case studies presented, the model predicted a small increase in the prevalence of contaminated lymph nodes after transport, with large variation in prevalence across batches. The authors concluded that factors such as stress can have large effects on individual batch prevalence, but that this occurs relatively rarely across all batches processed, and therefore does not have large effect on mean national *Salmonella* prevalence between farm and slaughter.

Hotes, Traulsen and Krieter (2012) also described a cost model for estimating the potential costs associated with segregated transport of slaughter pigs to the abattoir, based upon-farm *Salmonella* risk. The authors reported that costs incurred in such a programme would be driven by the proportion of changed shipments (as determined by the relationship between low and high prevalence farms), as well as selected cut-point for categorization, and the additional distance travelled per restructured shipment.

Krebs and Belloc (2012) described a model for *Salmonella* contamination in pork carcasses and potential effects of interventions applied at transport, lairage and slaughter. Interventions applied at transport had the least impact on prevalence in the three areas studied.

The Irish Agriculture and Food Development Authority (Teagasc) reported a stochastic model studying the process from slaughter to boned-out cuts, supported by abattoir and retail surveys in the Republic of Ireland and Northern Island (Teagasc, 2010); in the Republic of Ireland, pigs were also sampled for *Salmonella* from farm, through transport and lairage, to slaughter. These surveys demonstrated that transport of *Salmonella*-negative pigs in contaminated trucks could be a means of *Salmonella* infection. Cold power washing of transport trucks was inadequate to remove *Salmonella*, and some trucks were more widely contaminated after washing.

SC2.4 LAIRAGE

The effect of lairage and potential interventions to mitigate its effect were investigated in 10 primary research studies: seven controlled trials, one quasi-experiment, and two cross-sectional studies, as well as three stochastic models. Additionally, two challenge studies were conducted in conditions designed to simulate lairage.

Five controlled trials investigated the effect of lairage relative to no lairage on individual animals' *Salmonella* prevalence in caecal content (Fravalo *et al.*, 2003; Hurd *et al.*, 2001, 2002; Larsen *et al.*, 2004; Rostagno, Hurd and McKean, 2005), with six of seven individual trials reporting a protective treatment effect (median OR = 0.47, range (0.11–1.06)) for 'no lairage', with low confidence (Table SC2.1). Three controlled trials investigated the effect of lairage relative to none (Hurd *et al.*, 2001, 2002; Larsen *et al.*, 2004) with prevalence of carcass lymph nodes as the outcome measure; opposing results were reported (Table SC2.1). Low confidence in both summary estimates reflects significant heterogeneity of effects across trials.

Two studies investigated carcass *Salmonella* prevalence following animal exposure to contaminated or non-contaminated lairage. Boes *et al.* (2001) described a controlled trial in which pigs from a *Salmonella*-negative herd were kept either in clean lairage or in lairage contaminated by faeces containing *Salmonella,* for two to four hours; there was a non-significant difference in carcass *Salmonella* prevalence between the two groups. Arguello *et al.* (2014) studied the effect of slaughter pig exposure to previously contaminated lairage on individual *Salmonella* carcass prevalence in pigs originating from herds categorized as one of three (low, medium or high) risk levels for *Salmonella* infection. Similarly, a non-significant treatment ($P > 0.05$) effect was observed.

Schmidt *et al.* (2004) described four trials in which pigs arriving at an abattoir were randomly assigned to pens , either treated (cleaned and disinfected) or non-treated, prior to slaughter; carcasses were subsequently cultured for *Salmonella* (faeces, caecal contents, and ileo-caecal lymph nodes). Inconsistent results were reported, with one trial yielding a non-significant treatment effect for cleaning-disinfection on *Salmonella*, two trials yielding a significantly harmful treatment effect, and one trial yielding a significantly protective effect. Pen swabs taken after cleaning and disinfection in the treated pens but prior to animal entry confirmed that cleaning had been effective, and environmental *Salmonella* prevalence was significantly reduced in treated pens relative to control pens.

Two challenge trials were conducted in conditions designed to simulate lairage. Wall *et al.* (2010) investigated the use of a phage cocktail to reduce *Salmonella* in slaughter pigs, reporting no significant treatment effect on prevalence, assaying either ileo-caecal lymph nodes or faecal samples. However, caecal *Salmonella* concentrations were significantly reduced in the treatment group. Callaway *et al.* (2003) reported a non-significant reduction, after four hours of exposure to feed containing sodium chlorate, on *Salmonella* prevalence in the tonsils of market pigs, and no difference in lymph node or intestinal content *Salmonella* status between treatment and control groups. In contrast, in a quasi-experiment using 10 naturally infected pigs, the same authors report that chlorate significantly reduced caecal *Salmonella* colonization.

In a cross-sectional study, Milnes *et al.* (2009) studied potential effect of lairage on *Salmonella* contamination of intestinal contents in slaughter pigs, reporting that pigs held in lairage for more than 12 hours had significantly greater odds (OR = 2.83, 95% CI (1.33–6.01)) of *Salmonella* contamination relative to those held for 12 hours or less. In contrast, Kim and Song (2009) reported that increased duration of lairage (in hours) was protective of *Salmonella* carcass contamination.

Bollaerts *et al.* (2010) describe a stochastic pork farm to fork risk model (METZOON) and its use in modelling the effects of 14 intervention scenarios on reducing human salmonellosis cases. Scenarios include reduction of *Salmonella* prevalence at lairage, as measured by both internal and external contamination, by 10%, 25%, 50% or 75%. The authors categorized the scenarios involving reduction of internally contaminated animals as 'less effective' and the scenarios involving reduction of externally contaminated animals as 'not effective', relative to other scenarios considered (e.g. during polishing, evisceration or chilling).

Delhalle *et al.* (2009) described a quantitative risk model of *Salmonella* prevalence in Belgian minced pork from lairage to consumption; scenarios include effect of changes in prevalence at various stages of the pork production chain. Reduction of *Salmonella* prevalence in pigs entering lairage by 10%, 25% and 75% was modelled, yielding a predicted relative reduction in human salmonellosis cases ranging from 24.7% to 84.2%.

Summary-of-findings table for lairage interventions

TABLE SC2.1. Controlled trials investigating the effect of presence or absence of lairage

Intervention	Outcome sample	No. trials/ studies	Odds ratio for intervention effect(1)	% Samples expected to be *Salmonella* positive in study population(2)		GRADE rating(3)	References
				Without treatment	With treatment		
No lairage	Carcasses (Caecal content)	8/5	Median = 0.19 (Range: 0.12-1.06)	20.7%	4.8% (Range: 2.4-24.7)	Low	Fravalo *et al.*, 2003a; Hurd *et al.*, 2001, 2002; Larsen *et al.*, 2004; Rostagno, Hurd and McKean, 2005;
No lairage	Carcasses (Lymph nodes)	4/4	Median = 0.79 (Range: 0.37-2.19)	35.6%	30.4% (Range: 14.5-54.3)	Low	Fravalo *et al.*, 2003a; Hurd *et al.*, 2001, 2002; Larsen *et al.*, 2004

NOTES:

(1) For interventions with multiple trials, refers to the average meta-analysis estimate and 95% CI from random-effects meta-analysis if heterogeneity was not significant (I^2<60%). If heterogeneity was significant, the median and range of odds ratios from individual studies are presented instead. If only one trial was available, refers to the odds ratio and 95% confidence interval from the individual study.

(2) The percentage of samples positive for *Salmonella* in the comparison group (assumed control risk, ACR) refers to the median prevalence in the comparison group trials. The percentage in the intervention/treatment group was calculated using the following formula for the odds ratio (OR) measure of effect: (OR × ACR / 1-ACR + [OR × ACR]), and provides an indication of the effect of the intervention on changing a given baseline level of *Salmonella* prevalence as reported in the included studies. The odds ratio used in this formula corresponds to the odds ratio presented in the intervention effect column as described under (1) above. If heterogeneity was not significant (I^2<60%), the estimate is presented with a 95% CI in brackets. If heterogeneity was significant, the median estimate is presented with the range of individual study effects in brackets to illustrate the variability in results across studies.

(3) GRADE rating descriptions: very low = the true effect is likely to be substantially different from the measured estimate; low = the true effect may be substantially different from the measured estimate; moderate = the true effect is likely to be close to the measured estimate, but there is a possibility that it is substantially different; high = there is strong confidence that the true effect lies close to that of the measured estimate.

Similar to the findings described above for effect of transport, Hill *et al.* (2011) reported that their stochastic model of transport and lairage predicts a small increase in the prevalence of contaminated lymph nodes after lairage, with large variation in prevalence across batches.

The farm to abattoir surveys reported by Teagasc (2010) as inputs to the stochastic model described, identified that lairage was a source of infection for pigs (and carcasses) from herds categorized as 'low risk', and pigs from 'high risk' herds were a significant source of lairage contamination. Expert elicitation performed in conjunction with the field surveys identified logistical difficulties and 'just in time' delivery difficulties as barriers to minimizing time in lairage; inadequate facilities and resources were suggested as barriers to improved cleaning of lairage.

SC2.5 LOGISTIC SLAUGHTER

Logistic slaughter was investigated in three primary research studies (one controlled trial, one quasi-experiment and one cohort study), and five stochastic models.

Boes *et al.* (2001) mixed 'control' pigs from *Salmonella*-negative herds, with *Salmonella*-positive pigs in lairage; all were slaughtered together. In contrast, 'intervention' pigs from *Salmonella*-negative herds were kept separately and slaughtered first at the abattoir. *Salmonella* carcass prevalence was slightly greater for the intervention group relative to the control group.

Arguello *et al.* (2014) described a study in which 42 pig herds were categorized as low, medium, or high risk for *Salmonella* infection based on serum ELISA of finishers, then all of the low risk herds were slaughtered at the same abattoir on day one, high risk herds on day two, and medium risk herds on day three. The authors report no significant treatment effect on carcass *Salmonella* prevalence, with the greatest absolute number of contaminated carcasses occurring on day three (i.e. in the carcasses originating from herds categorized as 'medium' risk).

Swanenburg *et al.* (2001) described a cohort study in which farms were categorized as low, medium or high risk for *Salmonella* infection based on finisher serum ELISA. Then slaughter pigs were processed at one abattoir, with low risk herds going first in the afternoon, followed by high-risk herds, and then more low risk herds. In a second trial, 200 pigs from *Salmonella*-free farms were slaughtered after 200 pigs from *Salmonella*-positive farms. In both trials, prevalence of *Salmonella* was greater in herds considered positive relative to those considered negative. However, in trial two, more *Salmonella*-positive samples were cultured from carcasses originating from low risk herds, slaughtered at the beginning of the abattoir shift, than at the end. The authors cite Snijders (1976) in hypothesizing that equipment at the end of a shift was contaminated even after it was cleaned, and the first animals through the line on the next shift physically removed the residual contamination.

Alban and Stärk (2005) describe a stochastic model of *Salmonella* prevalence from farm to final carcass, which examined logistic slaughter as a potential intervention. A minor (3.9% vs 3.8%) difference in prevalence of human salmonellosis was achieved only if there was no mixing and the stay in lairage was short (<3 hours).

Baptista *et al.* (2011) described a Danish stochastic model estimating carcass *Salmonella* prevalence in the pork production chain. The authors report that sanitary slaughter did not reduce carcass contamination to less than 1%, in 90% of model iterations. When only level 1/2/3 (i.e. higher risk) pigs were subjected to logistic slaughter, and level 0 pigs to routine abattoir procedures, the model did not achieve a (targeted) national *Salmonella* prevalence of 1%.

EFSA (2010) reported a quantitative risk assessment of *Salmonella* in slaughter and breeder pigs. The authors concluded that unhygienic practices enabling direct or indirect, or both, faecal contamination during transport and lairage increase the risk of carcass contamination with *Salmonella*. However, due to insufficient quantitative data on the microbial load of *Salmonella* on the skin of slaughter pigs, it was not possible to quantify the effect of cross-contamination during transport and lairage.

Goldbach and Alban (2006) described a stochastic cost-benefit model of the Danish pork production industry, comparing the effects of feeding home-prepared feed, feeding acid, hot water decontamination, and sanitary slaughter. Hot water decontamination was found to be the only economically viable option.

Hotes, Traulsen and Krieter (2011) described a stochastic model for *Salmonella* transmission in a pork production chain. All scenarios involving logistic slaughter produced a prevalence of carcass contamination not significantly different from the baseline (i.e. no logistic slaughter) scenario.

Teagasc (2010), reporting on the Irish farm to slaughter surveys of pigs and pork carcasses for *Salmonella*, concluded that, on an individual level, there was little association between serological status and culture of rectal or caecal contents at slaughter; therefore, logistic slaughter based on historical serological data was unlikely to be effective in reducing *Salmonella* prevalence on carcasses. They also identified inadequate lairage design as a potential barrier to successful implementation of logistic slaughter.

SC2.6 REFERENCES CITED IN SUMMARY CARD 2

Alban, L. & Stärk, K.D.C. 2005. Where should the effort be put to reduce the *Salmonella* prevalence in the slaughtered swine carcass effectively? *Preventive Veterinary Medicine,* 68(1): 63–79.

Arguello, H., Carvajal, A., Alvarez-Ordonez, A., Jaramillo-Torres, H. & Rubio, P. 2014. Effect of logistic slaughter on *Salmonella* contamination on pig carcasses. *Food Research International,* 55: 77–82.

Baptista, F.M., Halasa, T., Alban, L. & Nielsen, L.R. 2011. Modelling food safety and economic consequences of surveillance and control strategies for *Salmonella* in pigs and pork. *Epidemiology and Infection,* 139(5): 754–764.

Boes, J., Dahl, J., Nielsen, B. & Krog, H.H. 2001. Effect of separate transport, lairage, and slaughter on occurrence of *Salmonella* Typhimurium on slaughter carcasses. *Berliner und Munchener Tierarztliche Wochenschrift,* 114: 363–365.

Bollaerts, K., Messens, W., Aerts, M., Dewulf, J., Maes, D., Grijspeerdt, K. & Stede, Y. 2010. Evaluation of scenarios for reducing human salmonellosis through household consumption of fresh minced pork meat. *Risk Analysis,* 30: 853–865.

Callaway, T.R., Anderson, R.C., Edrington, T.S., Genovese,K.J., Stahl,C.H., Jung, Y.S., Bischoff, K.M., Poole, T.L. Harvey, R.B. & D. J. Nisbet. 2003. Experimental chlorate product treatment reduces *Salmonella* populations in swine during lairage [Abstract]. *Journal of Animal Science,* 81(Supp 1): No. 529.

Delhalle, L., Saegerman, C., Messens, W., Farnir, F., Korsak, N., Van Der Stede, Y. & Daube, G. 2009. Assessing interventions by quantitative risk assessment tools to reduce the risk of human salmonellosis from fresh minced pork meat in Belgium. *Journal of Food Protection,* 72: 2252–2263.

Dorr, P.M., Lowman, H. & Gebreyes, W. 2005. The role of truck wash practices in the dissemination of *Salmonella* and Campylobacter in commercial swine production. pp. 161–163, *in:* Proceedings of 6th International Symposium on the Epidemiology and Control of Foodborne Pathogens in Pork. Rohnert Park, CA, USA.

EFSA [European Food Safety Authority]. 2010. Scientific opinion on a quantitative mi crobiological risk assessment of Salmonella in slaughter and breeder pigs. *EFSA Journal,* 8(4): 1547 [90 p.].

Fravalo, P., Cariolet, R., Quegiuner, M. & Salvat, G. 2003. Individual effect of the steps preceding slaughtering on *Salmonella* contamination of pigs. pp. 61–64, *in:* Proceedings of the 5th International Symposium on the Epidemiology and Control of Foodborne Pathogens in Pork. Heraklion-Crete, Greece.

Goldbach, S.G., Alban, L., 2006. A cost-benefit analysis of *Salmonella*-control strategies in Danish pork production. *Preventive Veterinary Medicine,* 77(1-2): 1–14.

Gotter, V., Klein, G., Koesters, S., Kreienbrock, L., Blaha, T. & Campe, A. 2012. Main risk factors for *Salmonella*-infections in pigs in north-western Germany. *Preventive Veterinary Medicine,* 106: 301–307.

Hill, A., Simons, R., Swart, A. & Snary, E. 2011. A transport and lairage model for *Salmonella* transmission in pigs for individual EU member states. p. 385, *in:* Proceedings of 9th International Conference on the Epidemiology and Control of Biological, Chemical and Physical Hazards in Pigs and Pork. Maastricht, The Netherlands.

Hotes, S., Traulsen, I. & Krieter, J., 2011. *Salmonella* control measures with special focus on vaccination and logistic slaughter procedures. *Transboundary and Emerging Diseases*, 58(5): 434–444.

Hotes, S., Traulsen, I. & Krieter, J. 2012. The additional costs of segregated transport to slaughter to decrease *Salmonella* prevalence in pork – a simulation study. *Preventive Veterinary Medicine*, 104: 174–178.

Hurd, H.S., McKean, J.D., Wesley, I.V. & Karriker, L.A. 2001. The effect of lairage on *Salmonella* isolation from market swine. *Journal of Food Protection,* 64: 939-944.

Hurd, H.S., McKean, J.D., Griffith, R.W., Wesley, I.V. & Rostagno, M.H. 2002. *Salmonella enterica* infections in market swine with and without transport and holding. *Applied and Environmental Microbiology,* 68: 2376–2372.

Kim, Y.-J. & Song, Y. 2009. An analysis of the factors for microbial contamination risk for pork at slaughterhouses in Korea using the logit model. *Journal of Toxicology and Environmental Health,* 72: 1470–1474.

Larsen, S.T., Hurd, H.S., McKean, J.D., Griffith, R.W. & Wesley, I.V. 2004. Effect of short-term lairage on the prevalence of *Salmonella enterica* in cull sows. *Journal of Food Protection,* 67: 1489–1493.

Krebs, S. & Belloc, C. 2012. Maîtrise des Salmonelles en production porcine: a quel(s) maillon(s) de la filière intervenir? [Control of *Salmonella* in pig supply chain: which step to target?] 44e Journées de la Recherche Porcine en France, Paris, France, 7–8 February 2012. *Journées de la Recherche Porcine en France,* 44: 31–36.

Mannion, C., Egan, J., Lynch, P.B., Leonard, F.C., 2007. An investigation into the efficacy of washing trucks following the transportation of pigs – a *Salmonella* perspective. pp. 253–255, *in:* Proceedings of the 7th International Symposium on the Epidemiology and Control of Foodborne Pathogens in Pork. Verona, Italy.

Milnes, A.S., Sayers, A.R., Stewart, I., Clifton-Hadley, F., Davies, R.H., Newell, D.G., Cook, A.J.C., Evans, S.J., Smith, R.P. & Paiba, G.A. 2009. Factors related to the carriage of Verocytotoxigenic *E. coli*, *Salmonella*, thermophilic *Campylobacter* and *Yersinia enterocolitica* in cattle, sheep and pigs at slaughter. *Epidemiology and Infection,* 137: 1135–1148.

Rajkowski, K.T., Eblen, S. & Laubauch, C. 1998. Efficacy of washing and sanitizing trailers used for swine transport in reduction of *Salmonella* and *Escherichia coli*. *Journal of Food Protection,* 61: 31–35.

Rostagno, M.H., Hurd, H.S. & McKean, J.D. 2005. Resting pigs on transport trailers as an intervention strategy to reduce *Salmonella enterica* prevalence at slaughter. *Journal of Food Protection,* 68: 1720–1723.

Rostagno, M.H., Eicher, S.D. & Lay, D.C. Jr. 2009. Effect of feed withdrawal and transportation on *Salmonella enterica* infection in market weight pigs. p. 92, *in:* Proceedings of the 8th International Symposium on the Epidemiology and Control of Foodborne Pathogens in Pork. Quebec City, Canada.

Rostagno, M.H. & Lay, D.C. Jr. 2011. Effect of transportation and mixing with unfamiliar pig on *Salmonella* susceptibility in market weight pigs. pp. 308–309, Proceedings of 9th International Conference on the Epidemiology and Control of Biological, Chemical and Physical Hazards in Pigs and Pork. Maastricht, The Netherlands.

Schmidt, P.L., O'Connor, A.M., McKean, J.D. & Hurd, H.S. 2004. The association between cleaning and disinfection of lairage pens and the prevalence of *Salmonella enterica* in swine at harvest. *Journal of Food Protection,* 67: 1384–1388.

Snijders, J.M.A., 1976. Pig slaughtering hygiene. Thesis, Faculty of Veterinary Medicine. University of Utrecht, The Netherlands.

Swanenburg, M., Van der Wolf, P., Urlings, H.A.P., Snijders, J.M.A., Van Knapen, F., 2001. *Salmonella* in slaughter pigs: The effect of logistic slaughter procedures of pigs on the prevalence of *Salmonella* in pork. *International Journal of Food Microbiology,* 70, 231-242.

Teagasc [The Irish Agriculture and Food Development Authority]. 2010. *Salmonella* in pork on the island of Ireland: A microbial risk assessment. http://www.safefood.eu/Publications/Research-reports/Salmonella-in-Pork.aspx.

Wall, S.K., Zhang, J.Y., Rostagno, M.H. & Ebner, P.D. 2010. Phage therapy to reduce preprocessing *Salmonella* infections in market-weight swine. *Applied and Environmental Microbiology,* 76: 48–53.

SUMMARY CARD 3:
Pork Chain – Processing Interventions

SC3.1 SUMMARY OF KEY FINDINGS

This Summary Card covers the evidence for interventions implemented at various stages of pork processing on *Salmonella* control.

SC3.1.1 Scalding and singeing

- Results from a previous systematic review and several studies of different designs (quasi-experimental, challenge trial and cross-sectional) indicate that these steps can provide a large reduction in *Salmonella* prevalence on pork carcasses when operating under good hygienic practices, with low (scalding) to moderate (singeing) confidence in the specific magnitudes of effect from quasi-experimental field studies.

SC3.1.2 Post-slaughter and pre-chill

- Results from three controlled trial studies conducted under commercial conditions indicate that organic acid washes can significantly reduce *Salmonella* prevalence on pork carcasses prior to chilling, with high confidence in the specific magnitudes of effect; while three studies suggest a similar trend for thermal treatments, but with an overall non-significant effect. One study each found that an acidified sodium chlorite wash, hot water/organic acid wash combinations, steam vacuuming, and knife trimming may also be effective. Results with low to very low confidence from six quasi-experimental studies also found that various pre-chill washes were effective to reduce *Salmonella* prevalence.
- Several challenge trials conducted under laboratory and pilot plant conditions found that several different wash treatments can reduce inoculated *Salmonella* levels on pork carcasses. However, enhanced formulations (e.g. containing organic acids or other chemicals) did not always provide significant benefits compared with ambient water washes alone. There is low to very low confidence in results of these studies due to the artificial study conditions, small sample sizes, and, in some cases, large variation in effect between studies.

SC3.1.3 Chg

- The primary purpose of chilling is to inhibit microbial growth; however, the process also can have an 'intervention' effect on reducing *Salmonella*. One previous systematic review that included 13 quasi-experimental studies found that chilling tends to reduce *Salmonella* prevalence on pork carcasses, but some studies found higher prevalence values post- vs pre-chilling, indicating potential for cross-contamination during this step. The confidence in the specific quantitative estimate of *Salmonella* reduction due to chilling is very low due to the large variation in results among studies and possible risks of bias in the available studies.

- One controlled trial conducted under commercial conditions found, with low confidence, that spray chilling (intermittent water sprays during the first 8 h) reduced *Salmonella* on carcasses compared with conventional air chilling at 4°C.

SC3.1.4 Packaging

- Various laboratory challenge trials indicate that modified packaging and preservation with various chemicals and extracts have potential to reduce *Salmonella* concentrations and growth in pork products during distribution and storage.

SC3.1.5 Multiple processing stages

- One systematic review found evidence with low confidence from three quasi-experimental studies that HACCP implementation on its own is not linked to reductions in *Salmonella* prevalence on pork carcasses.
- Two quasi-experimental studies found that, with low confidence, cleaning and disinfection practices in processing plants can lead to reductions in *Salmonella* prevalence on environmental samples. Two cross-sectional studies also found relationships between good hygienic practices and lower *Salmonella* prevalence.
- Results from eight other quasi-experimental studies suggest a clear trend for *Salmonella* to be reduced to very low levels by implementing multiple interventions throughout processing from slaughter to chilling, but with very low confidence in the specific magnitude of effect due to possible confounding factors and variation among studies.
- Six challenge trials conducted under laboratory conditions found, with low to very low confidence, that various non-thermal interventions (UV irradiation, plasma, and supercritical carbon dioxide) could be promising to further control *Salmonella* in pork.
- Various quantitative microbial risk assessment models (QMRAs) and other simulation models suggest that multiple intervention combinations are probably needed to achieve the largest reductions in *Salmonella* contamination and human illness, but the extent of external generalizability of these models to other settings and contexts is unclear.

SC3.1.6 Overall implications

- Moderate to high confidence was identified across multiple studies for the effect estimates of singeing and pre-chill thermal and organic acid washes to reduce *Salmonella* prevalence on pork carcasses.
- Other intervention effect estimates were of lower confidence due to one or more of the following factors: potential for confounding bias, small sample sizes, variability in effects across studies, and laboratory conditions that might not reflect real-world settings.
- In challenge trials, enhanced washes (e.g. containing organic acids or other chemicals) did not always achieve more effective reductions compared with ambient temperature water washes alone.
- Greatest reductions in *Salmonella* are likely to be achieved when multiple interventions are implemented in combination, as part of a "multiple-hurdle" strategy.
- Potential for increases in Salmonella prevalence at various points in the chain (e.g. evisceration) highlight the importance of preventing cross-contamination and maintaining good hygienic practices.

SC3.2 INTERVENTION DESCRIPTION

This Summary Card summarizes the evidence for a range of interventions that can be implemented at the processing level to reduce *Salmonella* contamination of pork. Specific categories of interventions where evidence was identified in the literature on *Salmonella* control and that are covered in this Summary Card include:

- **Organic acid washes:** refers to washes with antimicrobials such as lactic, acetic or citric acid that affect microbial growth through disruptions to nutrient transport and energy generation, and can cause injury to microbial cells through their low pH (Wheeler, Kalchayanand and Bosilevac, 2014).
- **Washes containing other chemicals and oxidizers:** includes washes containing other *miscellaneous products* that destroy bacteria through various actions, such as oxidation and disruption of cellular functions (Wheeler, Kalchayanand and Bosilevac, 2014), or that prevent bacterial attachment to meat. Examples include chlorine solutions, trisodium phosphate, acidified sodium chlorite, electrolysed oxidized water, and ozone.
- **Water washes:** refers to an ambient, warm or cold temperature wash to physically remove contamination from carcasses.
- **Thermal washes:** refers to various heat treatment washes to destroy microbial cells. Examples include hot water and steam pasteurization.
- **Steam vacuuming** refers to steam treatment of carcasses to loosen soil and destroy bacteria, followed by application of a vacuum to remove contaminants.

- **Non-thermal interventions:** refers to alternative, non-chemical and non-thermal interventions that aim to reduce microbial contamination while preserving product quality and nutrients that can be affected by thermal treatments (Wheeler, Kalchayanand and Bosilevac, 2014). Examples include ionizing irradiation, ultraviolet (UV) light, plasma gases, and high pressure processing.
- **Modified packaging and preservation techniques:** refers to a range of interventions that can be applied to prevent spoilage and inhibit microbial growth during final product distribution and storage.
- **Good hygienic and production practices:** includes a range of different practices that are implemented in processing plants and that may have a pathogen-reduction effect. Examples with evidence included in this summary are:
  - *Scalding* is the process of immersing pigs in hot water tanks with the primary purpose of softening hairs on the pig skin to make them easier to remove.
  - *Singeing* is the process of removing remaining hairs from a pig carcass.
  - *Chilling* is the step at the end of the slaughter process and before fabrication, for rapidly reducing the carcass temperature to prevent microbial growth, and preserve product quality.
  - *Hazard analysis critical control point (HACCP)* is a system that identifies, evaluates, and controls hazards which are significant for food safety (Codex Alimentarius Commission, 2003).

These interventions can be applied at various stages of the processing line, with some steps themselves having a pathogen-reduction effect. Evidence was identified for interventions at the following stages of processing, each of which is presented and discussed separately below:

- Scalding and singeing
- Post-slaughter and pre-chill
- Chilling
- Packaging
- Multiple and unspecified processing stages

SC3.3 SCALDING AND SINGEING

One previous systematic review of primary research studies describing the prevalence of *Salmonella* in pork from slaughter to chilling found that both the scalding and singeing steps were associated with a decrease in prevalence (O'Connor *et al.*, 2012). However, the review also noted that following these steps, evisceration, splitting and post-mortem inspection were associated with an increase in *Salmonella* prevalence, highlighting the importance of preventing cross-contamination following scalding and singeing (O'Connor *et al.*, 2012). The review met 10/11 recommended reporting reliability criteria using the AMSTAR quality assessment tool.

Six quasi-experimental studies conducted under commercial conditions were identified that evaluated the effect of either scalding or singeing in pre- vs post-samples, five of which reported extractable data and are summarized in Table SC3.1. The studies found that both scalding and singeing significantly reduced *Salmonella* prevalence on pork carcasses (Davies, McLaren and Bedford, 1999; Bolton *et al.*, 2002; Pearce *et al.*, 2004; Da Silva *et al.*, 2012; Cocora *et al.*, 2013; Hernandez *et al.*, 2013). The confidence in the estimate of reduction for singeing is moderate, while for scalding it is low due to significant heterogeneity in the findings among studies.

One challenge trial conducted under pilot plant conditions (Table SC3.2) found that singeing reduced inoculated *Salmonella* levels on pork carcasses by more than 2 logs compared with no treatment (Clayton, 2002).

One cross-sectional study in Canada found that contamination of the scald tank water with *Salmonella* was significantly associated with a higher *Salmonella* prevalence in carcass lots (Letellier *et al.*, 2009), while another cross-sectional study in Thailand found that re-use of the scalding tank water in multiple processing batches was associated with a higher *Salmonella* prevalence on carcass swabs (Tadee, Boonkhot and Patchanee, 2014). A cross-sectional study of 10 Belgian slaughterhouses found that scalding with steam was associated with a lower odds of identifying *Salmonella* on post-chill pork carcasses (Delhalle *et al.*, 2008). Another cross-sectional study found that increasing time spent in the singeing unit was associated with lower *Salmonella* prevalence in carcass swabs (Marier *et al.*, 2014).

SC3.4 POST-SLAUGHTER AND PRE-CHILLING

Seven controlled trials conducted under commercial conditions were identified that evaluated the effect of pre-chill washes on *Salmonella* reduction in pork (Table SC3.3). Three studies investigated hot water (74°C for 5 sec, and 76.5–81°C for 15 sec) or steam (82–85°C for 60 sec), finding a trend for reduction in *Salmonella* prevalence, but an overall non-significant effect, with high confidence (Eggenberger-Solorzano *et al.*, 2002; Trivedi, Reynolds and Chen, 2007; Hamilton *et al.*, 2010). Four studies found that organic acid sprays (2–5% lactic acid for up to 60 sec and 1.8–2% acetic acid for 3–25 sec) also significantly reduced *Salmonella* prevalence on carcasses, with high confidence in the results when compared with no treatment (Epling, Carpenter and Blankenship, 1993; Frederick *et al.*, 1994; Eggenberger-Solorzano *et al.*, 2002). One study found that an acidified sodium chlorite

Summary-of-findings tables for the effects of scalding and singeing

TABLE SC3.1 Quasi-experimental studies

Intervention	Intervention/ outcome sample	No. trials or studies	Odds ratio for intervention effect(1)	% Samples *Salmonella* positive in study population(2)		GRADE rating(3)	References
				Pre-treatment	Post-treatment		
Scalding	Carcass	4/4	Median = 0.03 (Range: 0.02-0.17)	34.8%	1.5% (Range: 0.9-8.3)	Low	Davies, McLaren and Bedford, 1999; Bolton *et al.*, 2002; Pearce *et al.*, 2004; Hernandez *et al.*, 2013
Singeing	Carcass	4/4	MA = 0.26 (95% CI: 0.15-0.43)	18.1%	5.4% (95% CI: 3.3-8.8)	Moderate	Davies, McLaren and Bedford, 1999; Pearce *et al.*, 2004; Da Silva *et al.*, 2012; Hernandez *et al.*, 2013

NOTES: MA = meta-analysis average estimate from random-effects model; CI = confidence interval.

(1) For interventions with multiple trials, refers to the average odds ratio estimate and 95% CI from random-effects meta-analysis if heterogeneity was not significant (I^2<60%). If heterogeneity was significant, the median and range of odds ratios from individual studies is presented instead. If only one trial was available, refers to the odds ratio and 95% CI from the individual study.
For challenge trials, meta-analysis was not conducted due to limiting data reporting. Instead, results are presented as a median (range) log reduction for multiple trials, and as the single estimate and 95% CI for individual studies.

(2) The percentage of samples expected to be positive for *Salmonella* in the comparison group (assumed control risk, ACR) refers to the median prevalence in the comparison group trials. The expected percentage in the intervention/treatment group was calculated using the following formula: (OR × ACR / 1 -ACR + [OR × ACR]). The odds ratio used in this formula corresponds to the odds ratio presented in the intervention effect column as described under (1) above. If heterogeneity was not significant (I^2<60%), the estimate is presented with a 95% CI in brackets. If heterogeneity was significant, the median estimate is presented with the range of individual study effects in brackets to illustrate the variability in results across studies.

(3) GRADE rating descriptions: very low = the true effect is likely to be substantially different from the measured estimate; low = the true effect may be substantially different from the measured estimate; moderate = the true effect is likely to be close to the measured estimate, but there is a possibility that it is substantially different; high = there is strong confidence that the true effect lies close to that of the measured estimate.

TABLE SC3.2 Challenge trial studies

Intervention	Intervention/ outcome sample	Comparison group	No. trials/ studies	Log CFU/cm² reduction(1)	GRADE rating(2)	References
Singeing	Carcass	No treatment	1/1	Single study = 2.20 (95% CI: 2.10-2.30)	Low	Clayton, 2002

NOTES: CI = confidence interval.

(1) For interventions with multiple trials, refers to the average odds ratio estimate and 95% CI from random-effects meta-analysis if heterogeneity was not significant (I^2<60%). If heterogeneity was significant, the median and range of odds ratios from individual studies is presented instead. If only one trial was available, refers to the odds ratio and 95% CI from the individual study.
For challenge trials, meta-analysis was not conducted due to limiting data reporting. Instead, results are presented as a median (range) log reduction for multiple trials, and as the single estimate and 95% CI for individual studies.

(2) GRADE rating descriptions: very low = the true effect is likely to be substantially different from the measured estimate; low = the true effect may be substantially different from the measured estimate; moderate = the true effect is likely to be close to the measured estimate, but there is a possibility that it is substantially different; high = there is strong confidence that the true effect lies close to that of the measured estimate.

spray (900–1100 ppm for 15 sec) reduced *Salmonella* prevalence on pork carcasses compared with the control group (Hamilton *et al.*, 2010), and one study found that a combination of hot water and organic acid washes also tended to reduce *Salmonella* prevalence on pork carcasses compared with a control group (Eggenberger-Solorzano *et al.*, 2002). Le Roux, Minvielle and Gault (2008) found that steam vacuuming (at 90°C for 15 sec) was equally as effective as knife trimming to reduce *Salmonella* prevalence on pork carcasses (Table SC3.3).

Four other controlled trials conducted under commercial conditions were found, but none identified *Salmonella*, precluding evaluation of intervention efficacy on this pathogen (Prasai *et al.*, 1992; Fu, Sebranek and Murano, 1994; Morris *et al.*, 1997; Carr *et al.*, 1998). One of these studies evaluated the effect of hot fat trimming against *Salmonella* in pork, finding that it resulted in significantly lower coliform counts on pork carcasses compared with those that were not trimmed of fat, but had no effect on *Staphylococcus* spp., lactic acid bacteria or aerobic plate counts (Carr *et al.*, 1998). The other three trials found mixed evidence for the effect of pre-chill wash interventions: one study found that a trisodium phosphate wash, dip or spray was not effective to reduce aerobic plate counts on pork carcasses and loins (Morris *et al.*, 1997); another found that organic acid sprays had limited effect on aerobic plate counts, coliforms, *E. coli* and other pathogens on pork loins (Fu, Sebranek and Murano, 1994); and the third found that a hot (55°C) water spray with 1% lactic acid did not significantly reduce aerobic plate counts on pork carcasses (Prasai *et al.*, 1992).

Six quasi-experiments conducted under commercial conditions were identified evaluating the effect of pre-chill washes on *Salmonella* control in pork (Table SC3.4). Five of these studies were conducted under natural contamination conditions and found that a lactic acid wash (Larsen *et al.*, 2003), unspecified pre-chill washes (Mannion *et al.*, 2012; Hernandez *et al.*, 2013), and multiple interventions applied post-evisceration and pre-chilling (Creus, Pérez and Mateu, 2005; Da Silva *et al.*, 2012) tended to reduce *Salmonella* prevalence pre- vs post-treatment, with moderate to low confidence in the specific estimates of reduction. The other study found that a warm (22–23°C) water wash at high pressure (8 bar) reduced *Salmonella* prevalence on carcasses that were artificially inoculated with faecal contamination (Brustolin *et al.*, 2014).

A total of 14 challenge trials were identified that investigated the efficacy of post-slaughter and pre-chill interventions on pork carcasses in laboratory and pilot plant settings, eight of which reported extractable data and are summarized in Table SC3.5. The available evidence indicates that ambient water washes and washes containing organic acids or other chemicals tend to be effective in reducing inoculated *Salmonella* concentrations on various pork carcass tissues (Frederick *et al.*, 1994; van Netten *et al.*, 1994; Morris *et al.*, 1997; Fabrizio and Cutter, 2004; Choi *et al.*, 2009; Carpenter, Smith and Broadbent, 2011), but enhanced washes did not always provide significant benefits compared with simple ambient water washes alone (Fabrizio and Cutter, 2004; Carpenter, Smith and Broadbent, 2011; Machado *et al.*, 2013). The studies in Table SC3.5, as well as others that did not report extractable data, found that combinations of carcass washes and other processing interventions tended to achieve the largest reductions in inoculated *Salmonella* levels (Choi *et al.*, 2009; Christiansen, Krag and Aabo, 2009; Kich *et al.*, 2011; Morild *et al.*, 2011; King *et al.*, 2012; Machado *et al.*, 2013; Mansur *et al.*, 2015). There is low to very low confidence in these specific estimates of reduction due to the artificial study conditions, small sample sizes, and in some cases heterogeneity among the studies.

Preliminary results from another systematic review and network meta-analysis of water, thermal, and organic acid/chlorine-based wash interventions to control *Salmonella* on pork carcasses or carcass parts with skin found that there was no strong evidence for the superiority of any one of the three treatment types (Totton *et al.*, unpublished data). This review found that the most consistently observed association was a positive effect of organic acid/chlorine-based washes on the reduction in *Salmonella* prevalence.

One cross-sectional study found that use of chlorine in the wash water was associated with a lower *Salmonella* prevalence on carcass swabs (Tadee, Boonkhot and Patchanee, 2014).

A European Union QMRA found that implementation of pre-chill interventions to consistently reduce *Salmonella* concentrations by two logs could reduce the number of salmonellosis cases by over 90% in member states (EFSA, 2010). The study found that processing interventions are currently best placed to consistently reduce the number of human cases due to *Salmonella* from pork, but that combinations of pre- and post-harvest interventions can achieve larger reductions (EFSA, 2010).

Summary-of-findings tables for the effects of post-slaughter and pre-chill interventions

TABLE SC3.3 Controlled trial studies

Intervention	Intervention/ outcome sample	Comparison group	No. trials/ studies	Odds ratio for intervention effect(1)	% Samples *Salmonella* positive in study population(2)		GRADE rating(3)	References
					Without treatment	With treatment		
Thermal wash	Carcass (immediately after)	No treatment	4/3	MA = 0.37 (95% CI: 0.09-1.43)	4.3%	1.6% (95% CI: 0.4-6.0)	High	Eggenberger-Solorzano *et al.*, 2002; Trivedi, Reynolds and Chen, 2007; Hamilton *et al.*, 2010
Organic acid washes	Carcass (immediately after)	No treatment	4/3	MA = 0.29 (95% CI: 0.12-0.75)	7.7%	2.4% (95% CI: 0.9-5.9)	High	Epling, Carpenter and Blankenship, 1993; Frederick *et al.*, 1994; Eggenberger-Solorzano *et al.*, 2002
	Carcass (immediately after)	Water wash	2/1	MA = 0.12 (95% CI: 0.03-0.49)	100.0%	19.2% (95% CI: 1.6-77.7)	Low	van Netten *et al.*, 1995
	Carcass (24 hours after)	No treatment	1/1	Single study = 0.003 (95% CI: 0.00-0.05)	10.7%	0.7% (95% CI: 0.002-0.6)	Low	Epling, Carpenter and Blankenship, 1993
Other chemical/ oxidizer wash	Carcass (immediately after)	No treatment	1/1	Single study = 0.40 (95% CI: 0.16-0.96)	16.0%	7.0% (95% CI: 3.0-15.4)	Low	Hamilton *et al.*, 2010
Steam vacuuming	Carcass (immediately after)	Knife trimming	1/1	Single study = 1.12 (95% CI: 0.12-77.32)	2.2%	6.5% (95% CI: 0.3-63.7)	Low	Le Roux, Minvielle and Gault 2008
Multiple Interventions	Carcass (immediately after)	No treatment	2/1	MA = 0.53 (95% CI: 0.08-3.34)	4.2%	2.3% (95% CI: 0.4-12.7)	Low	Eggenberger-Solorzano *et al.*, 2002

NOTES: MA = meta-analysis average estimate from random-effects model; CI = confidence interval.

(1) For interventions with multiple trials, refers to the average meta-analysis estimate and 95% CI from random-effects meta-analysis if heterogeneity was not significant (I^2<60%). If heterogeneity was significant, the median and range of odds ratios or log reduction values from individual studies are presented instead. If only one trial was available, refers to the odds ratio or log reduction value and 95% CI from the individual study. For challenge trials, meta-analysis was not conducted due to limiting data reporting. Instead, results are presented as a median (range) log reduction for multiple trials, and as the single estimate and 95% CI for individual studies.

(2) The percentage of samples expected to be positive for *Salmonella* in the comparison group (assumed control risk, ACR) refers to the median prevalence in the comparison group trials. The expected percentage in the intervention/treatment group was calculated using the following formula for the odds ratio (OR) measure of effect: (OR × ACR / 1 - ACR + [OR × ACR]). The odds ratio used in this formula corresponds to the odds ratio presented in the intervention effect column as described under (1) above. If heterogeneity was not significant (I^2<60%), the estimate is presented with a 95% CI in brackets. If heterogeneity was significant, the median estimate is presented with the range of individual study effects in brackets to illustrate the variability in results across studies.

(3) GRADE rating descriptions: very low = the true effect is likely to be substantially different from the measured estimate; low = the true effect may be substantially different from the measured estimate; moderate = the true effect is likely to be close to the measured estimate, but there is a possibility that it is substantially different; high = there is strong confidence that the true effect lies close to that of the measured estimate.

TABLE SC3.4 Quasi-experimental studies

Intervention	Intervention/ outcome sample	No. trials/ studies	Odds ratio for intervention effect(1)	% Samples *Salmonella* positive in study population(2)		GRADE rating(3)	References
				Pre-treatment	Post-treatment		
Water wash	Carcass	1/1	Single study = 0.02 (95% CI: 0.001-0.23)	91.7%	16.7% (95% CI: 1.5-71.9)	Very low	Brustolin *et al.*, 2014
Organic acid wash	Carcass	1/1	Single study = 0.45 (95% CI: 0.22-0.92)	13.9%	6.7% (95% CI: 3.4-12.9)	Low	Larsen *et al.*, 2003
Unspecified washes	Carcass	2/2	MA = 0.60 (95% CI: 0.10-3.52)	5.1%	3.1% (95% CI: 0.5-15.9)	Low	Mannion *et al.*, 2012; Hernandez *et al.*, 2013
Multiple Interventions	Carcass	2/2	MA = 0.94 (95% CI: 0.47-1.88)	8.6%	8.1% (95% CI: 4.2-15.0)	Moderate	Creus, Pérez and Mateu, 2005; Da Silva *et al.*, 2012

NOTES: MA = meta-analysis average estimate from random-effects model; CI = confidence interval.

(1) For interventions with multiple trials, refers to the average meta-analysis estimate and 95% CI from random-effects meta-analysis if heterogeneity was not significant (I^2<60%). If heterogeneity was significant, the median and range of odds ratios or log reduction values from individual studies are presented instead. If only one trial was available, refers to the odds ratio or log reduction value and 95% CI from the individual study.
For challenge trials, meta-analysis was not conducted due to limiting data reporting. Instead, results are presented as a median (range) log reduction for multiple trials, and as the single estimate and 95% CI for individual studies.

(2) The percentage of samples expected to be positive for *Salmonella* in the comparison group (assumed control risk, ACR) refers to the median prevalence in the comparison group trials. The expected percentage in the intervention/treatment group was calculated using the following formula for the odds ratio (OR) measure of effect: (OR × ACR / 1 -ACR + [OR × ACR]). The odds ratio used in this formula corresponds to the odds ratio presented in the intervention effect column as described under (1) above. If heterogeneity was not significant (I^2<60%), the estimate is presented with a 95% CI in brackets. If heterogeneity was significant, the median estimate is presented with the range of individual study effects in brackets to illustrate the variability in results across studies.

(3) GRADE rating descriptions: very low = the true effect is likely to be substantially different from the measured estimate; low = the true effect may be substantially different from the measured estimate; moderate = the true effect is likely to be close to the measured estimate, but there is a possibility that it is substantially different; high = there is strong confidence that the true effect lies close to that of the measured estimate.

TABLE SC3.5 Challenge trial studies

Intervention	Intervention/ outcome sample	Comparison group	No. trials/ studies	Log CFU/cm² reduction[(1)]	GRADE rating[(2)]	References
Water washes	Carcass tissues	No treatment	3/3	Median = 1.03 (Range: 0.75-1.36)	Very low	Frederick *et al.*, 1994; Fabrizio and Cutter, 2004; Carpenter, Smith and Broadbent, 2011
Thermal wash	Carcass tissues	No treatment	2/2	Median = 1.76 (Range: -0.20-3.71)	Very low	Clayton, 2002; Morild *et al.*, 2011
Organic acid washes	Carcass tissues	Water wash	4/4	Median = 0.79 (Range: 0.31-1.81)	Very low	Frederick *et al.*, 1994; van Netten *et al.*, 1994; Fabrizio and Cutter, 2004; Carpenter, Smith and Broadbent, 2011
	Carcass tissues	No treatment	6/6	Median = 1.43 (Range: -0.10-2.83)	Very low	Clayton, 2002; Frederick *et al.*, 1994; Fabrizio and Cutter, 2004; Choi *et al.*, 2009; Carpenter, Smith and Broadbent, 2011; Morild *et al.*, 2011
Other chemical/ oxidizer washes	Carcass tissues	Water wash	1/1	Single study = 0.17 (95% CI: -0.31-0.66)	Low	Fabrizio and Cutter, 2004
	Carcass tissues	No treatment	3/3	Median = 1.58 (Range: 1.53-2.28)	Very low	Clayton, 2002; Fabrizio and Cutter, 2004; Morild *et al.*, 2011
Multiple Interventions	Carcass tissues	Chilling only	1/1	Single study = 0.38 (No measure of variability reported)	Low	King *et al.*, 2012
	Carcass tissues (measured post-treatment)	Pre-treatment	1/1	Single study = 1.78 (95% CI: 0.69-2.87)	Very low	Morild *et al.*, 2011

NOTES: MA = meta-analysis average estimate from random-effects model; CI = confidence interval.

(1) For interventions with multiple trials, refers to the average meta-analysis estimate and 95% CI from random-effects meta-analysis if heterogeneity was not significant (I^2<60%). If heterogeneity was significant, the median and range of odds ratios or log reduction values from individual studies are presented instead. If only one trial was available, refers to the odds ratio or log reduction value and 95% CI from the individual study.
For challenge trials, meta-analysis was not conducted due to limiting data reporting. Instead, results are presented as a median (range) log reduction for multiple trials, and as the single estimate and 95% CI for individual studies.

(2) GRADE rating descriptions: very low = the true effect is likely to be substantially different from the measured estimate; low = the true effect may be substantially different from the measured estimate; moderate = the true effect is likely to be close to the measured estimate, but there is a possibility that it is substantially different; high = there is strong confidence that the true effect lies close to that of the measured estimate.

SC3.5 CHILLING

The primary purpose of carcass chilling is to inhibit microbial growth; however, the process also can have an 'intervention' effect on reducing *Salmonella*. One previous systematic review was identified (including a recent update) evaluating the effects of chilling to reduce *Salmonella* prevalence on pork carcasses (Barron, Bergin and Butler, 2008; Barron *et al.*, 2013). The review identified 13 quasi-experimental studies that measured changes in *Salmonella* pre vs post chilling and meta-analysis indicated a significant average reduction (Barron *et al.*, 2013). Reported in Table SC3.6 below is a median estimate of reduction for chilling (significant heterogeneity among studies was noted). The confidence in these estimates is very low due to the large variation in results among studies (with some showing higher prevalence values post-chilling), possibility for publication bias (lack of small studies showing no effect), and unclear risk of bias of the included studies. The authors found that the total sample size, number of batches sampled in an abattoir, and size of the carcass swab area had a significant impact on the effect of chilling, indicating that the sampling methods contributed to the heterogeneity between studies (Barron *et al.*, 2013). Data from this analysis were used as inputs into a QMRA model that found that the chilling, as well as pre-chill washing, were the processing stages that contributed most significantly to reducing *Salmonella* prevalence in pork joints (Barron *et al.*, 2009; Teagasc, 2010). The review did not meet several recommended reporting criteria for systematic reviews using the AMSTAR quality assessment tool.

Two controlled trials conducted under commercial conditions were identified that evaluated the effect of different chilling methods to reduce *Salmonella* in pork (Epling, Carpenter and Blankenship, 1993; Carr *et al.*, 1998). Epling, Carpenter and Blankenship (1993) found that spray chilling (intermittent water at 4°C every 20 min for 11 sec during the first 8 hours) reduced *Salmonella* on carcasses compared with conventional air chilling at 4°C (Table SC3.7). The other study isolated *Salmonella* from only one sample, precluding evaluation of intervention efficacy on this pathogen (Carr *et al.*, 1998). The authors compared blast chilling (-10 to -25°C for 45 min followed by chilling at 2°C for 23 hours) to standard chilling (2°C for 24 hours) and found that the former resulted in significantly lower coliform counts and *Staphylococcus* spp. on pork carcasses that were not trimmed of fat, but had no effect on fat-trimmed carcasses and on aerobic plate counts (Carr *et al.*, 1998). Significant reductions were also noted for lactic acid bacteria counts (Carr *et al.*, 1998).

One challenge trial conducted under laboratory conditions was identified that found that blast chilling was more effective to reduce *Salmonella* on pork belly surfaces compared with conventional chilling, with log reduction values ranging from 0.09 to 0.18 (Table SC3.8).

Summary-of-findings tables for the effects of chilling

TABLE SC3.6 Systematic review

Intervention	Study design	Intervention/ outcome sample	No. trials/ studies	Odds ratio for intervention effect[1]	% Samples *Salmonella* positive in study population[2]		GRADE rating[3]	References
					Pre treatment	Post treatment		
Chilling	Quasi-experiments	Carcass	13/13	Median = 0.41 (Range: 0.14-2.93)	10.3%	4.5% (Range: 1.6-25.2)	Very low	Barron, Bergin and Butler, 2008; Barron *et al.*, 2013

NOTES: CI = confidence interval.

(1) For interventions with multiple trials, refers to the average odds ratio estimate and 95% CI from random-effects meta-analysis if heterogeneity was not significant (I^2<60%). If heterogeneity was significant, the median and range of odds ratios from individual studies is presented instead. If only one trial was available, refers to the odds ratio and 95% CI from the individual study. For challenge trials, meta-analysis was not conducted due to limiting data reporting. Instead, results are presented as a median (range) log reduction for multiple trials, and as the single estimate and 95% CI for individual studies.

(2) The percentage of samples expected to be positive for *Salmonella* in the comparison group (assumed control risk, ACR) refers to the median prevalence in the comparison group trials. The expected percentage in the intervention/treatment group was calculated using the following formula for the odds ratio (OR) measure of effect: (OR × ACR / 1 - ACR + [OR × ACR]). The odds ratio used in this formula corresponds to the odds ratio presented in the intervention effect column as described under (1) above. If heterogeneity was not significant (I^2<60%), the estimate is presented with a 95% CI in brackets. If heterogeneity was significant, the median estimate is presented with the range of individual study effects in brackets to illustrate the variability in results across studies.

(3) GRADE rating descriptions: very low = the true effect is likely to be substantially different from the measured estimate; low = the true effect may be substantially different from the measured estimate; moderate = the true effect is likely to be close to the measured estimate, but there is a possibility that it is substantially different; high = there is strong confidence that the true effect lies close to that of the measured estimate.

TABLE SC3.7 Controlled trial studies

Intervention	Intervention/ outcome sample	Comparison group	No. trials/ studies	Odds ratio for intervention effect(1)	% Samples *Salmonella* positive in study population(2)		GRADE rating(3)	References
					Without treatment	With treatment		
Spray chilling	Carcass (immediately after)	Conventional (air) chilling	1/1	Single study = 0.67, (95% CI: 0.32-1.38)	13.3%	9.3% (95% CI: 4.8-17.5)	Low	Epling, Carpenter and Blankenship, 1993
	Carcass (24 h after)	Conventional (air) chilling	1/1	Single study = 0.66, (95% CI: 0.33-1.30)	15.3%	10.7% (95% CI: 5.7-19.1)	Low	Epling, Carpenter and Blankenship, 1993

NOTES: CI = confidence interval.

(1) For interventions with multiple trials, refers to the average odds ratio estimate and 95% CI from random-effects meta-analysis if heterogeneity was not significant (I^2<60%). If heterogeneity was significant, the median and range of odds ratios from individual studies is presented instead. If only one trial was available, refers to the odds ratio and 95% CI from the individual study. For challenge trials, meta-analysis was not conducted due to limiting data reporting. Instead, results are presented as a median (range) log reduction for multiple trials, and as the single estimate and 95% CI for individual studies.

(2) The percentage of samples expected to be positive for *Salmonella* in the comparison group (assumed control risk, ACR) refers to the median prevalence in the comparison group trials. The expected percentage in the intervention/treatment group was calculated using the following formula for the odds ratio (OR) measure of effect: (OR × ACR / 1 -ACR + [OR × ACR]). The odds ratio used in this formula corresponds to the odds ratio presented in the intervention effect column as described under (1) above. If heterogeneity was not significant (I^2<60%), the estimate is presented with a 95% CI in brackets. If heterogeneity was significant, the median estimate is presented with the range of individual study effects in brackets to illustrate the variability in results across studies.

(3) GRADE rating descriptions: very low = the true effect is likely to be substantially different from the measured estimate; low = the true effect may be substantially different from the measured estimate; moderate = the true effect is likely to be close to the measured estimate, but there is a possibility that it is substantially different; high = there is strong confidence that the true effect lies close to that of the measured estimate.

TABLE SC3.8 Challenge trial studies

Intervention	Intervention/ outcome sample	Comparison group	No. trials/ studies	Log CFU/cm² reduction(1)	GRADE rating(3)	References
Blast chilling	Carcass belly surfaces	Conventional (air) chilling	4/1	Median = 0.13 (0.09-0.18)	Low	Chang *et al.*, 2003

NOTES: CI = confidence interval.

(1) For interventions with multiple trials, refers to the average odds ratio estimate and 95% CI from random-effects meta-analysis if heterogeneity was not significant (I^2<60%). If heterogeneity was significant, the median and range of odds ratios from individual studies is presented instead. If only one trial was available, refers to the odds ratio and 95% CI from the individual study. For challenge trials, meta-analysis was not conducted due to limiting data reporting. Instead, results are presented as a median (range) log reduction for multiple trials, and as the single estimate and 95% CI for individual studies.

(3) GRADE rating descriptions: very low = the true effect is likely to be substantially different from the measured estimate; low = the true effect may be substantially different from the measured estimate; moderate = the true effect is likely to be close to the measured estimate, but there is a possibility that it is substantially different; high = there is strong confidence that the true effect lies close to that of the measured estimate.

SC3.6 PACKAGING

The effects of various packaging and final product preservation interventions were studied in eight laboratory challenge trials identified in this review (Van Laack *et al.*, 1993; Nam *et al.*, 2006; Latha *et al.*, 2009; Nanasombat and Chooprang, 2009; Shan *et al.*, 2009; Piachin and Trachoo, 2011; Chen *et al.*, 2013; Chantarasataporn *et al.*, 2014).

Three studies investigated the effects of various natural extracts as product preservatives/antimicrobials (Nam *et al.*, 2006; Shan *et al.*, 2009; Chen *et al.*, 2013). One study found that significant log reductions in *Salmonella* on ground pork over 7 days storage were achieved with cinnamon oil and olive extracts, while oregano oil and apple extracts showed limited effectiveness (Chen *et al.*, 2013). One study investigated the inhibitory effects of cinnamon stick, oregano, clove, pomegranate peel, and grape seed extracts on *Salmonella* in raw pork, finding that all extracts were effective to reduce *Salmonella* over 9 days of storage by ~1–2 logs compared with the control group, with clove extracts being the most effective (Shan *et al.*, 2009). Another study found that while a rosemary-tocopherol combination was effective to prevent quality changes in irradiated restructured pork loins, it had no effect on *Salmonella* (Nam *et al.*, 2006).

Two studies investigated salt preservatives (Latha *et al.*, 2009; Nanasombat and Chooprang, 2009). One study found that potassium sorbate and sodium lactate each lowered the survival rate of *Salmonella* on raw pork during freezing for 72 hours (Nanasombat and Chooprang, 2009), while the other study found that 5% potassium sorbate and a combination of 5% sodium chloride and 2.5% each of sodium acetate, sodium citrate, sodium lactate and potassium sorbate applied to pork carcasses before de-boning resulted in non-detectable levels of *Salmonella*, compared with a nearly 3 log contamination in the control group after storage until spoilage (Latha *et al.*, 2009).

One study found that vacuum packaging of pork loins prior to chilling for 24 hours at 2°C resulted in enhanced reductions of *Salmonella* compared with vacuuming packing after chilling and compared with unpacked loins (Van Laack *et al.*, 1993). One study found that 0.2–0.4% water-based oligochitosan reduced *Salmonella* to undetectable levels in ground pork after 1 or 2 days of storage (Chantarasataporn *et al.*, 2014). Another study found that fresh pork treated with 2-4% potassium lactate and packaged with and without ozone injected at 200–1000 mg/h reduced *Salmonella* levels by up to 0.8 logs over 15 days of storage compared with the untreated, packaged control group (Piachin and Trachoo, 2011).

SC3.7 MULTIPLE AND UNSPECIFIED PROCESSING STAGES

One previous systematic review evaluated the efficacy of HACCP to reduce *Salmonella* contamination of pork carcasses during processing (Wilhelm *et al.*, 2011). A meta-analysis was conducted on three eligible quasi-experimental studies measuring changes in *Salmonella* prevalence before and after implementation of HACCP programmes. Results indicate low confidence of no overall effect of HACCP on *Salmonella* reduction in pork (Table SC3.9). Significant heterogeneity between study findings was noted, which the authors noted could be due to the wide variability in how HACCP is defined and implemented in different settings (Wilhelm *et al.*, 2011). Although HACCP may be effective in some settings and likely depends on the specific critical control points and other implementation factors, these findings suggest that there is insufficient evidence that HACCP itself is effective in reducing *Salmonella* prevalence on pork during processing. The review met 9/11 recommended reporting reliability criteria using the AMSTAR quality assessment tool.

One controlled trial conducted under commercial conditions was identified that evaluated the effect of a modified pluck removal practice (leaving tongue in the intact head during plucking, no head splitting, removal of head pre-chilling and pre-inspection) on *Salmonella* reduction in pork (Olsen *et al.*, 2001). The study found no significant difference in *Salmonella* prevalence on pork between the modified and conventional plucking practice (Olsen *et al.*, 2001).

Ten quasi-experiments conducted under commercial conditions were identified evaluating the effect of interventions at multiple processing stages on *Salmonella* control in pork (Table SC3.10). Two studies found that cleaning and disinfection of the abattoir equipment and environment resulted in significant reductions in *Salmonella* prevalence in environmental samples (Conter *et al.*, 2006; Gantzhorn *et al.*, 2014). There is low confidence in this estimate in terms of a possible reduction on pork carcasses, since the studies only measured environmental samples. Two cross-sectional studies also found relationships between cleaning, disinfection, and good hygienic practices and a reduced *Salmonella* prevalence on pork carcasses (Kim and Song, 2009; Letellier *et al.*, 2009).

Eight other quasi-experimental studies measured the effects of multiple and unspecified interventions applied over various processing stages (Saide-Albornoz *et al.*, 1995; Tamplin *et al.*, 2001; Keenliside *et al.*, 2005; Algino *et al.*, 2009; De Busser *et al.*, 2011; Schmidt *et al.*, 2012; Keelara *et al.*, 2013; Williams, Ebel and Allender, 2015). The studies generally found a very large effect on *Salmonella* reduction (Table SC3.10). However, the overall confidence in the effect estimates is very low due to potential for confounding in these studies and because of large variations between studies in the magnitude of the effect, which could be due to differences in the specific interventions applied or other factors. While the overall confidence in identifying a specific estimate of reduction is very low due to these factors, the studies highlight a clear trend for *Salmonella* to be reduced to very low levels by implementing multiple interventions throughout processing. However, potential for cross-contamination in one study was noted, finding a higher prevalence of *Salmonella* post-chilling compared with pre-washing (Algino *et al.*, 2009).

Six challenge trials were identified that investigated the efficacy of various non-thermal interventions to reduce *Salmonella* contamination of pork carcasses in laboratory settings, four of which reported extractable data and are summarized in Table SC3.11. Two studies investigated UV light irradiation (Wong, Linton and Gerrard, 1998; Sommers, Sites and Musgrove, 2010) and two studies investigated plasma gas interventions (Kim *et al.*, 2013; Jayasena *et al.*, 2015), achieving up to 1 to 2 log reductions. Two other trials investigated the effect of supercritical carbon dioxide and found that at high doses it resulted in multiple log reductions of *Salmonella* in pork loins and ground pork (Bae *et al.*, 2011a, b).

Two cross-sectional studies found that a faster processing time was associated with a lower *Salmonella* prevalence on carcasses (Kim and Song, 2009; Marier *et al.*, 2014).

A QMRA to assess the risks for human salmonellosis due to consumption of fresh minced pork meat in Belgium evaluated different hypothetical *Salmonella* mitigation strategies at different stages of production and found that those implemented at the polishing, evisceration, chilling and post-processing stages were most effective, whereas pre-harvest interventions and those at the beginning of the slaughter process had a more limited effect (Bollaerts *et al.*, 2010).

An Irish expert elicitation study to rank interventions to control *Salmonella* in the pork chain found that, at processing, experts ranked (1) careful evisceration, (2) bagging the bung and (3) logistic slaughter as the interventions with the most potential to reduce the risk of carcass/cross contamination (Teagasc, 2010). Sufficient employee training was noted as a critical factor affecting the success of the first two interventions (Teagasc, 2010). The experts noted that a combination of interventions along the farm-to-fork chain is essential (Teagasc, 2010). QMRA models assessing the risk of human salmonellosis cases due to pork consumption in the UK and EU also found that a combination of interventions implemented along the food chain were most effective to reduce the illness burden (Hill *et al.*, 2009, 2011).

Four Danish stochastic simulation models were identified (Alban and Stärk, 2005; Goldbach and Alban, 2006; Lawson *et al.*, 2009; Baptista *et al.*, 2011). Alban and

Stärk (2005) found that herd prevalence, singeing efficiency, and cross-contamination during slaughter and handling were the most important factors affecting pork carcass *Salmonella* prevalence, and that a combination of interventions achieved the largest reductions in prevalence. A cost-benefit analysis of four interventions (hot water decontamination of carcasses, logistic slaughter, use of home-mixed feed, and acidified feed) found that hot water decontamination was the only economically viable option (Goldbach and Alban, 2006). Another cost-effectiveness analysis found that among four processing intervention options (hot water, steam ultrasound, steam vacuum and lactic acid decontamination), steam ultrasound was the most cost-effective, following by hot water decontamination (Lawson *et al.*, 2009). An economic model comparing alternative processing intervention scenarios (logistic slaughter, hot-water decontamination, steam ultrasound and steam vacuum) found that steam vacuuming and steam ultrasound were the most cost-effective, and that cost-effectiveness varied by abattoir size (Baptista *et al.*, 2011).

A cross-sectional study by Bollerslev *et al.* (2013) found a correlation between the level of *E. coli* and *Salmonella* prevalence on pork carcasses during processing, with a risk model finding that the number of human cases could have been reduced by approximately 50% if the *E. coli* levels at slaughter had not exceeded 3 to 4 log CFU/38 cm^2.

Summary-of-findings tables for the effects of interventions applied at multiple or unspecified processing stages

TABLE SC3.9 Systematic review

Intervention	Study design	Intervention/ outcome sample	No. trials/ studies	Odds ratio for intervention effect(1)	% Samples *Salmonella* positive in study population(2)		GRADE rating(3)	References
					Pre treatment	Post treatment		
HACCP	Quasi-experiments	Carcass	3/3	Median = 0.94 (Range: 0.51-1.29)	6.9%	6.5% (Range: 3.7-8.7)	Low	Wilhelm *et al.*, 2011

NOTES: CI = confidence interval.

(1) For interventions with multiple trials, refers to the average meta-analysis estimate and 95% CI from random-effects meta-analysis if heterogeneity was not significant (I^2<60%). If heterogeneity was significant, the median and range of odds ratios or log reduction values from individual studies are presented instead. If only one trial was available, refers to the odds ratio or log reduction value and 95% CI from the individual study. For challenge trials, meta-analysis was not conducted due to limiting data reporting. Instead, results are presented as a median (range) log reduction for multiple trials, and as the single estimate and 95% CI for individual studies.

(2) The percentage of samples expected to be positive for *Salmonella* in the comparison group (assumed control risk, ACR) refers to the median prevalence in the comparison group trials. The expected percentage in the intervention/treatment group was calculated using the following formula for the odds ratio (OR) measure of effect: (OR × ACR / 1 -ACR + [OR × ACR]). The odds ratio used in this formula corresponds to the odds ratio presented in the intervention effect column as described under (1) above. If heterogeneity was not significant (I^2<60%), the estimate is presented with a 95% CI in brackets. If heterogeneity was significant, the median estimate is presented with the range of individual study effects in brackets to illustrate the variability in results across studies.

(3) GRADE rating descriptions: very low = the true effect is likely to be substantially different from the measured estimate; low = the true effect may be substantially different from the measured estimate; moderate = the true effect is likely to be close to the measured estimate, but there is a possibility that it is substantially different; high = there is strong confidence that the true effect lies close to that of the measured estimate.

TABLE SC3.10 Quasi-experimental studies

Intervention	Intervention/ outcome sample	No. trials/ studies	Odds ratio for intervention effect[1]	% Samples *Salmonella* positive in study population[2]		GRADE rating[3]	References
				Pre treatment	Post treatment		
Cleaning and disinfection of abattoir equipment/ environment	Various environmental samples	2/2	MA = 0.27 (95% CI: 0.16-0.46)	22.8%	7.5% (95% CI: 4.6-12.1)	Low	Conter *et al.*, 2006; Gantzhorn *et al.*, 2014
Multiple interventions (post-exsanguin-ation to post-chill)	Carcass	1/1	Single study = 0.003 (95% CI: 0.00-0.02)	73.0%	0.8% (95% CI: 0.1-5.8)	Very low	Tamplin *et al.*, 2001
Multiple interventions (pre-evisceration to post-chill)	Carcass	4/4	Median = 0.06 (Range: 0.01-0.16)	14.9%	1.0% (Range: 0.2-2.8)	Very low	Saide-Albornoz *et al.*, 1995; Keenliside *et al.*, 2005; Schmidt *et al.*, 2012; Williams, Ebel and Allender, 2015
Multiple interventions (post-evisceration to post-chill)	Carcass	3/3	Median = 0.97 (Range: 0.14-2.23)	9.4%	9.1% (Range: 1.5-18.8)	Very low	Alginoet *al.*, 2009; De Busser *et al.*, 2011; Keelara *et al.*, 2013

NOTES: MA = meta-analysis average estimate; CI = confidence interval.

(1) For interventions with multiple trials, refers to the average meta-analysis estimate and 95% CI from random-effects meta-analysis if heterogeneity was not significant (I^2<60%). If heterogeneity was significant, the median and range of odds ratios or log reduction values from individual studies are presented instead. If only one trial was available, refers to the odds ratio or log reduction value and 95% CI from the individual study.
For challenge trials, meta-analysis was not conducted due to limiting data reporting. Instead, results are presented as a median (range) log reduction for multiple trials, and as the single estimate and 95% CI for individual studies.

(2) The percentage of samples expected to be positive for *Salmonella* in the comparison group (assumed control risk, ACR) refers to the median prevalence in the comparison group trials. The expected percentage in the intervention/treatment group was calculated using the following formula for the odds ratio (OR) measure of effect: (OR × ACR / 1 - ACR + [OR × ACR]). The odds ratio used in this formula corresponds to the odds ratio presented in the intervention effect column as described under (1) above. If heterogeneity was not significant (I^2<60%), the estimate is presented with a 95% CI in brackets. If heterogeneity was significant, the median estimate is presented with the range of individual study effects in brackets to illustrate the variability in results across studies.

(3) GRADE rating descriptions: very low = the true effect is likely to be substantially different from the measured estimate; low = the true effect may be substantially different from the measured estimate; moderate = the true effect is likely to be close to the measured estimate, but there is a possibility that it is substantially different; high = there is strong confidence that the true effect lies close to that of the measured estimate.

TABLE SC3.11 Challenge trial studies

Intervention	Intervention/ outcome sample	Comparison group	Units	No. trials/ studies	Log reduction[(1)]	GRADE rating[(3)]	References
UV irradiation	Carcass tissue	No treatment	CFU/ cm^2	1/1	Single study = 0.51 (95% CI: 0.34–0.67)	Very low	Sommers, Sites and Musgrove, 2010
	Pork chops	No treatment	CFU/g	1/1	Single study = 2.37 (no measure of variability reported)	Low	Wong, Linton and Gerrard, 1998
Non-thermal plasma	Pork loin/butts	No treatment	CFU/g	2/2	Median = 1.47 (Range: 1.38–1.56)	Low	Kim *et al.*, 2013; Jayasena *et al.*, 2015

NOTES: CI = confidence interval.

(1) For interventions with multiple trials, refers to the average meta-analysis estimate and 95% CI from random-effects meta-analysis if heterogeneity was not significant (I^2<60%). If heterogeneity was significant, the median and range of odds ratios or log reduction values from individual studies are presented instead. If only one trial was available, refers to the odds ratio or log reduction value and 95% CI from the individual study.
For challenge trials, meta-analysis was not conducted due to limiting data reporting. Instead, results are presented as a median (range) log reduction for multiple trials, and as the single estimate and 95% CI for individual studies.

(3) GRADE rating descriptions: very low = the true effect is likely to be substantially different from the measured estimate; low = the true effect may be substantially different from the measured estimate; moderate = the true effect is likely to be close to the measured estimate, but there is a possibility that it is substantially different; high = there is strong confidence that the true effect lies close to that of the measured estimate.

SC3.8 REFERENCES CITED IN SC3

Alban, L. & Stärk, K.D.C. 2005. Where should the effort be put to reduce the *Salmonella* prevalence in the slaughtered swine carcass effectively? *Preventive Veterinary Medicine*, 68(1): 63–79.

Algino, R.J., Badtram, G.A., Ingham, B.H. & Ingham, S.C. 2009. Factors associated with *Salmonella* prevalence on pork carcasses in very small abattoirs in Wisconsin. *Journal of Food Protection*, 72(4): 714–721.

Bae, Y.Y., Choi, Y.M., Kim, M.J., Kim, K.H., Kim, B.C. & Rhee, M.S. 2011a. Application of supercritical carbon dioxide for microorganism reductions in fresh pork. *Journal of Food Safety*, 31(4): 511–517.

Bae, Y.Y., Kim, N.H., Kim, K.H., Kim, B.C. & Rhee, M.S. 2011b. Supercritical carbon dioxide as a potential intervention for ground pork decontamination. *Journal of Food Safety*, 31(1): 48–53.

Baptista, F.M., Halasa, T., Alban, L., Nielsen, L.R., 2011. Modelling food safety and economic consequences of surveillance and control strategies for *Salmonella* in pigs and pork. *Epidemiology and Infection*, 139(5): 754–764.

Barron, U.G., Bergin, D. & Butler, F. 2008. A meta-analysis study of the effect of chilling on prevalence of *Salmonella* on pig carcasses. *Journal of Food Protection*, 71(7): 1330–1337.

Barron, U.G., Soumpasis, I., Butler, F., Prendergast, D., Duggan, S. & Duffy, G. 2009. Estimation of prevalence of *Salmonella* on pig carcasses and pork joints, using a quantitative risk assessment model aided by meta-analysis. *Journal of Food Protection*, 72(2): 274–285.

Barron, U.G., Cadavez, V., Sheridan, J.J. & Butler, F. 2013. Modelling the effect of chilling on the occurrence of *Salmonella* on pig carcasses at study, abattoir and batch levels by meta-analysis. *International Journal of Food Microbiology*, 163(2-3): 101–113.

Bollaerts, K., Messens, W., Aerts, M., Dewulf, J., Maes, D., Grijspeerdt, K. & Stede, Y. 2010. Evaluation of scenarios for reducing human salmonellosis through household consumption of fresh minced pork meat. *Risk Analysis*, 30(5): 853–865.

Bollerslev, A., Nauta, M., Hald, T., Hansen, T. & Aabo, S. 2013. Use of a slaughter hygiene indicator (*Escherichia coli*) to quantify the risk of human salmonellosis related to pork in Denmark - an approach for risk-based meat control? pp. 122–124, *in:* Proceedings of the 10th International Conference on the Epidemiology and Control of Biological, Chemical and Physical Hazards in Pigs and Pork. Portland, ME, USA.

Bolton, D.J., Pearce, R.A., Sheridan, J.J., Blair, I.S., McDowell, D.A. & Harrington, D. 2002. Washing and chilling as critical control points in pork slaughter hazard analysis and critical control point (HACCP) systems. *Journal of Applied Microbiology*, 92(5): 893–902.

Brustolin, J.C., Pisol, A., Steffens, J., Toniazzo, G., Valduga, E., di Luccio, M. & Cansian, R.L. 2014. Decontamination of pig carcasses using water pressure and lactic acid. *Brazilian Archives of Biology and Technology,* 57(6): 954–961.

Carpenter, C.E., Smith, J.V. & Broadbent, J.R. 2011. Efficacy of washing meat surfaces with 2% levulinic, acetic, or lactic acid for pathogen decontamination and residual growth inhibition. *Meat Science*, 88(2): 256–260.

Carr, M.A., Thompson, L.D., Miller, M.F., Boyd Ramsey, C. & Kaster, C.S. 1998. Chilling and trimming effects on the microbial populations of pork carcasses. *Journal of Food Protection,* 61: 487–489.

Chang, V.P., Mills, E.W. & Cutter, C.N. 2003. Reduction of bacteria on pork carcasses associated with chilling method. *Journal of Food Protection,* 66: 1019–1024.

Chantarasataporn, P., Tepkasikul, P., Kingcha, Y., Yoksan, R., Pichyangkura, R., Visessanguan, W. & Chirachanchai, S. 2014. Water-based oligochitosan and nanowhisker chitosan as potential food preservatives for shelf-life extension of minced pork. *Food Chemistry,* 159: 463–470.

Chen, C.H., Ravishankar, S., Marchello, J. & Friedman, M. 2013. Antimicrobial activity of plant compounds against *Salmonella* Typhimurium DT104 in ground pork and the influence of heat and storage on the antimicrobial activity. *Journal of Food Protection,* 76(7): 1264–1269.

Choi, Y.M., Kim, O.Y., Kim, K.H., Kim, B.C. & Rhee, M.S. 2009. Combined effect of organic acids and supercritical carbon dioxide treatments against non-pathogenic *Escherichia coli, Listeria monocytogenes, Salmonella* Typhimurium and *E. coli* O157:H7 in fresh pork. *Letters in Applied Microbiology*, 49(4): 510–515.

Christiansen, P., Krag, R. & Aabo, S. 2009. Effect of hot water and lactic acid decontamination on *Escherichia coli, Salmonella* Typhimurium and *Yersinia enterocolitica* on pork. p. 253, *in:* Proceedings of 8th International Symposium on the Epidemiology and Control of Foodborne Pathogens in Pork. Quebec City, Canada.

Clayton, N.C. 2002. The efficacy of various *Salmonella* intervention methods applied to pork carcasses during slaughter. Masters thesis. University of Kentucky, USA.

Cocora, Z.M., Dobai, G., Lazarescu, C. & Tibru, I. 2013. The impact of the slaughtering technological flow stages on the *Salmonella* spp. prevalence. *Lucrari Stiintifice - Universitatea de Stiinte Agricole a Banatului Timisoara, Medicina Veterinara*, 46: 37.

Codex Alimentarius Commission. 2003. Hazard analysis and critical control point (HACCP) system and guidelines for its application. Doc. CAC/RCP 1-1969, Rev. 4-2003.

Conter, M., Zanardi, E., Ghidini, S., Guidi, E. & Campanini, G. 2006. Microbiological condition of carcasses and equipment in a pig slaughterhouse and evaluation of a steam decontamination system. *Italian Journal of Animal Science,* 18(4): 387–396.

Creus, E., Pérez, J.F. & Mateu, E. 2005. *Salmonella* contamination on pork carcasses: A study of critical points. p. 304, *in:* Proceedings of the 6th International Symposium on the Epidemiology and Control of Foodborne Pathogens in Pork. Rohnert Park, CA, USA.

Da Silva, L.E., Dias, V., Ferronatto, A., Guerra, P., Berno, L., Triches, N., Kich, J.D., Corbellini, L.G. & Cardoso, M. 2012. Longitudinal dissemination of *Salmonella enterica* clonal groups through the slaughter process of *Salmonella*-positive pig batches. *Journal of Food Protection,* 75(9): 1580–1588.

Davies, R.H., McLaren, I.M. & Bedford, S. 1999. Distribution of *Salmonella* contamination in two pig abattoirs. p. 267, *in:* Proceedings of the 3rd International Symposium on the Epidemiology and Control of *Salmonella* in Pork. Washington, DC., USA.

De Busser, E.V., Maes, D., Houf, K., Dewulf, J., Imberechts, H., Bertrand, S. & De Zutter, L. 2011. Detection and characterization of *Salmonella* in lairage, on pig carcasses and intestines in five slaughterhouses. *International Journal of Food Microbiology,* 145(1): 279–286.

Delhalle, L., De Sadeleer, L., Bollaerts, K., Farnir, F., Saegerman, C., Korsak, N., Dewulf, J., De Zutter, L. & Daube, G. 2008. Risk factors for *Salmonella* and hygiene indicators in the 10 largest Belgian pig slaughterhouses. *Journal of Food Protection,* 71(7): 1320–1329.

EFSA [European Food Safety Authority]. 2010. Scientific opinion on a quantitative microbiological risk assessment of *Salmonella* in slaughter and breeder pigs. *EFSA Journal,* 8(4): 1547 [90 p.].

Eggenberger-Solorzano, L., Niebuhr, S.E., Acuff, G.R. & Dickson, J.S. 2002. Hot water and organic acid interventions to control microbiological contamination on hog carcasses during processing. *Journal of Food Protection,* 65: 1248–1252.

Epling, L.K., Carpenter, J.A. & Blankenship, L.C. 1993. Prevalence of *Campylobacter* spp. and *Salmonella* spp. on pork carcasses and the reduction effected by spraying with lactic acid. *Journal of Food Protection,* 56(6): 536–&.

Fabrizio, K.A. & Cutter, C.N., 2004. Comparison of electrolyzed oxidizing water with other antimicrobial interventions to reduce pathogens on fresh pork. *Meat Science,* 68(3): 463–468.

Frederick, T.L., Miller, M.F., Thompson, L.D. & Ramsey, C.B. 1994. Microbiological properties of pork cheek meat as affected by acetic acid and temperature. *Journal of Food Science,* 59(2): 300–302.

Fu, A.H., Sebranek, J.G. & Murano, E.A. 1994. Microbial and quality characteristics of pork cuts from carcasses treated with sanitizing sprays. *Journal of Food Science,* 59(2): 306–309.

Gantzhorn, M.R., Pedersen, K., Olsen, J.E. & Thomsen, L.E. 2014. Biocide and antibiotic susceptibility of *Salmonella* isolates obtained before and after cleaning at six Danish pig slaughterhouses. *International Journal of Food Microbiology*, 181: 53–59.

Goldbach, S.G. & Alban, L. 2006. A cost-benefit analysis of *Salmonella*-control strategies in Danish pork production. *Preventive Veterinary Medicine*, 77(1-2): 1–14.

Hamilton, D., Holds, G., Lorimer, M., Kiermeier, A., Kidd, C., Slade, J., Pointon, A., 2010. Slaughterfloor decontamination of pork carcases with hot water or acidified sodium chlorite - a comparison in two Australian abattoirs. *Zoonoses and Public Health*, 57(Suppl. 1): 16–22.

Hernandez, M., Gomez-Laguna, J., Luque, I., Herrera-Leon, S., Maldonado, A., Reguillo, L. & Astorga, R.J. 2013. *Salmonella* prevalence and characterization in a free-range pig processing plant: Tracking in trucks, lairage, slaughter line and quartering. *International Journal of Food Microbiology*, 162(1): 48–54.

Hill, A., Simons, R., O'Reilly, K., Kosmider, R. & Cook, A.J.C. 2009. A risk assessment for *Salmonella* in pigs in Great Britain. pp. 193–195, *in:* Proceedings of the 8th International Symposium on the Epidemiology and Control of Foodborne Pathogens in Pork. Quebec City, Canada.

Hill, A., Swart, A., Simons, R., Vigre, H., Domingues, A., Hald, T., Evers, E. & Snary, E. 2011. Assessing the effect of on-farm and abattoir interventions in reducing human salmonellosis from pig meat consumption in the EU. pp. 381–383, *in:* Proceedings of 9th International Conference on the Epidemiology and Control of Biological, Chemical and Physical Hazards in Pigs and Pork. Maastricht, The Netherlands.

Jayasena, D.D., Kim, H.J., Yong, H.I., Park, S., Kim, K., Choe, W. & Jo, C. 2015. Flexible thin-layer dielectric barrier discharge plasma treatment of pork butt and beef loin: Effects on pathogen inactivation and meat-quality attributes. *Food Microbiology*, 46: 51–57.

Keelara, S., Scott, H.M., Morrow, W.M., Gebreyes, W.A., Correa, M., Nayak, R., Stefanova, R. & Thakur, S. 2013. Longitudinal study of distributions of similar antimicrobial-resistant *Salmonella* serovars in pigs and their environment in two distinct swine production systems. *Applied and Environmental Microbiology*, 79(17): 5167–5178.

Keenliside, J., Gensler, G., McFall, M. & Goonewardene, L. 2005. Prevalence and relatedness of *Salmonella* spp. in a Canadian abattoir. pp. 38, *in:* Proceedings of the 6th International Symposium on the Epidemiology and Control of Foodborne Pathogens in Pork. Rohnert Park, CA, USA.

Kich, J., Pissetti, C., Cardoso, M., Coldebella, A., Nogueira, M. & Ferraz, S.M. 2011. Effect of different treatments on swine carcass surface contamination with *Salmonella* Typhimurium. p. 278, *in:* Proceedings of 9th International Conference on the Epidemiology and Control of Biological, Chemical and Physical Hazards in Pigs and Pork. Maastricht, The Netherlands.

Kim, H.J., Yong, H.I., Park, S., Kim, K., Bae, Y.S., Choe, W., Oh, M.H. & Jo, C. 2013. Effect of inactivating *Salmonella* Typhimurium in raw chicken breast and pork loin using an atmospheric pressure plasma jet. *Journal of Animal Science and Technology,* 55(6): 545–549.

Kim, Y.J. & Song, Y. 2009. An analysis of the factors for microbial contamination risk for pork at slaughterhouses in Korea using the logit model. *Journal of Toxicology and Environmental Health – Part A-Current Issues,* 72(21-22): 1470–1474.

King, A.M., Miller, R.K., Castillo, A., Griffin, D.B. & Hardin, M.D. 2012. Effects of lactic acid and commercial chilling processes on survival of *Salmonella, Yersinia enterocolitica,* and *Campylobacter coli* in pork variety meats. *Journal of Food Protection,* 75(9): 1589–1594.

Larsen, S.T., McKean, J.D., Hurd, H.S., Rostagno, M.H., Griffith, R.W. & Wesley, I.V. 2003. Impact of commercial preharvest transportation and holding on the prevalence of *Salmonella enterica* in cull sows. *Journal of Food Protection,* 66(7): 1134–1138.

Latha, C., Sherikar, A.T., Waskar, V.S., Dubai, Z.B. & Ahmed, S.N. 2009. Sanitizing effect of salts on experimentally inoculated organisms on pork carcasses. *Meat Science,* 83(4): 796–799.

Lawson, L.G., Jensen, J.D., Christiansen, P. & Lund, M. 2009. Cost-effectiveness of *Salmonella* reduction in Danish abattoirs. *International Journal of Food Microbiology,* 134(1-2): 126–132.

Le Roux, A., Minvielle, B. & Gault, E. 2008. Validation of steam-vacuum process as corrective measure for visible faecal contamination on carcasses: preliminary results. Paper 2A-12, *in:* Proceedings of the 54th International Congress of Meat Science and Technology. Cape Town, South Africa.

Letellier, A., Beauchamp, G., Guevremont, E., D'Allaire, S., Hurnik, D. & Quessy, S. 2009. Risk factors at slaughter associated with presence of *Salmonella* on hog carcasses in Canada. *Journal of Food Protection,* 72(11): 2326–2331.

Machado, A.R., Gouveia, F.C., Picinin, L.C.A., Kich, J.D., Cardoso, M.R. de I. & Ferraz, S.M. 2013. Avaliação microbiológica e físico-química de pernis suínos tratados com ácidos orgânicos e/ou vapor no controle da contaminação superficial por *Salmonella* Typhimurium [Microbiological and physico-chemical evaluation of pork leg treated with organic acids and/or steam under pressure in the control of surface contamination by *Salmonella* Typhimurium]. *Ciência Animal Brasileira,* 14(3): 345–351.

Mannion, C., Fanning, J., McLernon, J., Lendrum, L., Gutierrez, M., Duggan, S. & Egan, J. 2012. The role of transport, lairage and slaughter processes in the dissemination of *Salmonella* spp. in pigs in Ireland. *Food Research International,* 45(2): 871–879.

Mansur, A.R., Tango, C.N., Kim, G.H. & Oh, D.H. 2015. Combined effects of slightly acidic electrolyzed water and fumaric acid on the reduction of foodborne pathogens and shelf life extension of fresh pork. *Food Control,* 47: 277–284.

Marier, E.A., Snow, L.C., Floyd, T., McLaren, I.M., Bianchini, J., Cook, A.J.C. & Davies, R.H. 2014. Abattoir based survey of *Salmonella* in finishing pigs in the United Kingdom 2006-2007. *Preventive Veterinary Medicine,* 117(3-4): 542–553.

Morild, R.K., Christiansen, P., Sorensen, A.H., Nonboe, U. & Aabo, S. 2011. Inactivation of pathogens on pork by steam-ultrasound treatment. *Journal of Food Protection,* 74(5): 769–775.

Morris, C.A., Lucia, L.M., Savell, J.W. & Acuff, G.R. 1997. Trisodium phosphate treatment of pork carcasses. *Journal of Food Science,* 62(2): 402–403+405.

Nam, K.C., Ko, K.Y., Min, B.R., Ismail, H., Lee, E.J., Cordray, J. & Ahn, D.U. 2006. Influence of rosemary-tocopherol/packaging combination on meat quality and the survival of pathogens in restructured irradiated pork loins. *Meat Science,* 74(2): 380–387.

Nanasombat, S. & Chooprang, K. 2009. Control of pathogenic bacteria in raw pork using organic acid salts in combination with freezing and thawing. *Kasetsart Journal (Natural Sciences),* 43(3): 576–583.

O'Connor, A.M., Wang, B., Denagamage, T. & McKean, J. 2012. Process mapping the prevalence of *Salmonella* contamination on pork carcass from slaughter to chilling: A systematic review approach. *Foodborne Pathogens and Disease,* 9: 386–395.

Olsen, A., Jensen, T., Dahl, J. & Christensen, H. 2001. Reduction in level of *Salmonella* on swine carcasses after slaughter without splitting the head. p. 124, *in:* Proceedings of the 4th International Symposium on the Epidemiology and Control of *Salmonella* and Other Food Borne Pathogens in Pork. Leipzig, Germany.

Pearce, R.A., Bolton, D.J., Sheridan, J.J., McDowell, D.A., Blair, I.S. & Harrington, D. 2004. Studies to determine the critical control points in pork slaughter hazard analysis and critical control point systems. *International Journal of Food Microbiology,* 90(3): 331–339.

Piachin, T. & Trachoo, N. 2011. Effect of ozone and potassium lactate on lipid oxidation and survival of *Salmonella* Typhimurium on fresh pork. *Pakistan Journal of Biological Sciences,* 14(3): 236–240.

Prasai, R.K., Acuff, G.R., Lucia, L.M., Morgan, J.B., May, S.G. & Savell, J.W. 1992. Microbiological effects of acid decontamination of pork carcasses at various locations in processing. *Meat Science,* 32: 413–423.

Saide-Albornoz, J.J., Knipe, C.L., Murano, E.A. & Beran, G.W. 1995. Contamination of pork carcasses during slaughter, fabrication, and chilled storage. *Journal of Food Protection,* 58(9): 993–997.

Schmidt, J.W., Brichta-Harhay, D., Kalchayanand, N., Bosilevac, J.M., Shackelford, S.D., Wheeler, T.L. & Koohmaraie, M. 2012. Prevalence, enumeration, serotypes, and antimicrobial resistance phenotypes of *Salmonella enterica* isolates from carcasses at two large United States pork processing plants. *Applied and Environmental Microbiology,* 78(8): 2716–2726.

Shan, B., Cai, Y.Z., Brooks, J.D. & Corke, H. 2009. Antibacterial and antioxidant effects of five spice and herb extracts as natural preservatives of raw pork. *Journal of the Science of Food and Agriculture,* 89(11): 1879–1885.

Sommers, C.H., Sites, J.E. & Musgrove, M. 2010. Ultraviolet light (254 nm) inactivation of pathogens on foods and stainless steel surfaces. *Journal of Food Safety,* 30(2): 470–479.

Tadee, P., Boonkhot, P. & Patchanee, P. 2014. Quantification of contamination levels and particular risk of *Salmonella* spp. in pigs in slaughterhouses in Chiang Mai and Lamphun provinces, Thailand. *Japanese Journal of Veterinary Research,* 62(4): 171–179.

Tamplin, M.L., Feder, I., Palumbo, S.A., Oser, A., Yoder, L. & Luchansky, J.B. 2001. *Salmonella* spp. and *Escherichia coli* biotype I on swine carcasses processed under the hazard analysis and critical control point-based inspection models project. *Journal of Food Protection,* 64(9): 1305–1308.

Teagasc [The Irish Agriculture and Food Development Authority]. 2010. *Salmonella* in pork on the island of Ireland: A microbial risk assessment. http://www.safefood.eu/Publications/Research-reports/Salmonella-in-Pork.aspx.

Trivedi, S., Reynolds, A.E. & Chen, J. 2007. Use of a commercial household steam cleaning system to decontaminate beef and hog carcasses processed by four small or very small meat processing plants in Georgia. *Journal of Food Protection,* 70(3): 635–640.

Van Laack, R.L.J.M., Johnson, J.L., Der, P.V., Smulders, F.J.M. & Snijders, J.M.A. 1993. Survival of pathogenic bacteria on pork loins as influenced by hot processing and packaging. *Journal of Food Protection,* 56(10): 847–851.

van Netten, P., Huis in 't Veld, J. & Mossel, D.A. 1994. An *in vitro* meat model for the immediate bactericidal effect of lactic acid decontamination on meat surfaces. *Journal of Applied Bacteriology,* 76(1): 49–54.

van Netten, P., Mossel, D.A. & Huis In 't Veld, J. 1995. Lactic acid decontamination of fresh pork carcasses: A pilot plant study. *International Journal of Food Microbiology*, 25(1): 1–9.

Wheeler, T.L., Kalchayanand, N. & Bosilevac, J.M. 2014. Pre- and post-harvest interventions to reduce pathogen contamination in the U.S. beef industry. *Meat Science*, 98(3; Special Issue): 372–382.

Wilhelm, B.J., Rajic, A., Greig, J.D., Waddell, L. & Harris, J. 2011. The effect of hazard analysis critical control point programmes on microbial contamination of carcasses in abattoirs: A systematic review of published data. *Foodborne Pathogens and Disease*, 8(9): 949–960.

Williams, M.S., Ebel, E.D. & Allender, H.D. 2015. Industry-level changes in microbial contamination on market hog and broiler chicken carcasses between two locations in the slaughter process. *Food Control*, 51: 361–370.

Wong, E., Linton, R.H. & Gerrard, D.E. 1998. Reduction of *Escherichia coli* and *Salmonella* Senftenberg on pork skin and pork muscle using ultraviolet light. *Food Microbiology*, 15(4): 415–423.

SUMMARY CARD SC4:
Beef Chain – On-Farm Interventions

SC4.1 SUMMARY OF KEY FINDINGS

This Summary Card covers the evidence supporting on-farm interventions for reduction of *Salmonella* in beef. Overall, we found a relatively small volume of literature investigating similar outcomes in the population of greatest interest (i.e. feeder cattle and beef carcasses), which limited our ability to perform meta-analysis for estimation of pooled summary estimates of effect. Where possible, we selected data from controlled trials for meta-analysis to calculate a summary estimate of effect. However, for major topics which were not underpinned by evidence from controlled trials, we used data from challenge trials.

SC4.1.1 Antimicrobials

- Four controlled trials investigated antimicrobials. One described the feeding of a basal diet plus tylosin to feedlot heifers, reporting a non-significant ($P > 0.05$) treatment effect on faecal *Salmonella* prevalence. In contrast, two cross-sectional studies reported a significant ($P < 0.05$) protective association between feeding dairy calves medicated milk replacer and *Salmonella* isolation on-farm. The three other trials all reported a non-significant treatment effect of feeding ionophores to cattle, both in a feedlot and a pasture setting, with moderate confidence.

SC4.1.2 Biosecurity

- Biosecurity interventions were only investigated in observational studies and stochastic models. Three different observational study designs (cohort, case-control and cross-sectional studies) as well as stochastic models consistently identified maintaining an 'open' herd (e.g. allowing introduction of purchased animals) as a significant predictor of *Salmonella* isolation on-farm. One cross-sectional study reported that a proxy measure for owner attitude towards hygiene (clean parking lot) was significantly associated with reduced odds of *Salmonella* prevalence on-farm. A stochastic model of the Danish dairy industry predicted that enhanced external biosecurity practices could reduce the herd-level prevalence of *Salmonella* from 3.25% to 0.1% over 10 years.

SC4.1.3 Feed/water acidification

- One controlled trial reported a non-significant treatment effect when feeding micro-encapsulated acids to Friesian calves, with *Salmonella* shedding as the outcome measure. In contrast, one United States of America cross-sectional study reported a significant harmful effect of feeding anionic salts to pre-partum dairy cows on faecal *Salmonella* shedding.

SC4.1.4 Feed management

- Three controlled trials and three cross-sectional studies investigated feeding of distiller's grains to cattle. All controlled trials and one cross-sectional study reported a non-significant treatment effect, with the other two cross-sectional studies reporting a significant positive association between feeding distillers' grains and faecal *Salmonella* shedding. Two cross-sectional studies reported a significant positive association between feeding cottonseed hulls to feeder cattle and *Salmonella* shedding. Inconsistent findings were reported for the effect on *Salmonella* shedding of withholding feed.

SC4.1.5 Manipulation of gut microbiota

- A small dataset of studies investigated use of prebiotics, probiotics or synbiotics to reduce *Salmonella* in cattle, consistently reporting a non-significant treatment effect, with low confidence.

SC4.1.6 Vaccination

- The dataset underpinning this intervention consists of controlled trials, challenge trials and observational studies, largely conducted in a dairy setting. While three controlled trials yielded a significant protective pooled summary estimate of effect of *Salmonell*a vaccines on shedding in dairy cattle, with very low confidence, evidence for their effectiveness in reducing prevalence or load of shedding in feeder cattle is lacking. The effectiveness of a vaccine demonstrated to be effective in a dairy setting in reducing *Salmonella* shedding in beef herds, with potentially different age structures and population densities, is unclear.

SC4.1.7 Overall implications

- Use of biosecurity measures and vaccines are the intervention categories with the broadest dataset of supporting evidence, although confidence is 'low' to 'very low' meaning that the reported estimates are likely to change with further research. One risk assessment of *Salmonella* spp. in ground beef reported that implementing a generic package of on-farm interventions that reduced prevalence and load of deep tissue lymph node contamination in fed cattle could also significantly reduce total *Salmonella* load in ground beef from fed cattle (Li, Kundu and Holley, 2015).
- Much of the literature captured in this review describes research conducted in dairy cattle. The extent to which intervention effectiveness may be generalized to beef cattle is currently unknown, particularly for vaccine research, which uses specific serotypes in specific population structures that have different characteristics from those of beef cattle.

- In contrast with the pork dataset, which included 24 risk assessments or stochastic models studying on-farm interventions, we found only eight risk assessments or models investigating *Salmonella* interventions in cattle on-farm. Across these studies, the generalizability of data and assumptions from the target population to others was frequently unclear (n = 7 of 8 studies), making the wider applicability of these studies' findings uncertain. In further contrast with the pork dataset, we were not able to compile or compute summary estimates of effect for any interventions measuring lymph node contamination as the outcome measure.

SC4.2 INTERVENTION DESCRIPTION

This Summary Card summarizes the evidence for a range of interventions that can be implemented at the on-farm level to reduce *Salmonella* contamination of beef. Specific categories of interventions covered in this Summary Card include:

- **Antimicrobials**: any substance of natural, semi-synthetic, or synthetic origin, that kills or inhibits the growth of a microorganism but causes little or no damage to the host (Giguere *et al.*, 2006). Includes those administered parenterally and orally, for treatment or prophylaxis.
- **Biosecurity:** has been defined as the implementation of measures that reduce the risk of introduction and spread of disease agents (FAO/OIS/World Bank, 2010). Includes, but is not limited to: sanitation; biosafety; disinfection; hygiene and hygiene barriers; all-in-all-out production; depopulation; staff and the environment; litter testing and treatment; and pest control. Biosecurity may consist of external (targeting prevention of introduction of targeted pathogens to the farm or unit) or internal (aimed at reducing spread of pathogens on-farm) procedures.
- **Feed/water acidification:** addition of organic acids, such as lactic acid, to feed or water. This would include 'nutraceuticals' such as copper, chromium, zinc, betaine or carnitine. We also found investigations of the effects on *Salmonella* prevalence of feeding anionic salts to dairy cattle (for prevention of peri-parturient hypocalcaemia).
- **Feed management**: comparisons of coarse vs finely ground feed; specific feedstuffs; or additives. Also included in this category is deliberate withholding of feed in the hours immediately prior to transport to slaughter.
- **Manipulation of gut microbiota:** includes use of probiotics, prebiotics, and synbiotics. *Probiotics* are living microorganisms that are fed to animals to colonize the gut environment to encourage a better microbial balance. A *prebiotic* may be defined as a non-digestible food ingredient that beneficially affects the host by selectively stimulating the favourable growth and activity of

one or a limited number of bacteria in the colon. The term *synbiotic* describes a combination of probiotic and prebiotic approaches. It includes application of protective bacterial species or cultures to out-compete and prevent *Salmonella* colonization. It can include specific bacterial species or cultures, or caecal contents or other materials from animals or the environment that contain many different or unknown bacterial species.

- **Vaccination:** refers to immunization of the subject using either autogenous or commercial vaccines.

Evidence was identified for each of these six categories of interventions, which are presented and discussed separately below.

SC4.3 ANTIMICROBIALS

The effect of antimicrobial administration on *Salmonella* reduction in cattle was investigated in 11 primary research studies: four controlled trials, one cohort study and five cross-sectional studies.

Jacob *et al.* (2008) reported a controlled trial in which 370 feedlot heifers were fed a basal ration plus wet corn distiller's grains with solubles, with no antimicrobials, monensin, or with monensin (300 mg/animal/day) and tylosin (90 mg/animal/day). No significant treatment effect on *Salmonella* prevalence was reported for feeding antimicrobials.

Berge, Moore and Sischo (2006) investigated cohorts of calves 1 to 84 days old from 26 dairies and seven calf ranches, and reported a significantly ($P < 0.05$) protective association between antimicrobial supplementation of milk replacer and faecal *Salmonella* shedding in calves. Conversely, prophylactic antimicrobial treatment of day-old calves was associated with increased shedding.

Three cross-sectional studies investigated the effect of antimicrobial exposure on *Salmonella* in cattle. Fossler *et al.* (2005a) reported a significant harmful association on faecal *Salmonella* shedding of not feeding milk replacer containing antimicrobials to United States of America pre-weaned dairy calves (odds ratio (OR) = 2.8, 95% confidence interval (CI) (1.4–5.8)). Similarly, Losinger *et al.* (1995) reported a significant protective association between feeding medicated milk replacer to pre-weaned dairy heifers and culture of *Salmonella* spp. (OR = 0.35, 95% CI (0.20–0.61)). In a feedlot setting, Green *et al.* (2010) reported a significant protective treatment effect on faecal *Salmonella* shedding associated with feeding antimicrobials of the tetracycline class in the ration within 2 weeks before sampling

Summary-of-findings table for the effects of feeding antimicrobials

TABLE SC4.1 Controlled trials investigating feeding antimicrobials/ionophores

Intervention	Population (Sample)	No. trials/ studies	Odds ratio for intervention effect[1]	% Samples *Salmonella*-positive in study population[2]		GRADE rating[3]	References
				Without treatment	With treatment		
Ionophores: monensin/ lasalocid	Feedlot Bulls Stocker calves (Faecal/hide culture)	2/2	MA = 0.79 (95% CI: 0.20–3.15)	14.4%	11.7% (95% CI: 3.3–4.6)	Moderate	Devant *et al.*, 2009; Edrington *et al.*, 2006a

NOTES: MA = meta-analysis average estimate from random-effects model; CI = confidence interval.

(1) Refers to the average meta-analysis estimate and 95% CI from random-effects meta-analysis, as heterogeneity was not significant (I^2<60%).

(2) The percentage of samples positive for *Salmonella* in the comparison group (assumed control risk, ACR) refers to the median prevalence in the comparison group trials. The percentage in the intervention/treatment group was calculated using the following formula for the odds ratio (OR) measure of effect: (OR × ACR / 1 -ACR + [OR × ACR]), and provides an indication of the effect of the intervention on changing a given baseline level of *Salmonella* prevalence as reported in the included studies. The odds ratio used in this formula corresponds to the odds ratio presented in the intervention effect column as described under (1) above. If heterogeneity was not significant (I^2<60%), the estimate is presented with a 95% CI in brackets. If heterogeneity was significant, the median estimate is presented with the range of individual study effects in brackets to illustrate the variability in results across studies.

(3) GRADE rating descriptions: very low = the true effect is likely to be substantially different from the measured estimate; low = the true effect may be substantially different from the measured estimate; moderate = the true effect is likely to be close to the measured estimate, but there is a possibility that it is substantially different; high = there is strong confidence that the true effect lies close to that of the measured estimate.

(OR = 0.04–95% CI 0.02–0.09) and more than 2 weeks before sampling (OR = 0.23, 95% CI (0.06–0.80)) in United States of America feeder cattle.

Devant *et al.* (2009) investigated the effect of feeding monensin-supplemented or basal diet (straw and concentrate), to 90 Holstein bulls over the 108-day trial period; there was a non-significant difference in *Salmonella* prevalence on hides at slaughter between the control group and the group fed monensin. Edrington *et al.* (2006a) reported a non-significant treatment effect on *Salmonella* shedding of feeding a mineral supplement with ionophore (1.76 g lasalocid/kg) for 61 days to stocker calves on pasture, relative to control calves. A meta-analysis summary estimate from these two studies yielded a non-significant treatment effect, with moderate confidence (Table SC4.1). Similarly, Jacob *et al.* (2008) reported a non-significant treatment effect on *Salmonella* faecal shedding in feedlot heifers fed a basal diet plus either monensin, or monensin and tylosin, relative to controls.

Inconsistent findings were reported from cross-sectional studies investigating ionophore use and *Salmonella*. In a study of United States of America dairy cows, Fossler et al. (2005b) reported a significant association between not using monensin in weaned calf or bred heifer diets and *Salmonella* shedding (OR = 3.2, 95% CI (2.0–5.4)). In contrast, in another cross-sectional study of United States of America dairy cows, Habing et al. (2012) report a significant harmful association between feeding ionophores to dairy cows and *Salmonella* shedding (OR = 2.1, 95% CI (1.2–3.7)). Losinger et al. (1997), in a survey of 100 United States of America feedlots, reported a non-significant association between feeding ionophores in United States of America feeder cattle and *Salmonella* 'positive' pens.

SC4.4 BIOSECURITY

Twenty primary research studies investigated biosecurity measures for reduction of *Salmonella* in cattle, all of which were observational: three cohort studies, three case-control studies, and 14 cross-sectional studies, as well as four predictive models.

SC4.4.1 External biosecurity practices

External biosecurity practices were investigated in two cohort studies, and several cross-sectional studies. Berge, Moore and Sischo (2006) investigated risk factors for *Salmonella* shedding in cohorts of California calves (1 to 84 days of age). Herds which were 'open' to introductions had significantly greater odds of *Salmonella* shedding in calves (OR = 35.53, 95% CI (7.11–178.5)).

Similar findings were reported in the three case-control studies found. Evans (1996) reported that reporting the purchase of replacement stock from dealers was significantly associated with *Salmonella* Typhimurium identified on-farm in case vs control herds in the UK (OR = 3.9, 95% CI (1.62–9.36)). Similarly, Vaessen *et al.* (1998) reported significantly increased odds of farm-level *Salmonella* Dublin infection associated with the purchase of livestock in case vs control herds in the Netherlands (OR = 4.29, 95% CI (1.28–14.45)), although interestingly odds of infection decreased if cattle had contact with cattle from other farms (OR = 0.07, 95% CI (0.01–0.49)). In contrast, Vanselow *et al.* (2007) reported a significantly increased risk of *Salmonella* shedding in dairy cattle if they had access to new arrivals on the farm (risk ratio (RR) = 4.0, $P = 0.03$). A Dutch case-control study (Veling *et al.*, 2002) investigating risk factors for clinical infection with *Salmonella* Typhimurium on dairy farms reported that the purchase of manure significantly increased odds of infection (OR = 21.5, 95% CI (1.4–332)). Warnick *et al.* (2003), in a survey of 65 United States of America dairy herds, reported that movement of livestock off-premises and back on again within the past 12 months was a non-significant predictor of *Salmonella* serogroup B on-farm in the final multivariable model.

Davison *et al.* (2006) investigated risk factors for culturing environmental *Salmonella* on 449 dairy farms in England and Wales. Cattle introductions in the past 3 to 12 months were a significant predictor in multivariable analysis, with farms introducing only adult cattle over 24 months of age having greater odds of *Salmonella* than those farms not allowing introductions. Using Poisson multivariable regression, the authors unexpectedly demonstrated that farms that did not use common or rented cattle grazing had significantly greater risk of becoming *Salmonella*-positive (incidence risk ratio (IR) = 1.71–95% CI (0.98–2.97)). Green *et al.* (2010) reported that sourcing animals in a feedlot pen from multiple sources was associated with increased odds of *Salmonella* infection (OR = 5.17, 95% CI (2.32–11.31)). Nielsen *et al.* (2012a), in a study of 86 dairy herds diagnosed with endemic *Salmonella* Dublin, observed that purchasing cattle from known positive herds was strongly associated (OR = 14.5, 95% CI (3.1–67)) with unsuccessful control (successful control was defined as the 10 youngest calves above three months of age testing *Salmonella* Dublin antibody-negative).

Evans (1996) reported that a feral cat population on-farm increased odds of *Salmonella* on-farm (OR = 1.35, 95% CI ((1.09–1.65)). In contrast, Veling *et al.* (2002) reported that having cats on-farm was significantly associated with a reduced odds of detection of *Salmonella* Dublin (OR = 0.06, 95% CI (0.006–0.6)).

Carlson *et al.* (2011), in a survey of Texas dairy farms, reported that numbers of European starlings on-farm better explained *Salmonella enterica* contamination of cattle feed and water than other variables studied, including cattle stocking, facility management, and environmental variables.

SC4.4.2 Internal biosecurity practices

Davison *et al.* (2006) reported that farms that did not have a clean visitor parking area had significantly increased odds of *Salmonella* cultured on-farm (OR = 2.86, 95% CI (1.82–6.92)) relative to those that did. The authors speculated that this predictor could be a proxy measure for attitudes toward hygiene. Using Poisson multivariable regression to model risk factors associated with farms becoming positive for *Salmonella*, absence of a clean visitor parking lot was similarly a significant predictor (IR = 3.49, 95% CI (1.37–8.92)). Younis *et al.* (2009), in a study of 220 diarrhoeaic Egyptian calves, reported a significant protective association between hygiene on-farm (estimated as a categorical variable) and *Salmonella* culture (OR = 0.63, 95% CI (1.73–5.61)). In contrast, Dewell *et al.* (2008) studied 40 lots of feedlot cattle from the feedlot through to slaughter, and reported that presence of standing water in the pens and dirty feedlot water troughs were non-significant predictors in the final Poisson multivariable model for *Salmonella* hide contamination at slaughter.

Two cross-sectional studies (Fossler *et al.*, 2005b; Habing *et al.*, 2012) reported a significant association between manure management and *Salmonella* shedding in dairy cows, with Fossler *et al.* (2005b) reporting a harmful association between manure disposal as liquid slurry and *Salmonella* status (OR = 1.8, 95% CI (1.0–3.0)) and Habing *et al.* (2012) reporting a protective association with manure application by broadcast/solid spreader (OR = 0.26, 95% CI (0.11–0.63)).

Kabagambe *et al.* (2000) reported a significantly increased odds of faecal *Salmonella* shedding on dairy farms using flush water systems for removal of manure from alleys (OR = 3.5, 95% CI (0.9–14.7)). Vanselow *et al.* (2007) also reported that large numbers of flies in feedlot pens (RR = 9.0, $P = 0.02$), or around stored manure (RR = 6.5, $P = 0.04$), significantly increased the risk of *Salmonella* isolation in cattle on Australian feedlots. Huston *et al.* (2002), in a study of 105 United States of America dairy farms, reported that use of straw bedding was a non-significant predictor in the final model for *Salmonella* on-farm, relative to farms using shavings. Losinger *et al.* (1997) reported a non-significant association between use of sprinklers in the feedlot and *Salmonella* culture 'positive' pens. In contrast, Habing *et al.* (2012) reported significantly increased odds of *Salmonella* culture on United States of America dairy farms (OR = 2.8, 95% CI (1.6–4.9)) using sprinklers, relative to those not using them.

Veling *et al.* (2002) reported a significantly protective association between feeding only dam colostrum to calves and *Salmonella* infection on-farm (OR = 0.08, 95% CI (0.01–0.54)). Losinger *et al.* (1995) reported a significant protective association between calving in an individual area of a building and faecal *Salmonella* prevalence in pre-weaned United States of America dairy heifers (OR = 0.53, 95% CI (0.28–0.99)).

Bergevoet *et al.* (2009) described a stochastic compartmental model for herd-level spread of *Salmonella* in Dutch dairy herds, evaluating management measures including: no entry of infected animals; no entry of infected manure; hygienic measures; and culling chronically infected animals. Modelling suggested that culling chronically infected animals and not transporting potentially infected animals could significantly reduce the prevalence of *Salmonella*-positive herds, and was most cost-effective. Adding hygienic measures and banning the transportation of animal manure reduced prevalence slightly more, but with significantly greater costs.

Jordan, Nielsen and Warnick (2008) describe a stochastic compartment model for *Salmonella* kinetics within the Danish dairy industry using field survey inputs for parameterization, in which an enhanced biosecurity scenario model improved external biosecurity practices (fewer cattle trades per year, with fewer animals, with less risky herds). After 10 years, the effect of enhanced biosecurity reduced predicted median herd prevalence from 3.25% to 0.1%.

Nielsen, Kudahl and Ostergaard (2012) described a stochastic model (a modification of the existing SIMHERD model) for *Salmonella* spread in dairy herds, containing six age groups, five infectious states (susceptible, acutely infected, super-shedder, carrier, resistant), four hygiene levels, three herd sizes, and four herd susceptibility levels stratified by age group. The authors concluded that hygiene level was highly predictive of the probability of infection spread, length and size of epidemic.

Van Schaik, Nielen and Dijkhuizen (2001) described a deterministic model to investigate economic consequences of a 'more-closed system' of dairy cattle production. This system involved application of additional external biosecurity measures: no external purchases; provision of protective clothing to visitors and employees; and double perimeter fence. Positive economic outcomes were experienced by a 55-cow herd in the basal scenario, and also when a sanitary barrier was used as opposed to just protective clothing.

SC4.5 FEED AND WATER ACIDIFICATION

Two studies found in this review investigated the use of organic acids for reduction of *Salmonella* in cattle. Spanghero *et al.* (2007) reported a controlled trial investigating effects of feeding milk replacer supplemented with microencapsulated organic acids on faecal *Salmonella* Typhimurium shedding in male Friesian calves from birth to 32 days of age, and reported no significant treatment effect ($P > 0.05$). Habing *et al.* (2012) conducted a cross-sectional study of 265 dairy herds in 17 United States of America states, studying herd-level risk factors for *Salmonella* infection. Feeding anionic salts to close-up cows significantly increased odds of *Salmonella* on-farm prevalence (OR = 1.9, 95% CI (1.1–3.5)).

SC4.6 FEED MANAGEMENT

Twenty-one primary research studies investigated feed management or additives for reduction of *Salmonella* in beef: six controlled trials, one quasi-experiment, one cohort study, two case-control studies and 11 cross-sectional studies.

Jacob *et al.* (2009) investigated feeding a basal diet plus either dried distiller's grains with solubles or dry rolled corn to 700 yearling feedlot heifers, culturing pooled pen faecal samples as the outcome measure. No significant treatment effect was reported for pen-level *Salmonella* prevalence. Similarly, Jacob *et al.* (2008) investigated feeding basal diet plus wet corn distiller's grain with solubles to 370 yearling heifers, culturing pooled pen faecal samples as the outcome measure; no significant treatment effect was reported. Edrington *et al.* (2010) investigated the effect of feeding 20% wet distiller's grains with steam-flaked corn or dry-rolled corn to feedlot cattle on faecal prevalence of *Salmonella*; a significantly increased *Salmonella* prevalence was observed in the cattle fed dry-rolled corn relative to those fed dry-rolled corn plus distiller's grains on one sampling event (day 132 of the trial), but no difference was observed overall. Four cross-sectional studies investigated feeding cattle distiller's grains. Green *et al.* (2010) and Kabagambe *et al.* (2000) both reported a significant positive association between feeding distiller's grains and *Salmonella* culture, assaying feedlot cattle pen samples and dairy cattle faecal samples, respectively. In contrast, Habing *et al.* (2012) and Fossler *et al.* (2005b) reported a non-significant association between feeding distiller's grains and *Salmonella* culture in dairy cattle.

Huston *et al.* (2002) in a survey of 105 United States of America dairy herds reported that method of concentrate delivery (separate from forage vs total mixed ration) was a non-significant predictor of *Salmonella* on-farm. Losinger *et al.* (1995), in a survey of United States of America dairy herds, reported that feeding dairy heifers hay from one day of age to weaning had a significant protective association with *Salmonella* shedding on-farm (OR = 0.52, 95% CI (0.31–0.88)). Warnick *et al.* (2003), in a survey of 65 United States of America dairy herds, reported that source of water (well vs town) was a non-significant predictor for *Salmonella* presence on-farm in the final multivariable model. Berge, Moore and Sischo (2006), in a cohort study of pre-weaned dairy calves on United States of America dairy and calf ranches, reported that method of feeding (bottle vs bucket) was a non-significant predictor of faecal *Salmonella* shedding.

Looper *et al.* (2006) described a controlled trial comparing the effect of various forage diets (endophyte-infected or novel endophyte-infected fescue, or common Bermudagrass) on *Salmonella* shedding in yearling steers. The authors reported a non-significant treatment effect.

Devant *et al.* (2009) reported a controlled trial investigating feeding a basal diet or diet supplemented with monensin or plant extracts (Biostar™) containing artichoke (20 to 30%), Siberian ginseng (15 to 25%), and fenugreek (55 to 65%) to 90 Holstein bulls for a 108-day feeding period. The authors reported a significant protective treatment effect on *Salmonella* prevalence for the bulls in the plant extract treatment group.

Edrington *et al.* (2006b) described a controlled trial investigating the effect of feeding ractopamine on faecal *Salmonella* prevalence in feeder steers, reporting a non-significant ($P < 0.05$) treatment effect.

Edrington *et al.* (2012) reported a cross-sectional study investigating potential risk factors for *Salmonella* shedding in dairy calves. The authors reported that all 1-week-old calves fed non-pasteurized waste milk shed *Salmonella*; none of the calves receiving pasteurized waste milk did.

Two cross-sectional studies conducted in the United States of America investigated feeding cottonseed hulls to feeder cattle. Losinger *et al.* (1997) reported a significant association (OR = 3.5, 95% CI (1.0–11.9)) between feeding whole cottonseed or cottonseed hulls within 7 days before faecal sample collection, and *Salmonella* shedding. Similarly, Green *et al.* (2010) reported a significant association (OR = 8.34, 95% CI (3.58–19.42)) between *Salmonella* shedding and feeding cottonseed hulls. In contrast, Fossler *et al.* (2005b) reported a non-significant association between feeding cottonseed and isolation of *Salmonella* on 129 United States of America dairy farms.

Beach, Murano and Acuff (2002) used a quasi-experimental design to investigate risk factors for *Salmonella* shedding in United States of America feeder and cull adult cattle from farm to slaughter. The authors reported no significant association between withholding feed and odds of individual faecal, hide or carcass contamination. In contrast, Abouzeed *et al.* (2000) described a cross-sectional study of Canadian slaughter beef and dairy cattle in Prince Edward Island. Fasted animals (i.e. those experiencing 18 to 24 hours of lairage) had a significantly greater faecal *Salmonella* prevalence relative to non-fasted cattle (7.5% vs 0.94%, respectively). The authors do not report microbial sampling of the lairage area itself prior to cattle entry.

Two case-control studies investigated forage management in dairy cattle and potential effects on *Salmonella* shedding. Vaessen *et al.* (1998) reported that Dutch dairy cattle grazing grass only, as opposed to grazing supplemented with hay, had significantly increased odds of *Salmonella* Dublin shedding (OR = 14.16, 95% CI (2.15–93.3)). In contrast, Veling *et al.* (2002) reported that unrestricted grazing of lactating cows was significantly protective (OR = 0.07, 95% CI (0.006–0.70)) of *Salmonella* Typhimurium shedding.

SC4.7 MANIPULATION OF GUT MICROBIOTA

Six studies investigated manipulation of gut microbiota for *Salmonella* reduction in cattle: two controlled trials, two challenge trials and two cross-sectional studies.

Stephens *et al.* (2007) investigated feeding low (1×10^7), medium (1×10^8), and high (1×10^9) CFU/steer/day doses of *Lactobacillus acidophilus* as well as a high dose

Summary-of-findings table for the effects of manipulation of gut microbiota

TABLE SC4.2 Controlled trials investigating feeding probiotics

Intervention	Population (Sample)	No. trials/ studies	Odds ratio for intervention effect(1)	% Samples *Salmonella* positive in study population(2)		GRADE rating(3)	References
				Without treatment	With treatment		
Probiotics	Feedlot cattle (Faecal culture/ PCR)	4/2	MA = 0.68 (95% CI: 0.44-1.04)	8.3%	5.8% (95% CI: 3.8-8.6)	Low	Stephens *et al.*, 2007; Tabe *et al.*, 2008

NOTES: MA= meta-analysis average estimate from random-effects model; CI = confidence interval.

(1) Refers to the average meta-analysis estimate and 95% CI from random-effects meta-analysis, as heterogeneity was not significant (I^2<60%).

(2) The percentage of samples positive for *Salmonella* in the comparison group (assumed control risk, ACR) refers to the median prevalence in the comparison group trials. The percentage in the intervention/treatment group was calculated using the following formula for the odds ratio (OR) measure of effect: (OR × ACR / 1 -ACR + [OR × ACR]), and provides an indication of the effect of the intervention on changing a given baseline level of *Salmonella* prevalence as reported in the included studies. The odds ratio used in this formula corresponds to the odds ratio presented in the intervention effect column as described under (1) above. If heterogeneity was not significant (I^2<60%), the estimate is presented with a 95% CI in brackets. If heterogeneity was significant, the median estimate is presented with the range of individual study effects in brackets to illustrate the variability in results across studies.

(3) GRADE rating descriptions: very low = the true effect is likely to be substantially different from the measured estimate; low = the true effect may be substantially different from the measured estimate; moderate = the true effect is likely to be close to the measured estimate, but there is a possibility that it is substantially different; high = there is strong confidence that the true effect lies close to that of the measured estimate.

of *Propionobacterium freudenreichii* (NP24) to 269 feeder steers for an average 137 days, culturing faecal grab samples and hide sponge samples for *Salmonella*. No significant (P < 0.05) treatment effects were reported for any of the dosages, for either of the outcome measures. Similarly, Tabe *et al.* (2008) fed 1 ×10^9 CFU/ steer/day of *L. acidophilus* (LA 51 1) and *P. freudenreichii* (NP24) to steers for nine weeks, with culture of faecal grab samples as the outcome measure; the treatment effect was non-significant. An overall non-significant (P > 0.05) summary estimate of effect was calculated across both studies, with low confidence (Table SC4.2).

Supplementation with probiotics was also investigated in two challenge trials conducted in suckling calves (Brewer *et al.*, 2014; Frizzo *et al.*, 2012). Brewer *et al.* (2014) reported a significant reduction in faecal *Salmonella* load on day six post-infection in calves administered *Saccharomyces cerevisiae* fermentation products incorporated into both milk replacer and starter grain, relative to control calves. Frizzo *et al.* (2012) reported a non-significant treatment effect on detection of *Salmonella* in internal organs (liver, spleen, lung, and visceral lymph nodes) in calves administered *Lactobacillus casei* DSPV 318T, *L. salivarius* DSPV 315T, and *Pedio-*

coccus acidilactici DSPV 006T (10^9 CFU/kg body weight of each strain daily) with or without lactose (100 g/day) during the experiment (from 5 to 20 days of age).

Losinger *et al.* (1997), in a cross-sectional study of pen-level risk factors in United States of America feeder cattle, reported a non-significant association between feeding probiotics and *Salmonella* shedding. Similarly, Habing *et al.* (2012) reported a non-significant association between the use of probiotics and culture of *Salmonella* on United States of America dairy farms.

SC4.8 VACCINATION

Vaccination for reduction of *Salmonella* in cattle was investigated in 25 primary research studies: eight controlled trials, 11 challenge trials, two cohort studies and four cross-sectional studies, as well as three stochastic models.

Miller, Sprouse and Garner (1991) reported vaccinating 37 dairy cows in late gestation with either a *Salmonella* Typhimurium mutant bacterin-toxoid or placebo, and reported a significant treatment effect for dam seroconversion, colostral antibodies and calf seroconversion in the vaccinated group, relative to unvaccinated (placebo) controls. Smith *et al.* (2015) reported that administering a commercial *Salmonella* Dublin vaccine to 30 United States of America Holstein cows twice in late gestation had a significant treatment effect on cow seroconversion and calf acquisition of colostral antibodies relative to unvaccinated controls. Similarly, Smith *et al.* (2014) reported vaccinating 30 United States of America Holsteins at drying-off with a commercially available *Salmonella* Newport vaccine, and boosting 4 weeks later, reporting a significant treatment effect for cow seroconversion at calving, colostral antibodies, and calf seroconversion after colostral feeding ($P = 0.01$, $P = 0.011$, and $P = 0.003$, respectively). In contrast, Habing *et al.* (2011) administered a commercial modified live *Salmonella* Dublin vaccine orally to randomly selected Holstein calves at three and 10 days of age, reporting a non-significant treatment effect on *Salmonella*-specific morbidity rate from birth to weaning.

Dodd *et al.* (2011) reported a non-significant treatment effect in an investigation of *Salmonella* shedding in 10 pens of United States of America feeder cattle vaccinated with *Salmonella* Newport SRP vaccine relative to 10 pens of non-vaccinated controls.

Three controlled trials (Heider *et al.*, 2008; Hermesch *et al.*, 2008; House *et al.*, 2001) investigated the effect of vaccination using commercial *Salmonella* Dublin or autogenous *Salmonella* Montevideo bacterins, or a modified live *Salmonella* Choleraesuis vaccine, on individual dairy cow faecal *Salmonella* shedding, with a significant protective pooled summary estimate of effect (Table SC4.3).

Challenge trials investigated the potential effect of vaccination on *Salmonella* prevalence, measured from culture of various organs and load of bacterial shedding. Immunization of calves from 1-7 weeks of age with a modified live *Salmonella* Typhimurium vaccine was reported in three challenge trials (Mizuno, McLennan and Trott, 2008; Mohler *et al.*, 2006, 2008), all reporting significantly reduced frequency of *Salmonella* shedding and reduced colonization of mesenteric lymph nodes, relative to controls. Two studies investigated the use of modified live *Salmonella* Dublin (Mukkur *et al.*, 1991; Smith *et al.*, 1993) or *Salmonella* Typhimurium (Van der Walt *et al.*, 2001) vaccine in calves, reporting significantly improved clinical outcomes in the vaccinated group post-challenge; in contrast, Anderson, Smith and Ulrich (1991) reported a non-significant treatment effect of a killed *Salmonella* Typhimurium vaccine on calf mortality post-challenge. Fox *et al.* (1997) reported the use of a commercial modified live *Salmonella* Choleraesuis vaccine in calves 3 to 5 weeks of age; all treatment groups demonstrated a significant protective treatment effect on *Salmonella* prevalence and colonization of organs and lymph nodes. Mortola *et al.* (1992) described a challenge trial in which pregnant dairy cows were vaccinated with a killed *Salmonella* Dublin vaccine and their calves challenged orally with *Salmonella* Dublin; both clinical outcomes and faecal shedding were improved in the offspring of vaccinated cows.

Edrington *et al.* (2013) described a series of experiments investigating development of a *Salmonella* challenge model in cattle. In the first trial, weaned steers were vaccinated with a commercial *Salmonella* Newport vaccine and challenged with either *Salmonella* Newport or *Salmonella* Montevideo. The authors reported a significant reduction in prevalence of *Salmonella* Newport contaminated lymph nodes in the vaccinated group, at 21 days post-challenge. In a second, similar trial, using a larger challenge dose, no significant treatment effects were observed with the exception of one of the six lymph node sampling sites. In a trial investigating the use of a transdermal challenge, a significant treatment effect was reported (P = 0.03) on prevalence of *Salmonella* Newport contaminated lymph nodes.

Cummings *et al.* (2009) described a cohort study of 831 United States of America dairy herds, reporting that use of a commercial gram-negative vaccine was a non-significant predictor for *Salmonella* incidence on-farm. Davies (1997) describe a cohort study of 14 UK cattle farms, reporting a significant difference (P < 0.01) in the incidence of *Salmonella* in herds implementing a combined *Salmonella* Dublin/ Typhimurium vaccination programme during the study.

Four cross-sectional studies reported a non-significant association between use of *Salmonella* vaccines and presence of *Salmonella* on United States of America dairy farms (Fossler *et al.*, 2005b; Huston *et al.*, 2002; Habing *et al.*, 2012; Kabagambe *et al.*, 2000).

Lu *et al.* (2014) described a stochastic model of *Salmonella* kinetics in United States of America dairy herds, and the potential effects of imperfect vaccines. Parameters considered included reducing shedding, reducing the length of the infectious period, reducing the number of clinical cases, and reducing host susceptibility. The authors concluded that vaccines effective at reducing the length of the infectious period are best for reducing prevalence, and vaccines effective at reducing host susceptibility are best for reducing outbreak size. Vaccines moderately effective in multiple domains may be considered in the absence of availability of a vaccine more effective in a specific domain.

Lu *et al.* (2009) described a deterministic model of *Salmonella* kinetics in an endemically infected United States of America dairy herd, concluding that lifetime (continuous) vaccination was more effective in reducing *Salmonella* prevalence than cohort vaccination.

Nielsen, Kudahl and Ostergaard (2012) described a stochastic model for *Salmonella* spread in dairy herds, investigating the potential effects of several interventions. They report that within the model, manipulation of herd susceptibility by itself was insufficient to control infection within the herd.

Summary-of-findings table for the effects of vaccines

TABLE SC4.3 Controlled trials investigating *Salmonella* vaccines and shedding

Intervention	Population (Sample)	No. trials/ studies	Odds ratio for intervention effect[(1)]	% Samples *Salmonella* positive in study population[(2)]: Without treatment	% Samples *Salmonella* positive in study population[(2)]: With treatment	GRADE rating[(3)]	References
Vaccine	Adult dairy cows (Faecal culture)	7/3	MA = 0.81 (95% CI: 0.67-0.97)	50.4%	45.1% (95% CI: 28.9-49.6)	Very low	Heider *et al.*, 2008; Hermesch *et al.*, 2008; House *et al.*, 2001

NOTES: MA = meta-analysis average estimate from random-effects model; CI = confidence interval.

(1) Refers to the average meta-analysis estimate and 95% CI from random-effects meta-analysis, as heterogeneity was not significant (I^2<60%).

(2) The percentage of samples positive for *Salmonella* in the comparison group (assumed control risk, ACR) refers to the median prevalence in the comparison group trials. The percentage in the intervention/treatment group was calculated using the following formula for the odds ratio (OR) measure of effect: (OR × ACR / 1-ACR + [OR × ACR]), and provides an indication of the effect of the intervention on changing a given baseline level of *Salmonella* prevalence as reported in the included studies. The odds ratio used in this formula corresponds to the odds ratio presented in the intervention effect column as described under (1) above. If heterogeneity was not significant (I^2<60%), the estimate is presented with a 95% CI in brackets. If heterogeneity was significant, the median estimate is presented with the range of individual study effects in brackets to illustrate the variability in results across studies.

(3) GRADE rating descriptions: very low = the true effect is likely to be substantially different from the measured estimate; low = the true effect may be substantially different from the measured estimate; moderate = the true effect is likely to be close to the measured estimate, but there is a possibility that it is substantially different; high = there is strong confidence that the true effect lies close to that of the measured estimate.

SC4.9 MULTIPLE STRATEGIES CONCURRENTLY OR COMPARATIVELY

Li, Kundu and Holley (2015) described a stochastic simulation model covering the pre-to post-harvest stages of beef production to assess the relative contribution of deep tissue lymph nodes (DTLNs) as compared with carcass surface contamination to influence *Salmonella* contamination in ground beef. The potential impact of various pre- and post-harvest interventions was also evaluated. Increasing the effectiveness of generic pre-harvest interventions on prevalence and load of DTLNs was predicted to reduce overall load of the ground beef, but did not have a major effect on the relative contribution from DTLNs.

SC4.10 MISCELLANEOUS PREDICTORS

Observational studies have identified several factors which may not be directly amenable to an intervention but are important to understand as they influence other intervention strategies' effectiveness. Berge, Moore and Sischo (2006) investigated risk factors for *Salmonella* shedding in cohorts of Californian calves (1 to 84 days of age), and reported that calf age in days was negatively associated with shedding (OR = 0.97, 95% CI (0.95–0.98). Davison *et al.* (2006) reported a cross-sectional study of UK dairy farms in which farms with more than 100 cattle (of all ages) on-site had significantly greater odds of *Salmonella* presence relative to those with 100 cattle or less. Five cross-sectional studies (Fossler *et al.*, 2005a; Huston *et al.*, 2002; Kabagambe *et al.*, 2000; Losinger *et al.*, 1997; Warnick *et al.*, 2003) studied the effect of herd size on odds of *Salmonella* infection on-farm, with three studies (Huston *et al.*, 2002; Kabagambe *et al.*, 2000; Warnick *et al.*, 2003) reporting a significant association between larger farms and increased odds of *Salmonella* infection on-farm. This observation was supported by one United States of America cohort study (Cummings *et al.*, 2009). Vaessen *et al.* (1998) reported a significant association between presence of liver flukes on-farm and odds of *Salmonella* Dublin isolation in Dutch dairy cattle (OR = 14.16, 95% CI (2.15–93.3)).

SC4.11 REFERENCES CITED IN SC4

Abouzeed, Y.M., Hariharan, H., Poppe, C., Kibenge, F.S.B., 2000. Characterization of *Salmonella* isolates from beef cattle, broiler chickens and human sources on Prince Edward Island. *Comparative Immunology Microbiology and Infectious Diseases,* 23(4): 253–266.

Anderson, J., Smith, B.P. & Ulrich, J.T. 1991. Vaccination of calves with a modified bacterin or oil-in-water emulsion containing alkali-detoxified *Salmonella* Typhimurium lipopolysaccharide. *American Journal of Veterinary Research,* 52(4): 596–601.

Beach, J.C., Murano, E.A. & Acuff, G.R. 2002. Prevalence of *Salmonella* and *Campylobacter* in beef cattle from transport to slaughter. *Journal of Food Protection,* 65, 1687-1693.

Berge, A.C.B., Moore, D.A. & Sischo, W.M. 2006. Prevalence and antimicrobial resistance patterns of *Salmonella enterica* in pre-weaned calves from dairies and calf ranches. *American Journal of Veterinary Research,* 67(9): 1580–1588.

Bergevoet, R.H.M., van Schaik, G., Veling, J., Backus, G.B.C. & Franken, P. 2009. Economic and epidemiological evaluation of *Salmonella* control in Dutch dairy herds. *Preventive Veterinary Medicine,* 89(1-2): 1–7.

Brewer, M.T., Anderson, K.L., Yoon, I.K., Scott, M.F. & Carlson, S.A. 2014. Amelioration of salmonellosis in pre-weaned dairy calves fed *Saccharomyces cerevisiae* fermentation products in feed and milk replacer. *Veterinary Microbiology,* 172(1-2): 248–255.

Carlson, J.C., Engeman, R.M., Hyatt, D.R., Gilliland, R.L., Deliberto, T.J., Clark, L., Bodenchuk, M.J. & Linz, G.M. 2011. Efficacy of European starling control to reduce *Salmonella enterica* contamination in a concentrated animal feeding operation in the Texas panhandle. *BMC Veterinary Research,* 7: Article 9.

Cummings, K.J., Warnick, L.D., Alexander, K.A., Cripps, C.J., Grohn, Y.T., McDonough, P.L., Nydam, D.V. & Reed, K.E. 2009. The incidence of salmonellosis among dairy herds in the northeastern United States. *Journal of Dairy Science,* 92(8): 3766–3774.

Davies, R.H. 1997. A two year study of *Salmonella* Typhimurium DT 104 infection and contamination on cattle farms. *Cattle Practice,* 5(3): 189–194,

Davison, H.C., Sayers, A.R., Smith, R.P., Pascoe, S.J.S., Davies, R.H., Weaver, J.P. & Evans, S.J. 2006. Risk factors associated with the *Salmonella* status of dairy farms in England and Wales. *Veterinary Record,* 159(26): 871–880.

Devant, M., Adelantado, C., Anglada, A., Calvo, M.A. & Bach, A. 2009. Effect of plant extract supplementation on *Escherichia coli* O157:H7 and *Salmonella* carcass isolation in young Holstein bulls fed a high-concentrate diet. *Journal of Food Protection,* 72(1): 147–150.

Dewell, G.A., Simpson, C.A., Dewell, R.D., Hyatt, D.R., Belk, K.E., Scanga, J.A., Morley, P.S., Grandin, T., Smith, G.C., Dargatz, D.A., Wagner, B.A. & Salman, M.D. 2008. Risk associated with transportation and lairage on hide contamination with *Salmonella enterica* in finished beef cattle at slaughter. *Journal of Food Protection,* 71(11): 2228–2232.

Dodd, C.C., Renter, D.G., Thomson, D.U. & Nagaraja, T.G. 2011. Evaluation of the effects of a commercially available *Salmonella* Newport siderophore receptor and porin protein vaccine on faecal shedding of *Salmonella* bacteria and health and performance of feedlot cattle. *American Journal of Veterinary Research*, 72(2): 239–247.

Edrington, T.S., Looper, M.L., Duke, S.E., Callaway, T.R., Genovese, K.J., Anderson, R.C. & Nisbet, D.J. 2006a. Effect of ionophore supplementation on the incidence of *Escherichia coli* O157:H7 and *Salmonella* and antimicrobial susceptibility of faecal coliforms in stocker cattle. *Foodborne Pathogens and Disease*, 3(3): 284–291.

Edrington, T.S., Callaway, T.R., Ives, S.E., Engler, M.J., Welsh, T.H., Hallford, D.M., Genovese, K.J., Anderson, R.C. & Nisbet, D.J. 2006b. Effect of ractopamine HCl supplementation on faecal shedding of *Escherichia coli* O157:H7 and *Salmonella* in feedlot cattle. Current Microbiology, 53(4): 340–345.

Edrington, T.S., MacDonald, J.C., Farrow, R.L., Callaway, T.R., Anderson, R.C. & Nisbet, D.J. 2010. Influence of wet distiller's grains on prevalence of *Escherichia coli* O157:H7 and *Salmonella* in feedlot cattle and antimicrobial susceptibility of generic *Escherichia coli* isolates. *Foodborne Pathogens and Disease,* 7(5): 605–608.

Edrington, T.S., Dowd, S.E., Farrow, R.F., Hagevoort, G.R., Callaway, T.R., Anderson, R.C. & Nisbet, D.J. 2012. Development of colonic microflora as assessed by pyrosequencing in dairy calves fed waste milk. *Journal of Dairy Science,* 95(8): 4519–4525.

Edrington, T.S., Loneragan, G.H., Hill, J., Genovese, K.J., He, H., Callaway, T.R., Anderson, R.C., Brichta-Harhay, D.M. & Nisbet, D.J. 2013. Development of a transdermal *Salmonella* challenge model in calves. *Journal of Food Protection,* 76(7): 1259–1263.

Evans, S.J. 1996. A case control study of multiple resistant *Salmonella* Typhimurium DT 104 infection of cattle in Great Britain. *Cattle Practice,* 4: 259–266.

FAO [Food and Agriculture Organization of the United Nations]/OIE [World Organisation for Animal Health]/World Bank. 2010. Good practices for biosecurity in the pig sector: issues and options in developing and transition countries. *[FAO] Animal Production and Health Paper,* No. 169. Available at: http://www.fao.org/docrep/012/i1435e/i1435e00.pdf Accessed 2015-11-10.

Fossler, C.P., Wells, S.J., Kaneene, J.B., Ruegg, P.L., Warnick, L.D., Bender, J.B., Eberly, L.E., Godden, S.M. & Halbert, L.W. 2005a. Herd-level factors associated with isolation of *Salmonella* in a multi-state study of conventional and organic dairy farms: II. *Salmonella* shedding in calves. *Preventive Veterinary Medicine,* 70(3-4): 279–291.

Fossler, C.P., Wells, S.J., Kaneene, J.B., Ruegg, P.L., Warnick, L.D., Bender, J.B., Eberly, L.E., Godden, S.M. & Halbert, L.W. 2005b. Herd-level factors associated with isolation of *Salmonella* in a multi-state study of conventional and organic dairy farms: I. *Salmonella* shedding in cows. *Preventive Veterinary Medicine,* 70(3-4): 257–277.

Fox, B.C., Roof, M.B., Carter, D.P., Kesl, L.D. & Roth, J.A. 1997. Safety and efficacy of an avirulent live *Salmonella* Choleraesuis vaccine for protection of calves against *S.* Dublin infection. *American Journal of Veterinary Research,* 58(3): 265–271.

Frizzo, L.S., Zbrun, M.V., Soto, L.P., Bertozzi, E., Sequeira, G.J., Marti, L.E., Signorini, M.L., Armesto, R.R. & Rosmini, M.R. 2012. Pathogen translocation and histopathological lesions in an experimental model of *Salmonella* Dublin infection in calves receiving lactic acid bacteria and lactose supplements. *Journal of Veterinary Science,* 13(3): 261–270.

Giguere, S., Prescott, J.F., Baggot, J.D., Walker, R.D. & Dowling, P.M. 2006. *Antimicrobial therapy in veterinary medicine.* Blackwell Publishing, Ames, Iowa, USA.

Green, A.L., Dargatz, D.A., Wagner, B.A., Fedorka-Cray, P., Ladely, S.R. & Kopral, C.A. 2010. Analysis of risk factors associated with *Salmonella* spp. isolated from U.S. feedlot cattle. *Foodborne Pathogens and Disease,* 7(7): 825–833.

Habing, G.G., Neuder, L.M., Raphael, W., Piper-Youngs, H. & Kaneene, J.B. 2011. Efficacy of oral administration of a modified-live Salmonella Dublin vaccine in calves. *Journal of the American Veterinary Medical Association,* 238(9): 1184–1190.

Habing, G.G., Lombard, J.E., Kopral, C.A., Dargatz, D.A. & Kaneene, J.B. 2012. Farm-level associations with the shedding of *Salmonella* and antimicrobial-resistant *Salmonella* in U.S. dairy cattle. *Foodborne Pathogens and Disease,* 9(9): 815–821.

Heider, L.C., Meiring, R.W., Hoet, A.E., Gebreyes, W.A., Funk, J.A. & Wittum, T.E. 2008. Evaluation of vaccination with a commercial subunit vaccine on shedding of *Salmonella enterica* in subclinically infected dairy cows. *Journal of the American Veterinary Medical Association,* 233(3): 466–469.

Hermesch, D.R., Thomson, D.U., Loneragan, G.H., Renter, D.R. & White, B.J. 2008. Effects of a commercially available vaccine against *Salmonella enterica* serotype Newport on milk production, somatic cell count, and shedding of *Salmonella* organisms in female dairy cattle with no clinical signs of salmonellosis. *American Journal of Veterinary Research,* 69(9): 1229–1234.

House, J.K., Ontiveros, M.M., Blackmer, N.M., Dueger, E.L., Fitchhorn, J.B., McArthur, G.R. & Smith, B.P. 2001. Evaluation of an autogenous *Salmonella* bacterin and a modified live *Salmonella* serotype Choleraesuis vaccine on a commercial dairy farm. *American Journal of Veterinary Research,* 62(12): 1897–1902.

Huston, C.L., Wittum, T.E., Love, B.C. & Keen, J.E. 2002. Prevalence of faecal shedding of *Salmonella* spp. in dairy herds. *Journal of the American Veterinary Medical Association,* 220(5): 645–649.

Jacob, M.E., Fox, T., Narayanan, S.K., Drouillard, J.S., Renter, D.G. & Nagaraja, T.G. 2008. Effects of feeding wet corn distiller's grains with solubles with or without monensin and tylosin on the prevalence and antimicrobial susceptibilities of faecal foodborne pathogenic and commensal bacteria in feedlot cattle. *Journal of Animal Science,* 86(5): 1182–1190.

Jacob, M.E., Fox, J.T., Drouillard, J.S., Renter, D.G. & Nagaraja, T.G. 2009. Evaluation of feeding dried distiller's grains with solubles and dry-rolled corn on the faecal prevalence of *Escherichia coli* O157:H7 and *Salmonella* spp. in cattle. *Foodborne Pathogens and Disease,* 6(2): 145–153.

Jordan, D., Nielsen, L.R. & Warnick, L.D. 2008. Modelling a national programme for the control of foodborne pathogens in livestock: the case of *Salmonella* Dublin in the Danish cattle industry. *Epidemiology and Infection,* 136(11): 1521–1536.

Kabagambe, E.K., Wells, S.J., Garber, L.P., Salman, M.D., Wagner, B., Fedorka-Cray, P., 2000. Risk factors for faecal shedding of *Salmonella* in 91 US dairy herds in 1996. *Preventive Veterinary Medicine,* 43(3): 177–194.

Li, M., Malladi, S., Hurd, H.S., Goldsmith, T.J., Brichta-Harhay, D.M. & Loneragan, G.H. 2015. *Salmonella* spp. in lymph nodes of fed and cull cattle: Relative assessment of risk to ground beef. *Food Control,* 50: 423–434.

Looper, M.L., Edrington, T.S., Flores, R., Rosenkrans, C.F., Nihsen, M.E. & Aiken, G.E. 2006. Prevalence of *Escherichia coli* O157:H7 and *Salmonella* in beef steers consuming different forage diets. *Letters in Applied Microbiology,* 42(6): 583–588.

Losinger, W.C., Wells, S.J., Garber, L.P., Hurd, H.S. & Thomas, L.A. 1995. Management factors related to *Salmonella* shedding by dairy heifers. *Journal of Dairy Science,* 78(11): 2464–2472.

Losinger, W.C., Garber, L.P., Smith, M.A., Hurd, H.S., Biehl, L.G., Fedorka-Cray, P., Thomas, L.A. & Ferris, K. 1997. Management and nutritional factors associated with the detection of *Salmonella* sp. from cattle faecal specimens from feedlot operations in the United States. *Preventive Veterinary Medicine,* 31(3-4): 231–244.

Lu, Z., Grohn, Y.T., Smith, R.L., Wolfgang, D.R., van Kessel, J.A.S. & Schukken, Y.H. 2009. Assessing the potential impact of *Salmonella* vaccines in an endemically infected dairy herd. *Journal of Theoretical Biology,* 259(4): 770–784.

Lu, Z., Grohn, Y.T., Smith, R.L., Karns, J.S., Hovingh, E. & Schukken, Y.H., 2014. Stochastic modelling of imperfect *Salmonella* vaccines in an adult dairy herd. *Bulletin of Mathematical Biology,* 76(3): 541–565.

Miller, R.B., Sprouse, R.F. & Garner, H.E. 1991. Active immunization of cows with a *Salmonella* Typhimurium mutant bacterin-toxoid and the passive transfer of anti-core-antigen antibodies in colostrum. *Bovine Practice,* 26: 169–170.

Mizuno, T., McLennan, M. & Trott, D. 2008. Intramuscular vaccination of young calves with a *Salmonella* Dublin metabolic-drift mutant provides superior protection to oral delivery. *Veterinary Research,* 39(3): Article 26.

Mohler, V.L., Heithoff, D.M., Mahan, M.J., Walker, K.H., Hornitzky, M.A., McConnell, C.S., Shum, L.W.C. & House, J.K. 2006. Cross-protective immunity in calves conferred by a DNA adenine methylase deficient *Salmonella enterica* serovar Typhimurium vaccine. *Vaccine*, 24(9): 1339–1345.

Mohler, V.L., Heithoff, D.M., Mahan, M.J., Walker, K.H., Hornitzky, M.A., Shum, L.W.C., Makin, K.J. & House, J.K. 2008. Cross-protective immunity conferred by a DNA adenine methylase deficient *Salmonella enterica* serovar Typhimurium vaccine in calves challenged with *Salmonella* serovar Newport. *Vaccine*, 26(14): 1751–1758.

Mortola, M.E., Pennimpede, P.E., Arauz, P.M., Lazovich, L.S. & Naumovich, T.D. 1992. Calf salmonellosis: prophylaxis by maternal immunization. *Avances en Ciencias Veterinarias*, 7(2): 203–208.

Mukkur, T.K.S., Walker, K.H., Jones, D., Wronski, E. & Love, D.N. 1991. Immunizing efficacy of aromatic-dependent *Salmonella* Dublin in mice and calves. *Comparative Immunology Microbiology and Infectious Diseases*, 14(3): 243–256.

Nielsen, T.D., Vesterbaek, I.L., Kudahl, A.B., Borup, K.J. & Nielsen, L.R. 2012a. Effect of management on prevention of *Salmonella* Dublin exposure of calves during a one-year control programme in 84 Danish dairy herds. *Preventive Veterinary Medicine*, 105(1-2): 101–109.

Nielsen, L.R., Kudahl, A.B. & Ostergaard, S. 2012. Age-structured dynamic, stochastic and mechanistic simulation model of *Salmonella* Dublin infection within dairy herds. *Preventive Veterinary Medicine*, 105(1-2): 59–74.

Smith, B.P., Dilling, G.W., Roden, L.D. & Stocker, B.A.D. 1993. Vaccination of calves with orally administered aromatic-dependent *Salmonella* Dublin. *American Journal of Veterinary Research*, 54(8): 1249–1255.

Smith, G.W., Alley, M.L., Foster, D.M., Smith, F. & Wileman, B.W. 2014. Passive immunity stimulated by vaccination of dry cows with a *Salmonella* bacterial extract. *Journal of Veterinary Internal Medicine*, 28(5): 1602–1605.

Smith, G.W., Smith, F., Zuidhof, S. & Foster, D.M. 2015. Short communication: Characterization of the serologic response induced by vaccination of late-gestation cows with a *Salmonella* Dublin vaccine. *Journal of Dairy Science*, 98(4): 2529–2532.

Spanghero, M., Zanfi, C., Fabbro, E., Scicutella, N. & Camellini, C. 2007. Effect of milk replacers added with microencapsulated organic acids or essential oils on the performance of weaning calves. *Italian Journal of Animal Science*, 6(Suppl. 1): 366–366.

Stephens, P., Loneragan, G.H., Karunasena, E. & Brashears, M.M. 2007. Reduction of *Escherichia coli* O157 and *Salmonella* in faeces and on hides of feedlot cattle using various doses of a direct-fed microbial. *Journal of Food Protection*, 70(10): 2386–2391.

Tabe, E.S., Oloya, J., Doetkott, D.K., Bauer, M.L., Gibbs, P.S. & Khaitsa, M.L. 2008. Comparative effect of direct-fed microbials on faecal shedding of *Escherichia coli*

O157:H7 and *Salmonella* in naturally infected feedlot cattle. *Journal of Food Protection,* 71(3): 539–544.

Vaessen, M.A., Veling, J., Frankena, K., Graat, E.A.M. & Klunder, T. 1998. Risk factors for *Salmonella* Dublin infection on dairy farms. *Veterinary Quarterly,* 20(3): 97–99

Van Der Walt, M.L., Vorster, J.H., Steyn, H.C. & Greeff, A.S. 2001. Auxotrophic, plasmid-cured *Salmonella enterica* serovar Typhimurium for use as a live vaccine in calves. *Veterinary Microbiology,* 80(4): 373–381.

Van Schaik, G., Nielen, M. & Dijkhuizen, A.A. 2001. An economic model for on-farm decision support of management to prevent infectious disease introduction into dairy farms. *Preventive Veterinary Medicine,* 51(3-4): 289–305.

Vanselow, B.A., Hornitzky, M.A., Walker, K.H., Eamens, G.J., Bailey, G.D., Gill, P.A., Coates, K., Corney, B., Cronin, J.P. & Renilson, S. 2007. *Salmonella* and on-farm risk factors in healthy slaughter-age cattle and sheep in eastern Australia. *Australian Veterinary Journal,* 85(12): 498–502.

Veling, J., Wilpshaar, H., Frankena, K., Bartels, C. & Barkema, H.W. 2002. Risk factors for clinical *Salmonella enterica* subsp. *enterica* serovar Typhimurium infection on Dutch dairy farms. *Preventive Veterinary Medicine,* 54(2): 157–168.

Warnick, L.D., Kanistanon, K., McDonough, P.L. & Power, L. 2003. Effect of previous antimicrobial treatment on faecal shedding of *Salmonella enterica* subsp. *enterica* serogroup B in New York dairy herds with recent clinical salmonellosis. *Preventive Veterinary Medicine,* 56(4): 285–297.

Younis, E.E., Ahmed, A.M., El-Khodery, S., Osman, S.A. & El-Naker, Y., 2009. Molecular screening and risk factors of enterotoxigenic *Escherichia coli* and *Salmonella* spp. in diarrheic neonatal calves in Egypt. *Research in Veterinary Science,* 87(3): 373–379.

SUMMARY CARD 5:
Beef Chain – Transport and Lairage Interventions

SC5.1 SUMMARY OF KEY FINDINGS

This Summary Card covers the evidence supporting transport and lairage interventions for reduction of *Salmonella* in beef. Overall, we found few studies investigating reduction of *Salmonella* in cattle during transport and none investigating reduction of *Salmonella* during lairage, although the potential amplification of bacteria at this stage has been confirmed in multiple studies.

SC5.1.1 Transport to slaughter

- A quasi-experiment (before-and-after trial) demonstrated that hide contamination significantly increased in both feedlot and adult cattle during transport, as did rectal shedding in adult cattle. Risk factors identified for increased odds of contamination post-transport included transportation distance and cattle behaviour during loading, although cleanliness of the transport trucks was a non-significant predictor of *Salmonella* contamination in one study.

SC5.1.2 Lairage

- A study of *Salmonella* spp. isolates collected from United States of America cattle in lairage concluded that the majority of isolates from both hides and carcasses at slaughter genotypically matched those from abattoir lairage, and not those from the farm of origin. However, a cross-sectional study conducted in the UK concluded that, in contrast with pigs, time in lairage was not a significant predictor of odds of carcass *Salmonella* contamination. Cattle type was, however, a significant predictor, with dairy cattle having increased odds of carcass contamination relative to beef breeds.

SC5.1.3 Overall implications

- The small body of literature found suggests the potential for both transport and lairage to be areas of amplification and transmission of *Salmonella* in cattle. However, currently, investigations of interventions applied at these points for *Salmonella* reduction are lacking.

SC5.2 INTERVENTION DESCRIPTION

This Summary Card summarizes the evidence for interventions that can be implemented post-farm during transport and at lairage to reduce *Salmonella* contamination of beef. Specific categories of interventions covered in this Summary Card are:

- **Transport to slaughter:** refers to the transportation of market animals from the final farm production unit to the abattoir.
- **Lairage:** refers to holding facilities encountered by animals at the abattoir from the time of unloading from trucks, to the stunning event.

SC5.3 TRANSPORT TO SLAUGHTER

Beach, Murano and Acuff (2002) describe a quasi-experiment in which United States of America cattle were sampled before and after transport to abattoir; rectal, hide and carcass swabs were cultured for *Salmonella*. For feedlot cattle, the prevalence of rectal shedding was similar pre-and post-transport. For adult pasture cattle, the prevalence of rectal shedding increased significantly post-transport relative to pre-transport. Hide contamination increased significantly post-transport for both types of cattle. A significantly greater proportion of adult carcasses cultured positive for *Salmonella*, relative to feedlot cattle.

Dewell *et al.* (2008) describe a multistage cross-sectional study of hide prevalence of *Salmonella* in United States of America cattle at the feedlot, and after transport to the abattoir, using multi-level Poisson regression to identify risk factors. The authors reported that transportation distance (long (>160.9 km) vs short) was a significant predictor of risk of *Salmonella* (relative risk (RR) = 2.28, 95% confidence interval (CI) (1.41–3.69)). Cattle behaviour during loading (agitated vs calm) was also a significant predictor (RR = 2.19, 95% CI (1.32–3.62)).

Reicks *et al.* (2007) report a cross-sectional study of United States of America feedlot cattle, sampled before and after transport to the abattoir, as well as samples of the trucks before and after transport. Although there was a significant difference in *Salmonella* prevalence in dirty trucks relative to clean ones prior to transporting the cattle being studied, the *Salmonella* prevalence of dirty and clean trucks was similar after transporting a load of cattle. The authors sampled the animals at two sites (midline and withers) and reported significantly greater prevalence at the midline site relative to the withers, in both before- and after-transport settings. However, at each site, the prevalence at the abattoir did not differ significantly between cattle transported in clean relative to dirty vehicles.

SC5.4 LAIRAGE

Our search found no investigations of specific interventions applied at the lairage stage of the journey to slaughter. We did, however, capture several observational studies of lairage cleaning and disinfection practices, as well as the change in *Salmonella* prevalence in cattle between their pen on-farm and at the abattoir.

Small *et al.* (2006) captured cleaning practices in 21 UK ruminant abattoirs. Authors report that bedding was used in the majority of lairages and was changed either between animal batches, daily, weekly or monthly. Roughly one in four lairages investigated was washed daily. None were cleaned using detergents or disinfectants. The authors concluded that the cleaning and disinfection protocols employed, in general, were unlikely to eliminate the microbial burden.

Small *et al.* (2002) reported an increase in *Salmonella* prevalence in environmental samples in the lairage of UK abattoirs, from 1.1% before a shift began to 11.1%, after a shift was completed.

Arthur *et al.* (2008) sampled 581 United States of America feedlot cattle hides on-farm, and again at the abattoir just prior to the wash cabinet. Prevalence of *Salmonella* on hides was significantly greater at the abattoir relative to on-farm (74.2% vs 0.7%). Additionally, pulsed field gel electrophoresis demonstrated that none of the *Salmonella* genotypes identified on-farm matched any of the genotypes isolated from animals at the abattoir. However, 959 of the 1007 hide isolates identified at the abattoir matched those from lairage, and 42 of 50 carcass isolates identified matched genotypes from lairage, highlighting the potential importance of lairage in transmission of hide-level infection.

Dewell *et al.* (2008) reported that United States of America slaughter cattle spending time in dirty lairage had greater risk of *Salmonella* positive hides at slaughter relative to those in clean lairage (RR = 1.83, 95% CI (0.7–3.14)).

Milnes *et al.* (2009) reported a cross-sectional study of *Salmonella* in carcasses in UK abattoirs, using logistic regression to identify risk factors for contamination. In contrast to findings for pork, time in lairage was a non-significant predictor for *Salmonella* contamination of beef carcasses. Breed was significant, however, with dairy breeds having increased odds of *Salmonella* contamination relative to beef breeds (OR = 3.38, 95% CI (1.49–7.66)). This is consistent with a cross-sectional Australian study of slaughter cattle, which reported that dairy cattle were significantly more likely to be faecally shedding *Salmonella* relative to beef cattle kept on pasture (Vanselow *et al.*, 2007).

SC5.5 REFERENCES CITED IN SC5

Arthur, T.M., Bosilevac, J.M., Brichta-Harhay, D., Kalchayanand, N., King, D.A., Shackelford, S.D., Wheeler, T.L. & Koohmaraie, M. 2008. Source tracking of *Escherichia coli* O157:H7 and *Salmonella* contamination in the lairage environment at commercial U.S. beef processing plants and identification of an effective intervention. *Journal of Food Protection,* 71(9): 1752–1760.

Beach, J.C., Murano, E.A. & Acuff, G.R. 2002. Prevalence of *Salmonella* and *Campylobacter* in beef cattle from transport to slaughter. *Journal of Food Protection,* 65(11): 1687–1693.

Dewell, G.A., Simpson, C.A., Dewell, R.D., Hyatt, D.R., Belk, K.E., Scanga, J.A., Morley, P.S., Grandin, T., Smith, G.C., Dargatz, D.A., Wagner, B.A. & Salman, M.D. 2008. Risk associated with transportation and lairage on hide contamination with *Salmonella enterica* in finished beef cattle at slaughter. *Journal of Food Protection,* 71(11): 2228–2232.

Li, M., Malladi, S., Hurd, H.S., Goldsmith, T.J., Brichta-Harhay, D.M. & Loneragan, G.H. 2015. *Salmonella* spp. in lymph nodes of fed and cull cattle: Relative assessment of risk to ground beef. *Food Control,* 50: 423–434.

Milnes, A.S., Sayers, A.R., Stewart, I., Clifton-Hadley, F., Davies, R.H., Newell, D.G., Cook, A.J.C., Evans, S.J., Smith, R.P. & Paiba, G.A. 2009. Factors related to the carriage of Verocytotoxigenic *E. coli*, *Salmonella*, thermophilic *Campylobacter* and *Yersinia enterocolitica* in cattle, sheep and pigs at slaughter. *Epidemiology and Infection,* 137(8): 1135–1148.

Reicks, A.L., Brashears, M.M., Adams, K.D., Brooks, J.C., Blanton, J.R. & Miller, M.F. 2007. Impact of transportation of feedlot cattle to the harvest facility on the prevalence of *Escherichia coli* O157:H7, *Salmonella*, and total aerobic microorganisms on hides. *Journal of Food Protection,* 70(1): 17–21.

Small, A., Reid, C.A., Avery, S.M., Karabasil, N., Crowley, C., Buncic, S., 2002. Potential for the spread of *Escherichia coli* O157, *Salmonella*, and *Campylobacter* in the lairage environment at abattoirs. *Journal of Food Protection,* 65(5): 931–936.

Small, A., James, C., James, S., Davies, R., Liebana, E., Howell, M., Hutchison, M. & Buncic, S. 2006. Presence of *Salmonella* in the red meat abattoir lairage after routine cleansing and disinfection and on carcasses. *Journal of Food Protection,* 69(10): 2342–2351.

Vanselow, B.A., Hornitzky, M.A., Walker, K.H., Eamens, G.J., Bailey, G.D., Gill, P.A., Coates, K., Corney, B., Cronin, J.P. & Renilson, S. 2007. *Salmonella* and on-farm risk factors in healthy slaughter-age cattle and sheep in eastern Australia. *Australian Veterinary Journal,* 85(12): 498–502.

SUMMARY CARD SC6:
Beef Chain – Processing Interventions

SC6.1 SUMMARY OF KEY FINDINGS

This Summary Card covers the evidence for interventions implemented at various stages of beef processing for *Salmonella* control:

SC6.1.1 Hide interventions

- Limited evidence was found for the effect of washes applied to live animals.
- Four quasi-experiments found that other chemical washes (including hydrogen bromide, chlorine, or sodium hydroxide) applied post-exsanguination significantly reduced *Salmonella* prevalence on hides, with moderate confidence in the estimate of effect. One study found a similar effect for an organic acid wash, while conflicting results were shown for water washes.
- Two challenge trials indicated that de-hairing may be effective to control *Salmonella* on hides, but two controlled trials showed conflicting evidence of its effect on other microbial hazards and indicator bacteria.

SC6.1.2 Bunging

- One controlled trial study suggests that bunging conducted prior to evisceration washes may provide significant *Salmonella* reductions compared with not bunging prior to evisceration, with low confidence in the specific estimate of reduction.

SC6.1.3 Post-evisceration and pre-chill

- Two controlled trials conducted under commercial conditions found, with high confidence, that thermal pre-chill washes can significantly reduce *Salmonella* prevalence on beef carcasses. Estimates of intervention effect with moderate to low confidence from seven quasi-experimental studies suggest that water washes, thermal washes, and multiple intervention combinations tend to decrease *Salmonella* prevalence on carcasses prior to chilling.
- Similar results with low to very low confidence were noted in several challenge trials. These studies did not always find a consistent benefit of thermal and organic acid washes compared with ambient or cold temperature water washes. Multiple intervention combinations were the most consistently effective compared with no treatment and water washes.

SC6.1.4 Chilling

- The primary purpose of carcass chilling is to inhibit microbial growth. Five quasi-experimental studies found a very low *Salmonella* prevalence in both pre- and post-chill samples, but the identification of positive samples post-chill could suggest a potential for cross-contamination, which could possibly be due to differences in study sampling methodologies or other factors.

SC6.1.5 Post-chilling and pre-fabrication

- One quasi-experimental study with a very low post-chill *Salmonella* prevalence (1%) suggested, with low confidence in the estimate of effect, that additional reductions in prevalence might be achievable with a post-chill organic acid wash.

SC6.1.6 Post-fabrication

- Several challenge trials suggest that all intervention categories tend to reduce inoculated *Salmonella* levels on beef trimmings when compared with no treatment, with low to very low confidence in the range of possible effects. Organic acid, other chemical/oxidizer, lactic acid bacteria, and multiple treatment combination washes showed consistent benefits when compared with ambient or cold temperature water.

SC6.1.7 Packaging

- Several challenge trials have shown that modified packaging and preservation with various chemicals and extracts have potential to mitigate *Salmonella* in beef products during distribution and storage, although some may affect product sensory quality.

SC6.1.8 Multiple processing stages

- One systematic review found evidence from five quasi-experimental studies that HACCP implementation on its own is not linked to reductions in *Salmonella* prevalence, with moderate confidence in the estimate of effect; but was linked to a reduction in aerobic bacterial counts.
- Results from several quasi-experimental studies suggest a clear trend for *Salmonella* to be reduced to very low levels by implementing multiple interventions between pre-evisceration and chilling; however, there is very low confidence in the expected magnitude of effect due to possible confounding factors and variation among studies.

SC6.1.9 Overall implications

- Overall, evidence was limited to studies with mostly low to very low confidence in terms of how closely one could expect a similar estimate of intervention effect in practice due to one or more of the following: potential for confounding bias; small sample sizes; variability across studies; and laboratory conditions that might not reflect commercial settings.

- Nevertheless, the evidence suggests that *Salmonella* reductions can be expected at multiple stages of processing with different washes, alternative treatments (e.g. irradiation), and various combinations of interventions.
- Use of thermal, organic acid, and other chemical washes at multiple stages of processing should provide consistent reductions in *Salmonella* contamination of beef; however, these enhanced washes did not always consistently achieve more effective reductions compared with ambient or cold temperature water alone.
- Greatest reductions in *Salmonella* are likely to be achieved when multiple interventions are implemented in combination, together with chilling, to prevent the growth of any microbes present, as part of a "multiple-hurdle" strategy.

SC6.2 INTERVENTION DESCRIPTION

This Summary Card summarizes the evidence for a range of interventions that can be implemented at the processing level to reduce *Salmonella* contamination of beef. Specific categories of interventions where evidence was identified in the literature on *Salmonella* control and that are covered in this Summary Card are:

- **Organic acid washes:** refers to washes with antimicrobials such as lactic, acetic, and citric acid, that affect microbial growth through disruptions to nutrient transport and energy generation and can cause injury to microbial cells through their low pH (Wheeler, Kalchayanand and Bosilevac, 2014).
- **Washes containing other chemicals and oxidizers:** includes washes containing other miscellaneous products that destroy bacteria through various actions, such as oxidation and disruption of cellular functions (Wheeler, Kalchayanand and Bosilevac, 2014), or that prevent bacterial attachment to meat. Examples include chlorine solutions, trisodium phosphate, alcohols, peracetic acid, peroxyacetic acid, hypobromous acid, acidified sodium chlorite, electrolyzed oxidized water, and ozone. Includes chemical de-hairing, which is a patented process of applying successive water and chemical washes in a cabinet to remove hair and improve visible cleanliness and therefore reduce microbial loads on animal hides (Schnell *et al.*, 1995).
- **Protective bacterial culture washes:** refers to washes containing lactic acid bacteria to control pathogens through the production of antimicrobial compounds such as bacteriocins, hydrogen peroxide, and organic acids (Baer, Miller and Dilger, 2013).
- **Water washes:** refers to an ambient or cold-temperature wash to physically remove contamination from carcasses.
- **Thermal washes:** refers to various heat treatment washes to destroy microbial

cells. Examples include hot water and steam pasteurization.

- **Dry heat:** refers to non-hydrating thermal interventions such a forced-air heating.
- **Electricity:** refers to application of electricity, such as through electric stunners.
- **Natural extracts:** refers to plant and other extracts applied as a treatment.
- **Non-thermal interventions:** refers to alternative, non-chemical and non-thermal interventions that aim to reduce microbial contamination while preserving product quality and nutrients that can be affected by thermal treatments (Wheeler, Kalchayanand and Bosilevac, 2014). Examples include electron beam and gamma irradiation, ultraviolet (UV) light, cold atmospheric plasma and high pressure processing.
- **Modified packaging and preservation techniques:** refers to a range of interventions that can be applied to prevent spoilage and inhibit microbial growth during final product distribution and storage.
- **Good hygienic and production practices:** includes a range of different practices that are or can be implemented in processing plants and that may have a pathogen-reduction effect. Examples with evidence included in this summary:
    - *Bunging* = tying off the rectum with a plastic bag during removal of intact viscera at slaughter to minimize the spread of cross-contamination on a carcass (Stopforth *et al.*, 2006).
    - *Trimming* = the removal of visible contamination from beef carcasses prior to washing (Graves Delmore *et al.*, 1997).
    - *Chilling* = step at the end of the slaughter process and before fabrication of rapidly reducing the carcass temperature to prevent microbial growth and preserve product quality.
    - *HACCP (Hazard analysis critical control point)* = a system that identifies, evaluates, and controls hazards which are significant for food safety (Codex Alimentarius Commission, 2003).

These interventions can be applied at various stages of the processing line, from decontamination of animal hides prior to slaughter to product decontamination post-fabrication and prior to shipment. Evidence was identified for interventions at the following stages of processing, each of which is presented and discussed separately below:

- Pre-de-hiding
- Bunging
- Post-evisceration and pre-chill
- Chilling
- Post-chilling and pre-fabrication
- Post-fabrication
- Packaging
- Multiple processing stages

SC6.3 PRE-DE-HIDING

SC6.3.1 Live animal treatments

One controlled trial conducted under commercial conditions in Pakistan was identified that investigated the effect of a live animal hide wash (hot water, 65.6°C) to reduce *Salmonella* contamination of the hide (Aftab *et al.*, 2012), but the reporting of the results was unclear in terms of the intervention's impact on *Salmonella*. In addition, one quasi-experiment conducted under commercial conditions was identified evaluating the effect of live animal hide washes (Table SC6.1). The study found that a single or double water wash and a lactic acid wash increased *Salmonella* prevalence of the hide, while a 50 ppm chlorine solution slightly decreased prevalence (Mies *et al.*, 2004).

SC6.3.2 Post-exsanguination treatments

One controlled trial conducted under commercial conditions was identified that evaluated the effect of a post-exsanguination hide wash to reduce *Salmonella* contamination; however, *Salmonella* was not isolated from any sample, precluding evaluation of intervention efficacy on this pathogen (Schnell *et al.*, 1995). The study investigated a de-hairing treatment and found no significant difference in reduction of aerobic plate counts or *E. coli* counts between de-haired and non-de-haired (control) carcasses (Schnell *et al.*, 1995). In contrast, another controlled trial that did not measure *Salmonella* found that de-hairing hides significantly reduced *E. coli* 0157 prevalence and aerobic plate counts and enterobacteriaceae counts on pre-evisceration carcasses (Nou *et al.*, 2003).

Five quasi-experiments conducted under commercial conditions were identified evaluating the effect of post-exsanguination hide washes (Table SC6.3). Two studies reported conflicting evidence on water washes with very low confidence in the intervention effect estimates (Arthur *et al.*, 2008a; Scanga *et al.*, 2011). In contrast, one study found that an organic acid wash (6.0% lactic acid at 30°C for 15 sec) was effective to reduce *Salmonella* prevalence of hides (Scanga *et al.*, 2011), while four studies found that other chemical washes (including hydrogen bromide, chlorine, or sodium hydroxide) also significantly reduced *Salmonella* prevalence on hides, with moderate confidence in the estimate of effect (Arthur *et al.*, 2007; Bosilevac *et al.*, 2009; Scanga *et al.*, 2011; Schmidt *et al.*, 2012).

In addition, interim results (July, 2015) from the United States Department of Agriculture (USDA) Nationwide Beef and Veal Carcass Baseline Survey found that a total of 143/561 (25.5%) post-hide removal/pre-evisceration samples collected from commercial slaughter and processing plants tested positive for *Salmonella*, a value that was lower among establishments that used one or more interventions

Summary-of-findings tables for pre-slaughter interventions

TABLE SC6.1 Quasi-experimental studies

Intervention	Intervention/ outcome sample	No. trials/ studies	Odds ratio for intervention effect[1]	% Samples *Salmonella* positive in study population[2]		GRADE rating[3]	References
				Pre-treatment	Post-treatment		
Water washes	Live animal (hide)	2/1	MA = 1.51 (95% CI: 0.97, 2.34)	46.7%	56.8% (95% CI: 45.8-67.2)	Very low	Mies *et al.*, 2004
	Post-exsang. animal (hide)	2/2	Median = 1.75 (Range: 0.04-3.45)	78.1%	86.1% (Range: 13.4-92.5)	Very low	Arthur *et al.*, 2008a; Scanga *et al.*, 2011
Organic acid wash	Live animal (hide)	1/1	Single study = 1.09 (95% CI: 0.61-1.96)	50.0%	52.2% (95% CI: 37.9-66.2)	Very low	Mies *et al.*, 2004
	Post-exsang. animal (hide)	1/1	Single study = 0.35 (95% CI: 0.15-0.81)	74.0%	50.0% (95% CI: 30.1-69.9)	Low	Scanga *et al.*, 2011
Other chemical/ oxidizer washes	Live animal (hide)	1/1	Single study = 0.83 (95% CI: 0.46-1.51)	60.0%	55.6% (Range: 40.9-69.3)	Very low	Mies *et al.*, 2004
	Post-exsang. animal (hide)	5/4	MA = 0.21 (95% CI: 0.13-0.34)	62.3%	26.0% (Range: 18.1-35.8)	Moderate	Arthur *et al.*, 2007; Bosilevac *et al.*, 2009; Scanga *et al.*, 2011; Schmidt *et al.*, 2012

NOTES: MA = meta-analysis average estimate from random-effects model; CI = confidence interval.

(1) For interventions with multiple trials, refers to the average meta-analysis estimate and 95% CI from random-effects meta-analysis if heterogeneity was not significant (I^2<60%). If heterogeneity was significant, the median and range of odds ratios or log reduction values from individual studies are presented instead. If only one trial was available, refers to the odds ratio or log reduction value and 95% confidence interval from the individual study.
For challenge trials, meta-analysis was not conducted due to limiting data reporting. Instead, results are presented as a median (range) log reduction for multiple trials, and as the single estimate and 95% CI for individual studies.

(2) The percentage of samples positive for *Salmonella* in the comparison group (assumed control risk, ACR) refers to the median prevalence in the comparison group trials. The percentage in the intervention/treatment group was calculated using the following formula for the odds ratio (OR) measure of effect: (OR × ACR / 1 -ACR + [OR × ACR]), and provides an indication of the effect of the intervention on changing a given baseline level of *Salmonella* prevalence as reported in the included studies. The odds ratio used in this formula corresponds to the odds ratio presented in the intervention effect column as described under (1) above. If heterogeneity was not significant (I^2<60%), the estimate is presented with a 95% CI in brackets. If heterogeneity was significant, the median estimate is presented with the range of individual study effects in brackets to illustrate the variability in results across studies.

(3) GRADE rating descriptions: very low = the true effect is likely to be substantially different from the measured estimate; low = the true effect may be substantially different from the measured estimate; moderate = the true effect is likely to be close to the measured estimate, but there is a possibility that it is substantially different; high = there is strong confidence that the true effect lies close to that of the measured estimate.

TABLE SC6.2 Challenge trial studies

Intervention	Intervention/ outcome sample	Comparison group	No. trials/ studies	Log CFU/cm^2 reduction$^{(1)}$	GRADE rating$^{(3)}$	References
Organic acid washes	Hide	Water wash	2/2	Median = 1.73 (Range: 0.70–2.75)	Very low	Mies *et al.*, 2004; Carlson *et al.*, 2008
Other chemical/ oxidizer washes	Hide	Water wash	3/2	Median = 0.90 (Range: 0.57–2.35)	Very low	Mies *et al.*, 2004; Carlson *et al.*, 2008
De-hairing	Hide	No treatment	1/1	Single study = 4.6 (no measure of variability reported)	Very low	Castillo *et al.*, 1998a
	Hide	Water wash	1/1	Single study = 0.75 (95% CI: 0.02–1.48)	Low	Carlson *et al.*, 2008

NOTES: MA = meta-analysis average estimate from random-effects model; CI = confidence interval.

(1) For interventions with multiple trials, refers to the average meta-analysis estimate and 95% CI from random-effects meta-analysis if heterogeneity was not significant (I^2<60%). If heterogeneity was significant, the median and range of odds ratios or log reduction values from individual studies are presented instead. If only one trial was available, refers to the odds ratio or log reduction value and 95% confidence interval from the individual study.
For challenge trials, meta-analysis was not conducted due to limiting data reporting. Instead, results are presented as a median (range) log reduction for multiple trials, and as the single estimate and 95% CI for individual studies.

(3) GRADE rating descriptions: very low = the true effect is likely to be substantially different from the measured estimate; low = the true effect may be substantially different from the measured estimate; moderate = the true effect is likely to be close to the measured estimate, but there is a possibility that it is substantially different; high = there is strong confidence that the true effect lies close to that of the measured estimate

during hide removal (22.9%) compared with samples from establishments that didn't use any interventions at this step (28.4%) (USDA, 2015). The preliminary findings indicated that the types of interventions used included cold water washes, hot water washes, and/or chlorine-based interventions.

Five challenge trials conducted under laboratory conditions were also identified, three of which reported extractable data and are summarized in Table SC6.2. Four studies found that washes containing either organic acids or other chemicals were effective in reducing inoculated *Salmonella* levels on hides compared with water washes (Mies *et al.*, 2004; Carlson *et al.*, 2008; Baskaran *et al.*, 2012; Jadeja and Hung, 2014). Two studies investigated hide de-hairing treatments, finding a reduction in inoculated *Salmonella* levels compared with no treatment and water washes (Castillo *et al.*, 1998a; Carlson *et al.*, 2008). However, the confidence in these estimates of reduction is low to very low due to the artificial study conditions, small sample sizes, and/or inconsistency across studies.

SC6.4 BUNGING

One controlled trial was identified conducted under commercial conditions that evaluated the effect of bunging applied prior to pre-evisceration washes on *Salmonella* reduction on beef carcasses (Stopforth *et al.*, 2006). The authors found a significant reduction in *Salmonella* prevalence in intervention carcasses compared with those where no bunging was conducted before the pre-evisceration wash (Table SC6.3).

Summary-of-findings table for the effects of bunging

TABLE SC6.3 Controlled trial studies

Intervention	Intervention/ outcome sample	Comparison group	No. trials/ studies	Odds ratio for intervention effect(1)	% Samples *Salmonella* positive in study population(2)		GRADE rating(3)	References
					No treatment	With treatment		
Bunging applied before pre-evisceration wash	Carcass	No bunging before pre-evisceration wash	1/1	Single study = 0.08 (95% CI: 0.004-1.54)	8.3%	0.8% (95% CI: 0.04-12.3)	Low	Stopforth *et al., 2006*

NOTES: CI = confidence interval.

(1) Refers to the odds ratio and 95% CI from the individual study

(2) The percentage of samples positive for *Salmonella* in the comparison group (assumed control risk, ACR) refers to the median prevalence in the comparison group trials. The percentage in the intervention/treatment group was calculated using the following formula for the odds ratio (OR) measure of effect: (OR × ACR / 1 - ACR + [OR × ACR]), and provides an indication of the effect of the intervention on changing a given baseline level of *Salmonella* prevalence as reported in the included studies. The odds ratio used in this formula corresponds to the odds ratio presented in the intervention effect column as described under (1) above. If heterogeneity was not significant (I^2<60%), the estimate is presented with a 95% CI in brackets. If heterogeneity was significant, the median estimate is presented with the range of individual study effects in brackets to illustrate the variability in results across studies.

(3) GRADE rating descriptions: very low = the true effect is likely to be substantially different from the measured estimate; low = the true effect may be substantially different from the measured estimate; moderate = the true effect is likely to be close to the measured estimate, but there is a possibility that it is substantially different; high = there is strong confidence that the true effect lies close to that of the measured estimate.

SC6.5 POST-EVISCERATION AND PRE-CHILL

Three controlled trials were identified that evaluated the effect of post-evisceration and pre-chill interventions against *Salmonella* in beef (Table SC6.4). Two studies found that a hot water (74–87.8°C for 18–39 sec) or steam (82–85°C for 60 sec) spray significantly reduced *Salmonella* prevalence on beef carcasses compared with a no treatment control group, with high confidence in the estimates of reduction (Reagan *et al.*, 1996; Trivedi, Reynolds and Chen, 2007). One of the studies also found that warm water washes and knife trimming significantly reduced *Salmonella* prevalence on treated carcasses, while other chemicals (specifically hydrogen peroxide and ozone) were not effective (Reagan *et al.*, 1996). *Salmonella* was not isolated from any sample in the third study, precluding evaluation of intervention efficacy on this pathogen (Prasai *et al.*, 1991). The study found that lactic acid washes at multiple points during processing (both after de-hiding and after evisceration) resulted in the largest reductions in aerobic plate counts on carcasses immediately and 72 hours post-slaughter. However, this reduction was in most cases not significantly greater than that achieved by a single post-evisceration lactic acid wash (Prasai *et al.*, 1991).

Seven quasi-experiments conducted under commercial conditions were identified evaluating the effect of post-evisceration and pre-chill washes and wash combinations (Table SC6.5). Studies found that water (Hajmeer *et al.*, 1999; Trairatapiwan, Lertpatarakomol and Mitchaothai, 2011; Narváez-Bravo *et al.*, 2013; Dong *et al.*, 2014) and thermal (Nutsch *et al.*, 1997; Wright, 2009) washes tended to decrease *Salmonella* prevalence prior to chilling (Table SC6.5), with low or moderate confidence, respectively, in the specific estimates of reduction. Large reductions were noted in one study that investigated a sequence of ambient temperature water, hot water (82°C), then 4–5% lactic acid washes (Ruby, Zhu and Ingham, 2007).

A total of 35 challenge trials were identified that investigated the efficacy of post-evisceration and pre-chill interventions on beef carcasses (Dickson and Anderson, 1991; Greer and Dilts, 1992; Dickson, Cutter and Siragusa, 1994; Kim and Slavik, 1994; Bawcom *et al.*, 1995; Hardin *et al.*, 1995; Bell, Cutter and Sumner, 1997; Cutter, Dorsa and Siragusa, 1997; Cutter *et al.*, 1999a, 2000; Phebus *et al.*, 1997; Tinney *et al.*, 1997; Arthur *et al.*, 2008b; Cabrera-Diaz *et al.*, 2009; Castillo *et al.*, 1998b, c, 1999; Dorsa *et al.*, 1998a; Dorsa *et al.*, 1998b; Cutter and Rivera-Betancourt, 2000; Castillo *et al.*, 1998b, 2001, 2003; Reyes *et al.*, 2003; Retzlaff *et al.*, 2004; Ellebracht *et al.*, 2005; King *et al.*, 2005; Niebuhr *et al.*, 2008; Pearce and Bolton, 2008; Sawyer *et al.*, 2008; Kalchayanand *et al.*, 2009; Laury *et al.*, 2009; Njongmeta *et al.*, 2011; Wolf *et al.*, 2012; Yoder *et al.*, 2010, 2012). Thirty-three of these reported extractable log reduction data and are summarized in Table SC6.6. Compared with no treatment, regular water, thermal, and organic acid washes tended to reduce inoculated *Salmonella* levels on beef carcasses by one or more logs, but thermal and organic acid washes did not always provide significant benefits compared with ambient or cold temperature water (Table SC6.6). Multiple intervention combinations were the most consistently effective to reduce inoculated *Salmonella* levels compared with no treatment and ambient or cold temperature water washes (Table SC6.6). Conflicting evidence was found for the efficacy of washes with other chemicals and oxidizers, with some studies indicating a potential for multiple log reductions (Table SC6.6). Two studies investigating trimming of visible contamination found that it reduced inoculated *Salmonella* levels by 1.5–1.9 logs compared with a warm-water wash (Hardin *et al.*, 1995; Phebus *et al.*, 1997). Two studies found that electricity via stunners slightly reduced inoculated *Salmonella* levels on carcasses by <1 log (Bawcom *et al.*, 1995; Tinney *et al.*, 1997), while two other studies found potential for multiple log reductions from sprays with natural (specifically dairy and grapefruit) extracts (Reyes *et al.*, 2003; Pearce and Bolton, 2008). In two studies that measured outcomes in treated beef carcasses that were

Summary-of-findings tables for interventions applied post-evisceration and pre-chill

TABLE SC6.4 Controlled trial studies

Intervention	Intervention/ outcome sample	Comparison group	No. trials/ studies	Odds ratio for intervention effect[1]	% Samples *Salmonella* positive in study population[2]		GRADE rating[3]	References
					Without treatment	With treatment		
Water wash	Carcass	No treatment	1/1	Single study = 0.23 (95% CI: 0.12-0.45)	30.3%	9.0% (95% CI: 4.8-16.3)	Very low	Reagan *et al.*, 1996
Thermal wash	Carcass	No treatment	2/2	MA = 0.09 (95% CI: 0.02-0.48)	15.4%	1.6% (95% CI: 0.3-8.0)	High	Reagan *et al.*, 1996; Trivedi, Reynolds and Chen, 2007
Other chemical/ oxidizer wash	Carcass	No treatment	1/1	Single study = 1.28 (95% CI: 0.79-2.09)	30.3%	35.8% (95% CI: 25.5-47.6)	Very low	Reagan *et al.*, 1996
Trimming	Carcass	No treatment	1/1	Single study = 0.19 (95% CI: 0.09-0.39)	30.3%	7.7% (95% CI: 4.0-14.6)	Very low	Reagan *et al.*, 1996
Multiple Interventions	Carcass	No treatment	1/1	Single study = 0.03 (95% CI: 0.01-0.14)	30.3%	1.4% (95% CI: 0.3-5.6)	Very low	Reagan *et al.*, 1996

NOTES: MA = meta-analysis average estimate from random-effects model; CI = confidence interval.

(1) For interventions with multiple trials, refers to the average meta-analysis estimate and 95% CI from random-effects meta-analysis if heterogeneity was not significant (I^2<60%). If heterogeneity was significant, the median and range of odds ratios or log reduction values from individual studies are presented instead. If only one trial was available, refers to the odds ratio or log reduction value and 95% CI from the individual study.

(2) The percentage of samples positive for *Salmonella* in the comparison group (assumed control risk, ACR) refers to the median prevalence in the comparison group trials. The percentage in the intervention/treatment group was calculated using the following formula for the odds ratio (OR) measure of effect: (OR × ACR / 1 -ACR + [OR × ACR]), and provides an indication of the effect of the intervention on changing a given baseline level of *Salmonella* prevalence as noted in the included studies. The odds ratio used in this formula corresponds to the odds ratio presented in the intervention effect column as described under (1) above. If heterogeneity was not significant (I^2<60%), the estimate is presented with a 95% CI in brackets. If heterogeneity was significant, the median estimate is presented with the range of individual study effects in brackets to illustrate the variability in results across studies.

(3) GRADE rating descriptions: very low = the true effect is likely to be substantially different from the measured estimate; low = the true effect may be substantially different from the measured estimate; moderate = the true effect is likely to be close to the measured estimate, but there is a possibility that it is substantially different; high = there is strong confidence that the true effect lies close to that of the measured estimate.

subsequently ground, limited reductions (<1 log) were noted for all intervention categories (Dorsa *et al.*, 1998a, b). One study that measured a dry heat intervention (forced-air heater at 400°C) found that the intervention was inferior when used alone compared with a warm water wash (Cutter, Dorsa and Siragusa, 1997). Overall, there is low to very low confidence in the specific effect estimates reported in Table 6.6 due to the artificial study conditions, small sample sizes, and/or inconsistency in results across studies.

TABLE SC6.5 Quasi-experimental studies

Intervention	Intervention/ outcome sample	No. trials/ studies	Odds ratio for intervention effect[(1)]	% Samples *Salmonella* positive in study population[(2)]		GRADE rating[(3)]	References
				Pre treatment	Post treatment		
Water washes	Carcass	4/4	MA = 0.53 (95% CI: 0.20-1.38)	7.2%	3.9% (95% CI: 1.5-9.6)	Low	Hajmeer *et al.*, 1999; Trairatapiwan, Lertpatarakomol and Mitchaothai, 2011;Narváez-Bravo *et al.*, 2013; Dong *et al.*, 2014
Thermal wash	Carcass	2/2	MA = 0.17 (95% CI: 0.02-1.51)	7.0%	1.3% (95% CI: 0.1-10.2)	Moderate	Nutsch *et al.*, 1997; Wright, 2009
Multiple interventions	Carcass	1/1	Single study = 0.06 (95% CI: 0.04-0.09)	28.1%	2.3% (95% CI: 1.5-3.5)	Low	Ruby, Zhu and Ingham, 2007

NOTES: MA = meta-analysis average estimate from random-effects model; CI = confidence interval.

(1) For interventions with multiple trials, refers to the average meta-analysis estimate and 95% CI from random-effects meta-analysis if heterogeneity was not significant (I^2<60%). If heterogeneity was significant, the median and range of odds ratios or log reduction values from individual studies are presented instead. If only one trial was available, refers to the odds ratio or log reduction value and 95% CI from the individual study.

(2) The percentage of samples positive for *Salmonella* in the comparison group (assumed control risk, ACR) refers to the median prevalence in the comparison group trials. The percentage in the intervention/treatment group was calculated using the following formula for the odds ratio (OR) measure of effect: (OR × ACR / 1 - ACR + [OR × ACR]), and provides an indication of the effect of the intervention on changing a given baseline level of *Salmonella* prevalence as noted in the included studies. The odds ratio used in this formula corresponds to the odds ratio presented in the intervention effect column as described under (1) above. If heterogeneity was not significant (I^2<60%), the estimate is presented with a 95% CI in brackets. If heterogeneity was significant, the median estimate is presented with the range of individual study effects in brackets to illustrate the variability in results across studies.

(3) GRADE rating descriptions: very low = the true effect is likely to be substantially different from the measured estimate; low = the true effect may be substantially different from the measured estimate; moderate = the true effect is likely to be close to the measured estimate, but there is a possibility that it is substantially different; high = there is strong confidence that the true effect lies close to that of the measured estimate.

TABLE SC6.6 Challenge trial studies

Intervention	Intervention sample	Outcome sample	Comparison group	No. trials/ studies	Log CFU/cm² reduction(1)	GRADE rating(3)	References
Water washes	Carcass	Carcass	No treatment	10/10	Median = 1.31 (Range: 0.15-2.76)	Very low	Smith, 1992; Cutter, Dorsa and Siragusa, 1997; Dorsa *et al.*, 1998a, bCutter, 1999a; Cutter and Rivera-Betancourt, 2000; Cutter *et al.*, 2000; Reyes *et al.*, 2003; Sawyer *et al.*, 2008; Yoder *et al.*, 2010)
	Carcass	Ground beef (0 days storage)	No treatment	2/2	Median = -0.10 (Range: -0.20-0)	Low	Dorsa *et al.*, 1998a, b
Thermal washes	Carcass	Carcass	Water wash	11/11	Median = 1.60 (Range: -0.20-3.0)	Very low	Smith, 1992; Phebus *et al.*, 1997; Castillo *et al.*, 1998b, c; Dorsa *et al.*, 1998a, b; Cutter and Rivera-Betancourt, 2000; Retzlaff *et al.*, 2004; Arthur *et al.*, 2008b; Niebuhr *et al.*, 2008; Yoder *et al.*, 2010
	Carcass	Carcass	No treatment	5/5	Median = 2.50 (Range: 0.10-2.74)	Very low	Smith, 1992; Dorsa *et al.*, 1998a, b; Cutter and Rivera-Betancourt, 2000; Castillo *et al.*,2003
	Carcass	Ground beef (0 days storage)	Water wash	2/2	Median = 0.35 (Range: 0-0.70)	Low	Dorsa *et al.*, 1998a, b;
	Carcass	Ground beef (0 days storage)	No treatment	2/2	Median = 0.25 (Range: 0-0.50)	Low	Dorsa *et al.*, 1998a, b

Intervention	Intervention sample	Outcome sample	Comparison group	No. trials/ studies	Log CFU/cm² reduction[1]	GRADE rating[3]	References
Organic acid washes	Carcass	Carcass	Water wash	22/14	Median = 0.75 (Range: 0.01-3.05)	Low	Dickson and Anderson, 1991; Greer and Dilts, 1992; Hardin *et al.*, 1995; Bell, Cutter and Sumner, 1997; Castillo *et al.*, 1998b; Dorsa *et al.*, 1998a, b; Cutter, 1999a; Cutter and Rivera-Betancourt, 2000; Arthur *et al.*, 2008b; Niebuhr *et al.*, 2008; Sawyer *et al.*, 2008; Njongmeta *et al.*, 2011; Yoder *et al.*, 2012
	Carcass	Carcass	No treatment	7/7	Median = 2.50 (Range: 0.82-4.94)	Very low	Dorsa *et al.*, 1998a, b; Cutter, 1999a; Cutter and Rivera-Betancourt, 2000; Reyes *et al.*, 2003; Tinney *et al.*, 1997; Sawyer *et al.*, 2008;Laury *et al.*, 2009
	Carcass	Ground beef (0 days storage)	Water wash	2/2	Median = 0.60 (Range: 0.45-0.75)	Low	Dorsa *et al.*, 1998a, b
	Carcass	Ground beef (0 days storage)	No treatment	2/2	Median = 0.50 (Range: 0.45-0.55)	Low	Dorsa *et al.*, 1998a, b

Intervention	Intervention sample	Outcome sample	Comparison group	No. trials/ studies	Log CFU/cm² reduction[1]	GRADE rating[3]	References
Other chemical/ oxidizer washes	Carcass	Carcass	Water wash	22/17	Median = 0.59 (Range: -1.90-2.60)	Low	Kim and Slavik, 1994; Bell, Cutter and Sumner, 1997; Dorsa *et al.*, 1998a, b; Castlllo *et al.*, 1999; Cutter, 1999a; Cutter and Rivera-Betancourt, 2000; Cutter *et al.*,2000; Castillo *et al.*, 2003; Reyes *et al.*, 2003; King *et al.*, 2005; Arthur *et al.*, 2008b; Niebuhr *et al.*, 2008; Sawyer *et al.*, 2008; Kalchayanand *et al.*, 2009; Njongmeta *et al.*, 2011; Yoder *et al.*, 2012)
	Carcass	Carcass	No treatment	9/8	Median = 1.85 (Range: -0.40-4.35)	Very low	Dorsa *et al.*, 1998a, b;Cutter, 1999a; Cutter and Rivera-Betancourt, 2000; Cutter *et al.*,2000; Reyes *et al.*, 2003; Ellebracht *et al.*, 2005; Sawyer *et al.*, 2008
	Carcass	Ground beef (0 days storage)	Water wash	2/2	Median = 0.65 (Range: 0.50-0.80)	Low	Dorsa *et al.*, 1998a, b;
	Carcass	Ground beef (0 days storage)	No treatment	2/2	Median = 0.55 (Range: 0.50-0.60)	Low	Dorsa *et al.*, 1998a, b;
Dry heat	Carcass	Carcass	Water wash	1/1	Single study = -3.40 (No measure of variability reported)	Low	Cutter, Dorsa and Siragusa, 1997
Electricity	Carcass	Carcass	No treatment	3/2	Median = 0.64 (Range: 0.54-0.83)	Very low	Bawcom *et al.*, 1995; Tinney *et al.*, 1997
Natural extracts	Carcass	Carcass	No treatment	4/2	Median = 5.34 (Range: 2.12-6.22)	Very low	Reyes *et al.*, 2003; Pearce and Bolton, 2008

Intervention	Intervention sample	Outcome sample	Comparison group	No. trials/ studies	Log CFU/cm² reduction(1)	GRADE rating(3)	References
	Carcass	Carcass	Water wash	1/1	Single study = 1.01 (No measure of variability reported)	Very low	Reyes *et al.*, 2003
Trimming	Carcass	Carcass	Water wash	2/2	Median = 1.67 (Range: 1.49-1.86)	Low	Hardin *et al.*, 1995; Phebus *et al.*, 1997
Multiple Interventions	Carcass	Carcass	Water wash	16/11	Median = 1.62 (Range: 0.30-3.24)	Very low	Bell, Cutter and Sumner, 1997; Cabrera-Diaz *et al.*, 2009; Castillo *et al.*, 1998b, 1999, 2001; Cutter, Dorsa and Siragusa, 1997; Dickson and Anderson, 1991; Phebus *et al.*, 1997; Niebuhr *et al.*, 2008; Njongmeta *et al.*, 2011; Sawyer *et al.*, 2008
	Carcass	Carcass	No treatment	4/4	Median = 2.69 (Range: 1.12-3.25)	Very low	Castillo *et al.*, 2003; Cutter, 1999a; Sawyer *et al.*, 2008; Tinney *et al.*, 1997

NOTES: MA = meta-analysis average estimate from random-effects model; CI = confidence interval.

(1) For interventions with multiple trials, refers to the average meta-analysis estimate and 95% CI from random-effects meta-analysis if heterogeneity was not significant (I^2<60%). If heterogeneity was significant, the median and range of odds ratios or log reduction values from individual studies are presented instead. If only one trial was available, refers to the odds ratio or log reduction value and 95% CI from the individual study.

(2) The percentage of samples positive for *Salmonella* in the comparison group (assumed control risk, ACR) refers to the median prevalence in the comparison group trials. The percentage in the intervention/treatment group was calculated using the following formula for the odds ratio (OR) measure of effect: (OR × ACR / 1 - ACR + [OR × ACR]), and provides an indication of the effect of the intervention on changing a given baseline level of *Salmonella* prevalence as noted in the included studies. The odds ratio used in this formula corresponds to the odds ratio presented in the intervention effect column as described under (1) above. If heterogeneity was not significant (I^2<60%), the estimate is presented with a 95% CI in brackets. If heterogeneity was significant, the median estimate is presented with the range of individual study effects in brackets to illustrate the variability in results across studies.

(3) GRADE rating descriptions: very low = the true effect is likely to be substantially different from the measured estimate; low = the true effect may be substantially different from the measured estimate; moderate = the true effect is likely to be close to the measured estimate, but there is a possibility that it is substantially different; high = there is strong confidence that the true effect lies close to that of the measured estimate.

Summary-of-findings tables for the effects of chilling

TABLE SC6.7 Quasi-experimental studies

Intervention	Intervention/ outcome sample	No. trials/ studies	Odds ratio for intervention effect(1)	% Samples *Salmonella* positive in study population(2)		GRADE rating(3)	References
				Pre-treatment	Post-treatment		
Chilling	Carcass	5/5	MA = 1.60 (95% CI: 0.58–4.41)	2.0%	3.1% (95% CI: 1.1–8.1)	Low	Dong *et al.*, 2014; Hajmeer *et al.*, 1999; Sofos *et al.*, 1999; Fegan *et al.*, 2005; Trairatapiwan, Lertpatarakomol and Mitchaothai, 2011

NOTES: MA = meta-analysis average estimate from random-effects model; CI = confidence interval.

(1) For interventions with multiple trials, refers to the average meta-analysis estimate and 95% CI from random-effects meta-analysis if heterogeneity was not significant (I^2<60%). If heterogeneity was significant, the median and range of odds ratios or log reduction values from individual studies are presented instead. If only one trial was available, refers to the odds ratio or log reduction value and 95% CI from the individual study. For challenge trials, meta-analysis was not conducted due to limiting data reporting. Instead, results are presented as a median (range) log reduction for multiple trials, and as the single estimate and 95% CI for individual studies.

(2) The percentage of samples positive for *Salmonella* in the comparison group (assumed control risk, ACR) refers to the median prevalence in the comparison group trials. The percentage in the intervention/treatment group was calculated using the following formula for the odds ratio (OR) measure of effect: (OR × ACR / 1 -ACR + [OR × ACR]), and provides an indication of the effect of the intervention on changing a given baseline level of *Salmonella* prevalence as noted in the included studies. The odds ratio used in this formula corresponds to the odds ratio presented in the intervention effect column as described under (1) above. If heterogeneity was not significant (I^2<60%), the estimate is presented with a 95% CI in brackets. If heterogeneity was significant, the median estimate is presented with the range of individual study effects in brackets to illustrate the variability in results across studies.

(3) GRADE rating descriptions: very low = the true effect is likely to be substantially different from the measured estimate; low = the true effect may be substantially different from the measured estimate; moderate = the true effect is likely to be close to the measured estimate, but there is a possibility that it is substantially different; high = there is strong confidence that the true effect lies close to that of the measured estimate.

SC6.6 CHILLING

The primary purpose of carcass chilling is to inhibit microbial growth. Five quasi-experimental studies conducted under commercial conditions were identified evaluating the effects of chilling on *Salmonella* reduction (Hajmeer *et al.*, 1999; Sofos *et al.*, 1999; Fegan *et al.*, 2005; Trairatapiwan, Lertpatarakomol and Mitchaothai, 2011; Dong *et al.*, 2014). The studies found infrequent *Salmonella* contamination in both pre- and post-chilling samples (Table SC6.7). The identification of positive *Salmonella* samples post-chilling in some studies could potentially be due to cross-contamination during the chilling process and emphasizes the importance of good hygienic practices during all processing steps (Hajmeer *et al.*, 1999; Trairatapiwan, Lertpatarakomol and Mitchaothai, 2011). There is low confidence in the specific estimate of effect of chilling to reduce *Salmonella* prevalence on beef carcasses due to potential for confounding bias in the included studies, and due to other factors, such as differences in study sampling methodologies and chilling procedures.

TABLE SC6.8 Challenge trial studies

Intervention	Intervention sample	Outcome sample	Comparison group	No. trials/ studies	Log CFU/g reduction(1)	GRADE rating(2)	References
Spray (wet) chilling	Carcass	Carcass (24 h post-chill)	Dry chilling	1/1	Single study = 0.28 (95% CI: 0.06, 0.50)	Very low	Tittor *et al.*, 2011
	Carcass	Carcass (48 h post-chill)	Dry chilling	1/1	Single study = 0.36 (95% CI: 0.14-0.58)	Very low	Tittor *et al.*, 2011
	Carcass	Carcass (7 days post-chill)	Dry chilling	1/1	Single study = -0.20 (95% CI: -0.51-0.11)	Very low	Tittor *et al.*, 2011
	Carcass	Carcass (28 days post-chill)	Dry chilling	1/1	Single study = -2.42 (95% CI: -2.87--1.97)	Very low	Tittor *et al.*, 2011
Organic acid wash	Carcass	Carcass	Water spray chill	2/1	Median = 1.26 (Range: 0.80-1.72)	Very low	Dickson, 1991
Multiple Interventions	Carcass	Carcass	Water spray chill	2/1	Median = 0.67 (Range: 0.48-0.84)	Very low	Dickson, 1991

NOTES: MA = meta-analysis average estimate from random-effects model; CI = confidence interval.

(1) For interventions with multiple trials, refers to the average meta-analysis estimate and 95% CI from random-effects meta-analysis if heterogeneity was not significant (I^2<60%). If heterogeneity was significant, the median and range of odds ratios or log reduction values from individual studies are presented instead. If only one trial was available, refers to the odds ratio or log reduction value and 95% CI from the individual study.
For challenge trials, meta-analysis was not conducted due to limiting data reporting. Instead, results are presented as a median (range) log reduction for multiple trials, and as the single estimate and 95% CI for individual studies.

(2) GRADE rating descriptions: very low = the true effect is likely to be substantially different from the measured estimate; low = the true effect may be substantially different from the measured estimate; moderate = the true effect is likely to be close to the measured estimate, but there is a possibility that it is substantially different; high = there is strong confidence that the true effect lies close to that of the measured estimate.

One challenge trial was identified that investigated the efficacy of spray vs dry chilling under simulated commercial conditions (Tittor *et al.*, 2011). The authors found that spray chilling reduced inoculated *Salmonella* levels when sampled within 48 hours of chilling, but after extended storage of 7 to 28 days, *Salmonella* counts were lower on dry chilled carcasses (Table SC6.8). Another challenge trial found that various cycles of spray chilling carcass tissues with acetic acid (at 0.5%, 1.0%, or 2.0%) and water and acetic acid combinations reduced inoculated *Salmonella* levels compared with spray chilling with water alone (Dickson, 1991).

SC6.7 POST-CHILL AND PRE-FABRICATION

One quasi-experiment conducted under commercial conditions was identified evaluating the effect of a post-chill and pre-fabrication wash (Table SC6.9). A very low level of contamination was found (Ruby, Zhu and Ingham, 2007), but a slight reduction in *Salmonella* prevalence was noted due to an organic (5% lactic) acid wash (Table SC6.9).

Two challenge trials conducted under laboratory conditions were also identified (Table SC6.10). One study found that a steam vacuum treatment at 130°C reduced inoculated *Salmonella* levels on post-chill beef carcasses by <1 log compared with no treatment (Bacon *et al.*, 2002b). The other study found that a combination of lactic acid wash followed by 200 ppm peroxyacetic acid wash reduced inoculated *Salmonella* concentrations on carcasses compared with samples measured after a pre-chill water wash, but a 200 ppm peroxyacetic acid wash alone was not effective (King *et al.*, 2005).

Summary-of-findings tables for interventions applied post-chill and pre-fabrication

TABLE SC6.9 Quasi-experimental studies

Intervention	Intervention/ outcome sample	No. trials/ studies	Odds ratio for intervention effect[(1)]	% Samples *Salmonella* positive in study population[(2)]		GRADE rating[(3)]	References
				Pre-treatment	Post-treatment		
Organic acid wash	Carcass	1/1	Single study = 0.33 (95% CI: 0.01-8.19)	1.0%	0.3% (95% CI: 0.01-7.9)	Low	Ruby, Zhu and Ingham, 2007

NOTES: CI = confidence interval.

(1) For interventions with multiple trials, refers to the average meta-analysis estimate and 95% CI from random-effects meta-analysis if heterogeneity was not significant (I^2<60%). If heterogeneity was significant, the median and range of odds ratios or log reduction values from individual studies are presented instead. If only one trial was available, refers to the odds ratio or log reduction value and 95% CI from the individual study.
For challenge trials, meta-analysis was not conducted due to limiting data reporting. Instead, results are presented as a median (range) log reduction for multiple trials, and as the single estimate and 95% CI for individual studies.

(2) The percentage of samples positive for *Salmonella* in the comparison group (assumed control risk, ACR) refers to the median prevalence in the comparison group trials. The percentage in the intervention/treatment group was calculated using the following formula for the odds ratio (OR) measure of effect: (OR × ACR / 1 - ACR + [OR × ACR]), and provides an indication of the effect of the intervention on changing a given baseline level of *Salmonella* prevalence as noted in the included studies. The odds ratio used in this formula corresponds to the odds ratio presented in the intervention effect column as described under (1) above. If heterogeneity was not significant (I^2<60%), the estimate is presented with a 95% CI in brackets. If heterogeneity was significant, the median estimate is presented with the range of individual study effects in brackets to illustrate the variability in results across studies.

(3) GRADE rating descriptions: very low = the true effect is likely to be substantially different from the measured estimate; low = the true effect may be substantially different from the measured estimate; moderate = the true effect is likely to be close to the measured estimate, but there is a possibility that it is substantially different; high = there is strong confidence that the true effect lies close to that of the measured estimate.

TABLE SC6.10 Challenge trial studies

Intervention	Intervention/ outcome sample	Comparison group	No. trials/ studies	Log CFU/cm² reduction[(1)]	GRADE rating[(2)]	References
Steam vacuum	Carcass	No treatment	1/1	Single study = 0.6 (95% CI: 0.43-0.77)	Low	Bacon *et al.*, 2002b
Other chemical/ oxidizer wash	Carcass	Water wash	1/1	Single study = -0.3 (95% CI: -0.62-0.02)	Low	King *et al.*, 2005
Multiple Interventions	Carcass	Water wash	1/1	Single study = 0.2 (95% CI: -0.76-1.16)	Low	King *et al.*, 2005

NOTES: CI = confidence interval.

(1) For interventions with multiple trials, refers to the average meta-analysis estimate and 95% CI from random-effects meta-analysis if heterogeneity was not significant (I^2<60%). If heterogeneity was significant, the median and range of odds ratios or log reduction values from individual studies are presented instead. If only one trial was available, refers to the odds ratio or log reduction value and 95% CI from the individual study.
For challenge trials, meta-analysis was not conducted due to limiting data reporting. Instead, results are presented as a median (range) log reduction for multiple trials, and as the single estimate and 95% CI for individual studies.

(2) GRADE rating descriptions: very low = the true effect is likely to be substantially different from the measured estimate; low = the true effect may be substantially different from the measured estimate; moderate = the true effect is likely to be close to the measured estimate, but there is a possibility that it is substantially different; high = there is strong confidence that the true effect lies close to that of the measured estimate.

SC6.8 POST-FABRICATION

One controlled trial was identified that evaluated the effect of post-fabrication and pre- and post-storage washes on *Salmonella* reduction of beef cuts, but *Salmonella* was not isolated from any sample, precluding evaluation of intervention efficacy on this pathogen (Prasai *et al.*, 1997). The authors found that lactic acid washes significantly reduced levels of aerobic plate counts (APCs) and the prevalence of *Listeria* spp. on beef cuts, with greater reductions in APCs for washes implemented pre-storage compared with post-storage (Prasai *et al.*, 1997).

A total of 32 challenge trials were identified that investigated the efficacy of post-fabrication interventions on beef trimmings in laboratories under simulated commercial conditions (Podolak *et al.*, 1995; Ellebracht *et al.*, 1999; Chung, Ko and Kim, 2000; Pohlman *et al.*, 2002a, b, 2009; Stivarius *et al.*, 2002a, b, c; Ellebracht *et al.*, 2005; Patel and Solomon, 2005; Harris *et al.*, 2006; McCann *et al.*, 2006; Ozdemir *et al.*, 2006; Echeverry *et al.*, 2009, 2010; Hughes *et al.*, 2010; Quilo *et al.*, 2010; Fouladkhah *et al.*, 2012; Geornaras *et al.*, 2012; Harris *et al.*, 2012; McDaniel *et al.*, 2012; Mehall *et al.*, 2012a, b; Mohan *et al.*, 2012; Patel *et al.*, 2012; Dias-Morse *et al.*, 2014; Kundu *et al.*, 2014; Pohlman, Dias-Morse and Pinidiya, 2014; Tango *et al.*, 2014; Zhao *et al.*, 2014; Li, Kundu and Holley, 2015). Twenty-nine of these reported extractable log reduction data and are summarized in Table SC6.11. Compared with no treatment, all intervention categories tended to reduce inoculated *Salmonella* levels on treated beef trimmings when measured as subsequently produced ground beef or beef cuts, with organic acid washes, irradiation, and multiple treatment combinations showing the largest log reductions (Table SC6.11). Consistent benefits were also observed when some interventions (organic acids washes, other chemical and oxidizer washes, lactic acid bacteria washes, and multiple treatment combinations) were compared with an ambient or cold temperature water wash and measured in subsequent beef cuts (Table SC6.11). Two studies investigated hydrodynamic pressure processing, finding that it significantly reduced inoculated *Salmonella* levels by 0.25 to 1.1 logs in beef stew pieces (Patel and Solomon, 2005; Patel *et al.*, 2012). Another study investigated a dry heat intervention and found it reduced inoculated *Salmonella* levels by 1–2 logs at lower temperatures (60° and 75°C) and by up to 4–6 logs at higher temperatures (90° and 100°C) (McCann *et al.*, 2006). Overall, there is low to very low confidence in these estimates of effect due to the artificial study conditions, small sample sizes, and/or variations in magnitude of effect across studies.

One challenge trial conducted under laboratory conditions was identified that evaluated the effect of various treatments to reduce *Salmonella* contamination of beef cheek meat (Schmidt *et al.*, 2014). The authors found a significant reduction in inoculated *Salmonella* levels for thermal, organic acid, and other chemical washes, with the former two having the largest overall reduction effect (Table SC6.11).

Summary-of-findings table for interventions applied post-fabrication

TABLE SC6.11 Challenge trial studies

Intervention	Intervention sample	Outcome sample	Comparison group	No. trials/ studies	Log CFU reduction(1)	GRADE rating(2)	References
Water washes	Beef trimmings	Ground beef (0 days storage)	No treatment	3/3	Median = 0.27 (Range: 0.25–0.95)	Very low	Ellebracht *et al.*, 2005; Harris *et al.*, 2012; Pohlman, Dias- Morse and Pinidiya, 2014
	Beef trimmings	Steaks (0 days storage)	No treatment	1/1	Single study = 0.18 (95% CI: -0.17–0.53)	Low	Mehall *et al.*, 2012a
	Cheek meat	Cheek meat	No treatment	1/1	Single study = 0.13 (no measure of variability reported)	Very low	Schmidt *et al.*, 2014
Thermal washes	Beef trimmings	Ground beef (across 7–42 days of storage)	No treatment	2/2	Median = 0.41 (Range: 0.11–0.70)	Very low	Ellebracht *et al.*, 1999; Stivarius *et al.*, 2002c
	Beef trimmings	Beef cuts/ steaks (0 days storage)	No treatment	1/1	Single study = 0.54 (no measure of variability reported)	Very low	Ozdemir *et al.*, 2006
	Cheek meat	Cheek meat	No treatment	1/1	Single study = 1.95 (no measure of variability reported)	Very low	Schmidt *et al.*, 2014
Organic acid washes	Beef trimmings	Ground beef (0 days storage)	Water wash	1/1	Single study = 0.17 (95% CI: 0.12–0.22)	Very low	Harris *et al.*, 2012
	Beef trimmings	Ground beef (0 days storage)	No treatment	4/4	Median = 0.50 (Range: 0.22–1.68)	Very low	Stivarius *et al.*, 2002a; Stivarius *et al.*, 2002c; Pohlman *et al.*, 2009; Harris *et al.*, 2012
	Beef trimmings	Beef cuts/ steaks (0 days storage)	Water wash	12/5	Median = 1.39 (Range: -0.10–3.96)	Very low	Podolak *et al.*, 1995; Echeverry *et al.*, 2009, 2010; Mehall *et al.*, 2012a; Tango *et al.*, 2014

Intervention	Intervention sample	Outcome sample	Comparison group	No. trials/ studies	Log CFU reduction(1)	GRADE rating(2)	References
	Beef trimmings	Beef cuts/ steaks (0 days storage)	No treatment	8/4	Median = 1.45 (Range: 0.31-1.70)	Very low	Ozdemir *et al.*, 2006; Fouladkhah *et al.*, 2012; Mehall *et al.*, 2012a; Li, Kundu and Holley, 2015
	Cheek meat	Cheek meat	Water wash	1/1	Single study = 1.42 (no measure of variability reported)	Very low	Schmidt *et al.*, 2014
	Cheek meat	Cheek meat	No treatment	1/1	Single study = 1.56 (no measure of variability reported)	Very low	Schmidt *et al.*, 2014
Other chemical/ oxidizer washes	Beef trimmings	Ground beef (0 days storage)	Water wash	2/2	Median = 0.11 (Range: -0.13-0.34)	Very low	Harris *et al.*, 2012; Pohlman, Dias-Morse and Pinidiya, 2014
	Beef trimmings	Ground beef (0 days storage)	No treatment	9/8	Median = 0.70 (Range: 0-1.97)	Very low	Stivarius *et al.*, 2002b; Ellebracht *et al.*, 2005; Pohlman *et al.*, 2009; Harris *et al.*, 2012; Mehall *et al.*, 2012a; Mohan *et al.*, 2012; Dias-Morse *et al.*, 2014; Pohlman, Dias-Morse and Pinidiya, 2014
	Beef trimmings	Beef cuts/ steaks (0 days storage)	Water wash	11/4	Median = 0.97 (Range: 0.11-3.38)	Very low	Echeverry *et al.*, 2009; Echeverry *et al.*, 2010; Mehall *et al.*, 2012a; Tango *et al.*, 2014
	Beef trimmings	Beef cuts/ steaks (0 days storage)	No treatment	7/2	Median = 0.60 (Range: 0.36-2.31)	Very low	Geornaras *et al.*, 2012; McDaniel *et al.*, 2012
	Cheek meat	Cheek meat	Water wash	1/1	Single study = 0.54 (no measure of variability reported)	Very low	Schmidt *et al.*, 2014

Intervention	Intervention sample	Outcome sample	Comparison group	No. trials/ studies	Log CFU reduction[1]	GRADE rating[2]	References
	Cheek meat	Cheek meat	No treatment	1/1	Single study = 0.68 (no measure of variability reported)	Very low	Schmidt *et al.*, 2014
Irradiation	Beef trimmings	Beef trimmings	No treatment	5/3	Median = 2.0 (Range: 1.79-6.0)	Very low	Chung, Ko and Kim, 2000; Kundu *et al.*, 2014; Li, Kundu and Holley, 2015
Lactic acid bacteria washes	Beef trimmings	Tenderized steaks (0 days storage)	Water wash	9/2	Median = 1.22 (Range: 0.20-2.94)	Very low	Echeverry*et al*, 2009, 2010
Multiple Interventions	Beef trimmings	Ground beef (0 days storage)	Water wash	1/1	Single study = 0.29 (95% CI: 0.10-0.48)	Low	Pohlman, Dias-Morse and Pinidiya, 2014
	Beef trimmings	Ground beef (0 days storage)	No treatment	8/8	Median = 1.45 (Range: 0.40-1.87)	Very low	Ellebracht *et al.*, 1999, 2005; Pohlman *et al.*, 2002a, b; Quilo *et al.*, 2010; Mohan *et al.*, 2012; Dias-Morse e*et al.*, 2014; Pohlman, Dias-Morse and Pinidiya, 2014
	Beef trimmings	Beef cuts/ steaks (0 days storage)	Water wash	2/2	Median = 3.45 (Range: 1.90-5.0)	Very low	Tango*et al.*, 2014; Zhao *et al.*, 2014
	Beef trimmings	Beef cuts/ steaks (0 days storage)	No treatment	4/4	Median = 1.64 (Range: 0.45-2.85)	Very low	Ozdemir *et al.*, 2006; McDaniel *et al.*, 2012; Mehall *et al.*, 2012b; Li, Kundu and Holley, 2015

NOTES: CI = confidence interval.

(1) Due to limiting data reporting, meta-analysis was not conducted for challenge trials. Instead, log reduction results in this column are presented as a median (range) log reduction for multiple trials and as the single estimate and 95% CI for individual studies.

(2) GRADE rating descriptions: very low = the true effect is likely to be substantially different from the measured estimate; low = the true effect may be substantially different from the measured estimate; moderate = the true effect is likely to be close to the measured estimate, but there is a possibility that it is substantially different; high = there is strong confidence that the true effect lies close to that of the measured estimate.

SC6.9 PACKAGING

The effects of various packaging and final product preservation interventions were studied in 31 laboratory challenge trials identified in this review.

Six studies were identified that investigated the use of *Lactobacillus* spp. to decrease *Salmonella* contamination on beef products (Gomólka-Pawlicka and Uradzinski, 2003; Smith *et al.*, 2005; Hoyle *et al.*, 2009; Ruby and Ingham, 2009; Olaoye and Onilude, 2010; Chaillou *et al.*, 2014). All studies reported a decrease in *Salmonella* that depended on the applied *Lactobacillus*/*Salmonella* ratio. Higher doses of *Lactobacillus spp.* were reported to be more effective in reducing pathogenic load (Smith *et al.*, 2005; Ruby and Ingham, 2009; Chaillou *et al.*, 2014); however, when applied in very high quantities product spoilage was observed (Ruby and Ingham, 2009). Some studies reported a reduction of *Salmonella* until undetectable levels in 5 days (Smith *et al.*, 2005; Hoyle *et al.*, 2009). A study that investigated differences in the effectiveness of various strains of lactic acid bacteria reported that *L. acidophilus* and *L. platarum* were most effective to inhibit *Salmonella* in ground beef (Gomólka-Pawlicka and Uradzinski, 2003)

Thirteen studies investigated the effects of natural extracts to reduce *Salmonella* contamination in beef products (Cutter, 2000; Skandamis, Tsigarida and Nychas, 2002; Careaga *et al.*, 2003; Ahn, Grun and Mustapha, 2004; Uhart, Maks and Ravishankar, 2006; Qiu and Wu, 2007; Hayouni *et al.*, 2008; Turgis *et al.*, 2008; Chao and Yin, 2009; Shalaby, 2011; Tayel *et al.*, 2012; Cruz-Galvez *et al.*, 2013; De Oliveira, Soares and Piccoli, 2013). Compounds investigated included various spices (oregano, red pepper, lemon grass, garlic, turmeric, cinnamon, mustard), fruits (pomegranate, grape seed, cranberry), or other plant extracts (roselle, pine bark, *Artemisia absinthium, Salvia officinalis* and *Schinus molle*). The compounds were typically administered as essential oils or aqueous extracts. Most of the above mentioned compounds were shown to be effective in reducing inoculated *Salmonella* levels, sometimes in conjugation with irradiation (Turgis *et al.*, 2008) or modified packaging (Skandamis, Tsigarida and Nychas, 2002), although the magnitude of the effects varied among the compounds and experimental conditions. The use of essential oils has been shown to affect the sensory properties of the meat in some cases (Hayouni *et al.*, 2008; De Oliveira, Soares and Piccoli, 2013).

Seven studies reported on modified atmosphere packaging interventions (Gill and DeLacy, 1991; Bergis, Poumeyrol and Beaufort, 1994; Cutter, 1999b; Nissen *et al.*, 2000; Skandamis, Tsigarida and Nychas, 2002; Brooks *et al.*, 2008; Miya *et al.*, 2014). Modified packages with increased oxygen (80%) or with added carbon monoxide (0.4%) were shown to result in a significant reduction of *Salmonella* after 7 and 14 days of storage in one study (Brooks *et al.*, 2008). Other studies reported effective reduction of *Salmonella* by applying a vacuum or a CO_2 atmosphere at lower temperatures (Gill and DeLacy, 1991), combinations of CO_2, O_2 and/or N_2 (Bergis, Poumeyrol and Beaufort, 1994), combining an epsilon-polylysine treatment with a CO_2 atmosphere (Miya *et al.*, 2014), and a combination of a modified atmosphere or vacuum package with an oregano essential oil extract (Skandamis, Tsigarida and Nychas, 2002). Other studies found that packaging materials containing 1500 ppm triclosan (Cutter, 1999b) and a high CO_2 / low CO mixture (0.4% CO / 60% CO_2 / 40% N_2) at 10°C (Nissen *et al.*, 2000) had no significant pathogen-reduction effect.

Other preservation treatments such as various salts, organic acids or volatile compounds were investigated in seven studies (Cutter and Siragusa, 1995; Tu and Mustapha, 2002; Jensen, Saxena and Keener, 2009; Ryu and Fung, 2010; Shalaby, 2011; Stelzleni, Ponrajan and Harrison, 2013; Miya *et al.*, 2014; Faith *et al.*, 2015). The use of gaseous anhydrous ammonia (5100 ppm) was shown to result in an up to 7 log reduction of *Salmonella* in textured beef in one study, whereas liquid ammonia and ammonium hydroxide were not effective (Jensen, Saxena and Keener, 2009). The use of sodium salts was investigated in several studies with inconsistent results (Cutter and Siragusa, 1995; Ryu and Fung, 2010; Stelzleni, Ponrajan and Harrison, 2013). Treatment with nisin, a polypeptide, was reported to be not effective (Tu and Mustapha, 2002) or to result in only a small (0.4) log reduction in combination with lactate (Cutter and Siragusa, 1995) in two studies. One study found that application of a mixture of volatile compounds resulted in a 1.7–2.2 log reduction of *Salmonella* in ground beef during a 5-day storage period at 8°C (Faith *et al.*, 2015). Another study found that an epsilon-polylysine treatment reduced *Salmonella* levels by 1.5–2.4 logs in fresh beef over 7 days, depending on the storage conditions (Miya *et al.*, 2014).

SC6.10 MULTIPLE PROCESSING STAGES

One previous systematic review evaluated the efficacy of HACCP to reduce *Salmonella* contamination of beef carcasses during processing (Wilhelm *et al.*, 2011). A meta-analysis was conducted on five eligible quasi-experimental studies measuring changes in *Salmonella* prevalence before and after implementation of HACCP programmes. Results indicate no overall effect of HACCP on *Salmonella* reduction in beef (Table SC6.12), with moderate confidence in the estimate of effect. However, the review found that HACCP was linked to a reduction in aerobic bacterial counts (Wilhelm *et al.*, 2011). Although HACCP may be effective in some settings (and may depend on the specific critical control points and other implementation factors), these findings suggest that there is limited evidence that HACCP itself is effective in reducing *Salmonella* prevalence on beef during processing (Wilhelm *et al.*, 2011). The review met 9/11 recommended reporting reliability criteria using the AMSTAR quality assessment tool.

Seven quasi-experimental studies conducted under commercial conditions were identified that evaluated the effect of multiple interventions applied between pre-evisceration and chilling (Sofos *et al.*, 1999; Bacon *et al.*, 2002a; Barkocy-Gallagher *et al.*, 2003; Rivera-Betancourt *et al.*, 2004; Ruby, Zhu and Ingham, 2007; Brichta-Harhay *et al.*, 2008; Koohmaraie *et al.*, 2012). The studies found a very large effect on *Salmonella* reduction (Table SC6.13), but the overall confidence in the median effect estimate is very low due to potential for confounding in these studies and because of large variations between studies in the magnitude of the effect, which could be due to differences in the intervention combinations applied across studies as well as other confounding factors. While the overall confidence in identifying a specific estimate of effect is very low due to these factors, the studies highlight a clear trend for *Salmonella* to be reduced to very low levels by implementing multiple interventions throughout processing.

In addition, interim results (July, 2015) from the USDA Nationwide Beef and Veal Carcass Baseline Survey correspond with the above findings from Table SC6.13, indicating that 25.5% of carcasses were positive during post-hide removal/pre-evisceration testing, and only 3.9% of paired carcass samples were positive pre-chill after receiving one or more different combinations of interventions (USDA, 2015). Furthermore, of 77 positive carcass samples at the post-hide removal/pre-evisceration step that did not receive an intervention at or before hide removal, only eight paired carcass samples remained positive (8/77, 10.4%) when tested at the pre-chill step after receiving one or more different combinations of interventions (USDA, unpublished data). MPN/ml values for the 77 positive samples at post-hide removal/pre-evisceration ranged from 0.225 to 645, with a median of 0.225. Of the eight carcass samples that remained positive, all had MPN/ml values of 0.225 on both pre- and post-samples.

Four challenge trials were identified that compared multiple combinations of pre-and post-evisceration washes, and/or pre- and post-chill washes with spray chilling, with and without organic acids and other chemicals, compared with spray chilling alone to reduce inoculated *Salmonella* on beef carcasses (Dickson and Anderson, 1991; Dickson and Siragusa, 1994; Tinney *et al.*, 1997; Stopforth *et al.*, 2005). All three studies that reported extractable data (Table SC6.14) found reductions in *Salmonella* levels ranging up to 2.5 logs. The other study found that organic acid water washes with and without spray chilling resulted in significant reductions of inoculated *Salmonella* levels, up to 3 logs, in carcass tissue after extended storage at 10–20 days (Dickson and Siragusa, 1994). The confidence in these estimates of reduction is very low due to the artificial study conditions, small sample sizes, and variation in effect across studies.

One cross-sectional study from Ethiopia found that several good hygienic practices (e.g. wearing protective garments during slaughtering, hand-washing after separating intestinal contents, washing of the knife before slaughtering, and slaughtering on sanitized floor) were associated with a lower *Salmonella* prevalence in carcass swab samples (Muluneh and Kibret, 2015).

Summary-of-findings tables for the effects of interventions applied at multiple processing stages

TABLE SC6.12 Systematic review

Intervention	Study design	Intervention/ outcome sample	No. trials/ studies	Odds ratio for intervention effect[(1)]	% Samples *Salmonella* positive in study population[(2)]		GRADE rating[(3)]	References
					Pre-treatment	Post-treatment		
HACCP	Quasi-experiments	Carcass	5/5	MA = 0.89 (95% CI: 0.53-1.48)	2.1%	1.9% (95% CI: 1.1-3.1)	Moderate	Wilhelm *et al.*, 2011

NOTES: MA = meta-analysis average estimate from random-effects model; CI = confidence interval.

(1) For interventions with multiple trials, refers to the average meta-analysis estimate and 95% CI from random-effects meta-analysis if heterogeneity was not significant (I^2<60%). If heterogeneity was significant, the median and range of odds ratios or log reduction values from individual studies are presented instead. If only one trial was available, refers to the odds ratio or log reduction value and 95% CI from the individual study. For challenge trials, meta-analysis was not conducted due to limiting data reporting. Instead, results are presented as a median (range) log reduction for multiple trials, and as the single estimate and 95% CI for individual studies.

(2) The percentage of samples positive for *Salmonella* in the comparison group (assumed control risk, ACR) refers to the median prevalence in the comparison group trials. The percentage in the intervention/treatment group was calculated using the following formula for the odds ratio (OR) measure of effect: (OR × ACR / 1 - ACR + [OR × ACR]), and provides an indication of the effect of the intervention on changing a given baseline level of *Salmonella* prevalence as noted in the included studies. The odds ratio used in this formula corresponds to the odds ratio presented in the intervention effect column as described under (1) above. If heterogeneity was not significant (I^2<60%), the estimate is presented with a 95% CI in brackets. If heterogeneity was significant, the median estimate is presented with the range of individual study effects in brackets to illustrate the variability in results across studies.

(3) GRADE rating descriptions: very low = the true effect is likely to be substantially different from the measured estimate; low = the true effect may be substantially different from the measured estimate; moderate = the true effect is likely to be close to the measured estimate, but there is a possibility that it is substantially different; high = there is strong confidence that the true effect lies close to that of the measured estimate.

TABLE SC6-13 Quasi-experimental studies

Intervention	Intervention/ outcome sample	No. trials/ studies	Odds ratio for Intervention effect[1]	% Samples *Salmonella* positive in study population[2]		GRADE rating[3]	References
				Pre-treatment	Post-treatment		
Multiple Interventions (pre-evisceration to pre-chill)	Carcass	7/7	Median = 0.01 (Range: 0.006-0.48)	28.7%	0.5% (Range: 0.2-16.0)	Very low	Sofos *et al.*, 1999; Bacon *et al.*, 2002a; Barkocy-Gallagher *et al.*, 2003; Rivera-Betancourt *et al.*, 2004; Ruby, Zhu and Ingham, 2007; Brichta-Harhay *et al.*, 2008; Koohmaraie *et al.*, 2012

NOTES: MA = meta-analysis average estimate from random-effects model; CI = confidence interval.

(1) For interventions with multiple trials, refers to the average meta-analysis estimate and 95% CI from random-effects meta-analysis if heterogeneity was not significant (I^2<60%). If heterogeneity was significant, the median and range of odds ratios or log reduction values from individual studies are presented instead. If only one trial was available, refers to the odds ratio or log reduction value and 95% CI from the individual study.
For challenge trials, meta-analysis was not conducted due to limiting data reporting. Instead, results are presented as a median (range) log reduction for multiple trials, and as the single estimate and 95% CI for individual studies.

(2) The percentage of samples positive for *Salmonella* in the comparison group (assumed control risk, ACR) refers to the median prevalence in the comparison group trials. The percentage in the intervention/treatment group was calculated using the following formula for the odds ratio (OR) measure of effect: (OR × ACR / 1 -ACR + [OR × ACR]), and provides an indication of the effect of the intervention on changing a given baseline level of *Salmonella* prevalence as noted in the included studies. The odds ratio used in this formula corresponds to the odds ratio presented in the intervention effect column as described under (1) above. If heterogeneity was not significant (I^2<60%), the estimate is presented with a 95% CI in brackets. If heterogeneity was significant, the median estimate is presented with the range of individual study effects in brackets to illustrate the variability in results across studies.

(3) GRADE rating descriptions: very low = the true effect is likely to be substantially different from the measured estimate; low = the true effect may be substantially different from the measured estimate; moderate = the true effect is likely to be close to the measured estimate, but there is a possibility that it is substantially different; high = there is strong confidence that the true effect lies close to that of the measured estimate.

TABLE SC6.14 Challenge trial studies

Intervention	Intervention/ outcome sample	Comparison group	No. trials/ studies	Log CFU/cm^2 reduction[1]	GRADE rating[2]	References
Multiple Interventions	Carcass	Spray chill only	4/3	Median = 1.58 (Range: 0.78-2.52)	Very low	Dickson and Anderson, 1991; Stopforth *et al.*, 2005; Tinney *et al.*, 1997

NOTES: MA = meta-analysis average estimate from random-effects model; CI = confidence interval.

(1) For interventions with multiple trials, refers to the average meta-analysis estimate and 95% CI from random-effects meta-analysis if heterogeneity was not significant (I^2<60%). If heterogeneity was significant, the median and range of odds ratios or log reduction values from individual studies are presented instead. If only one trial was available, refers to the odds ratio or log reduction value and 95% CI from the individual study.

For challenge trials, meta-analysis was not conducted due to limiting data reporting. Instead, results are presented as a median (range) log reduction for multiple trials, and as the single estimate and 95% CI for individual studies.

(2) GRADE rating descriptions: very low = the true effect is likely to be substantially different from the measured estimate; low = the true effect may be substantially different from the measured estimate; moderate = the true effect is likely to be close to the measured estimate, but there is a possibility that it is substantially different; high = there is strong confidence that the true effect lies close to that of the measured estimate.

SC6.11 REFERENCES CITED IN SC6

Aftab, M., Rahman, A., Qureshi, M.S., Akhter, S., Sadique, U., Sajid, A. & Zaman, S. 2012. Level of *Salmonella* in beef of slaughtered cattle at Peshawar. *Journal of Animal and Plant Sciences,* 22(2): 24–27.

Ahn, J.H., Grun, I.U. & Mustapha, A. 2004. Antimicrobial and antioxidant activities of natural extracts *in vitro* and in ground beef. *Journal of Food Protection,* 67(1): 148–155.

Arthur, T.M., Bosilevac, J.M., Brichta-Harhay, D., Kalchayanand, N., Shackelford, S.D., Wheeler, T.L. & Koohmaraie, M. 2007. Effects of a minimal hide wash cabinet on the levels and prevalence of *Escherichia coli* O157:H7 and *Salmonella* on the hides of beef cattle at slaughter. *Journal of Food Protection,* 70(5): 1076–1079.

Arthur, T.M., Bosilevac, J.M., Brichta-Harhay, D., Kalchayanand, N., King, D.A., Shackelford, S.D., Wheeler, T.L. & Koohmaraie, M. 2008. Source tracking of *Escherichia coli* O157:H7 and *Salmonella* contamination in the lairage environment at commercial U.S. beef processing plants and identification of an effective intervention. *Journal of Food Protection,* 71(9): 1752–1760.

Arthur, T. M., Kalchayanand, N., Bosilevac, J. M., Brichta-Harhay, D. M., Shackelford, S. D., Bono, J. L. & Koohmaraie, M. 2008b. Comparison of effects of antimicrobial interventions on multidrug-resistant *Salmonella*, susceptible *Salmonella*, and *Escherichia coli* O157:H7. *Journal of Food Protection,* 71(11): 2177–2181.

Bacon, R.T., Sofos, J.N., Belk, K.E., Hyatt, D.R. & Smith, G.C. 2002a. Prevalence and antibiotic susceptibility of *Salmonella* isolated from beef animal hides and carcasses. *Journal of Food Protection,* 65(2): 284–290.

Bacon, R.T., Sofos, J.N., Belk, K.E. & Smith, G.C. 2002b. Application of a commercial steam vacuum unit to reduce inoculated *Salmonella* on chilled fresh beef adipose tissue. *Dairy, Food and Environmental Sanitation,* 22: 184–190.

Baer, A.A., Miller, M.J. & Dilger, A.C. 2013. Pathogens of interest to the pork industry: A review of research on interventions to assure food safety. *Comprehensive Reviews in Food Science and Food Safety,* 12(2): 183–217.

Barkocy-Gallagher, G., Arthur, T.M., Rivera-Betancourt, M., Nou, X., Shackelford, S.D., Wheeler, T.L. & Koohmaraie, M. 2003. Seasonal prevalence of Shiga toxin-producing *Escherichia coli,* including O157:H7 and non-O157 serotypes, and *Salmonella* in commercial beef processing plants. *Journal of Food Protection,* 66(11): 1978–1986.

Baskaran, S.A., Upadhyay, A., Upadhyaya, I., Bhattaram, V. & Venkitanarayanan, K. 2012. Efficacy of octenidine hydrochloride for reducing *Escherichia coli* O157:H7, *Salmonella* spp., and *Listeria monocytogenes* on cattle hides. *Applied and Environmental Microbiology,* 78(12): 4538–4541.

Bawcom, D.W., Thompson, L.D., Miller, M.F. & Ramsey, C.B. 1995. Reduction of microorganisms on beef surfaces utilizing electricity. *Journal of Food Protection,* 58(1): 35–38.

Bell, K.Y., Cutter, C.N. & Sumner, S.S. 1997. Reduction of foodborne micro-organisms on beef carcass tissue using acetic acid, sodium bicarbonate, and hydrogen peroxide spray washes. *Food Microbiology,* 14(5): 439–448.

Bergis, H., Poumeyrol, G. & Beaufort, A. 1994. Growth of saprophytic flora and *Salmonella* in minced beef meat stored under modified atmosphere. *Sciences des Aliments,* 14(2): 217–227

Bosilevac, J.M., Arthur, T.M., Bono, J.L., Brichta-Harhay, D., Kalchayanand, N., King, D.A., Shackelford, S.D., Wheeler, T.L. & Koohmaraie, M. 2009. Prevalence and enumeration of *Escherichia coli* O157:H7 and *Salmonella* in U.S. abattoirs that process fewer than 1,000 head of cattle per day. *Journal of Food Protection,* 72(6): 1272–1278.

Brichta-Harhay, D., Guerini, M.N., Arthur, T.M., Bosilevac, J.M., Kalchayanand, N., Shackelford, S.D., Wheeler, T.L. & Koohmaraie, M. 2008. *Salmonella* and *Escherichia coli* O157:H7 contamination on hides and carcasses of cull cattle presented for slaughter in the United States: An evaluation of prevalence and bacterial loads by immunomagnetic separation and direct plating methods. *Applied and Environmental Microbiology,* 74(20): 6289-6297.

Brooks, J.C., Alvarado, M., Stephens, T.P., Kellermeier, J.D., Tittor, A.W., Miller, M.F. & Brashears, M.M. 2008. Spoilage and safety characteristics of ground beef packaged in traditional and modified atmosphere packages. *Journal of Food Protection,* 71(2): 293–301.

Cabrera-Diaz, E., Moseley, T.M., Lucia, L.M., Dickson, J.S., Castillo, A. & Acuff, G.R. 2009. Fluorescent protein-marked *Escherichia coli* biotype I strains as surrogates for enteric pathogens in validation of beef carcass interventions. *Journal of Food Protection,* 72(2): 295–303.

Careaga, M., Fernández, E., Dorantes, L., Mota, L., Jaramillo, M.A. & Hernandez-Sanchez, H. 2003. Antibacterial activity of *Capsicum* extract against *Salmonella* Typhimurium and *Pseudomonas aeruginosa* inoculated in raw beef meat. *International Journal of Food Microbiology,* 83(3): 331–335.

Carlson, B.A., Ruby, J., Smith, G.C., Sofos, J.N., Bellinger, G.R., Warren-Serna, W., Centrella, B., Bowling, R.A. & Belk, K.E. 2008. Comparison of antimicrobial efficacy of multiple beef hide decontamination strategies to reduce levels of *Escherichia coli* O157:H7 and *Salmonella. Journal of Food Protection,* 71(11): 2223–2227.

Castillo, A., Dickson, J.S., Clayton, R.P., Lucia, L.M. & Acuff, G.R. 1998a. Chemical dehairing of bovine skin to reduce pathogenic bacteria and bacteria of faecal origin. *Journal of Food Protection,* 61(5): 623–625.

Castillo, A., Lucia, L.M., Goodson, K.J., Savell, J.W. & Acuff, G.R. 1998b. Comparison of water wash, trimming, and combined hot water and lactic acid treatments for reducing bacteria of faecal origin on beef carcasses. *Journal of Food Protection,* 61(7): 823–828.

Castillo, A., Lucia, L.M., Goodson, K.J., Savell, J.W. & Acuff, G.R. 1998c. Use of hot water for beef carcass decontamination. *Journal of Food Protection,* 61(1): 19–25.

Castillo, A., Lucia, L.M., Roberson, D.B., Stevenson, T.H., Mercado, I. & Acuff, G.R. 2001. Lactic acid sprays reduce bacterial pathogens on cold beef carcass surfaces and in subsequently produced ground beef. *Journal of Food Protection,* 64(1): 58–62.

Castillo, A., McKenzie, K.S., Lucia, L.M. & Acuff, G.R. 2003. Ozone treatment for reduction of *Escherichia coli* O157:H7 and *Salmonella* serotype Typhimurium on beef carcass surfaces. *Journal of Food Protection,* 66(5): 775–779.

Castlllo, A., Lucia, L.M., Kemp, G.K. & Acuff, G.R. 1999. Reduction of *Escherichia coli* O157:H7 and *Salmonella* Typhimurium on beef carcass surfaces using acidified sodium chlorite. *Journal of Food Protection,* 62(6): 580–584.

Chaillou, S., Christieans, S., Rivollier, M., Lucquin, I., Champomier-Verges, M. & Zagorec, M. 2014. Quantification and efficiency of *Lactobacillus sakei* strain mixtures used as protective cultures in ground beef. *Meat Science,* 97(3; Special issue): 332–338.

Chao, C.Y. & Yin, M.C. 2009. Antibacterial effects of roselle calyx extracts and protocatechuic acid in ground beef and apple juice. *Foodborne Pathogens and Disease,* 6(2): 201–206.

Chung, M.S., Ko, Y.T. & Kim, W.S. 2000. Survival of *Pseudomonas fluorescens* and *Salmonella* Typhimurium after electron beam and gamma irradiation of refrigerated beef. *Journal of Food Protection,* 63(2): 162–166.

Codex Alimentarius Commission. 2003. Hazard analysis and critical control point (HACCP) system and guidelines for its application. CAC/RCP 1-1969, Rev. 4-2003.

Cruz-Galvez, A.M, Gomez-Aldapa, C.A., Villagomez-Ibarra, J.R., Chavarria-Hernandez, N., Rodriguez-Banos, J., Rangel-Vargas, E. & Castro-Rosas, J. 2013. Antibacterial effect against foodborne bacteria of plants used in traditional medicine in central Mexico: Studies *in vitro* and in raw beef. *Food Control,* 32(1): 289–295.

Cutter, C.N. 1999a. Combination spray washes of saponin with water or acetic acid to reduce aerobic and pathogenic bacteria on lean beef surfaces. *Journal of Food Protection,* 62(3): 280–283.

Cutter, C.N. 1999b. The effectiveness of triclosan-incorporated plastic against bacteria on beef surfaces. *Journal of Food Protection,* 62(5): 474–479.

Cutter, C.N. 2000. Antimicrobial effect of herb extracts against *Escherichia coli* O157:H7, *Listeria monocytogenes*, and *Salmonella* Typhimurium associated with beef. *Journal of Food Protection,* 63(5): 601–607.

Cutter, C.N., Dorsa, W.J., Handie, A., Rodriguez-Morales, S., Xiang, Z., Breen, P.J. & Compadre, C.M. 2000. Antimicrobial activity of cetylpyridinium chloride washes against pathogenic bacteria on beef surfaces. *Journal of Food Protection,* 63(5): 593–600.

Cutter, C.N., Dorsa, W.J. & Siragusa, G.R. 1997. Rapid desiccation with heat in combination with water washing for reducing bacteria on beef carcass surfaces. *Food Microbiology,* 14(5): 493–303.

Cutter, C.N. & Rivera-Betancourt, M. 2000. Interventions for the reduction of *Salmonella* Typhimurium DT 104 and non-O157:H7 enterohemorrhagic *Escherichia coli* on beef surfaces. *Journal of Food Protection,* 63(10): 1326–1332.

Cutter, C.N. & Siragusa, G.R. 1995. Treatments with nisin and chelators to reduce *Salmonella* and *Escherichia coli* on beef. *Journal of Food Protection,* 58(9): 1028–1030.

De Oliveira, T.L.C., Soares, R.D.A. & Piccoli, R.H. 2013. A weibull model to describe antimicrobial kinetics of oregano and lemongrass essential oils against *Salmonella* Enteritidis in ground beef during refrigerated storage. *Meat Science,* 93(3): 645–651.

Dias-Morse, P., Pohlman, F.W., Williams, J. & Brown, A.H. 2014. Single or multiple decontamination interventions involving lauric arginate on beef trimmings to enhance microbial safety of ground beef. *Professional Animal Scientist,* 30(5): 477–484.

Dickson, J.S. 1991. Control of *Salmonella* Typhimurium, *Listeria monocytogenes*, and *Escherichia coli* 0157-H7 on beef in a model spray chilling system. *Journal of Food Science,* 56(1): 191–193.

Dickson, J.S. & Anderson, M.E. 1991. Control of *Salmonella* on beef tissue surfaces in a model system by pre- and post-evisceration washing and sanitizing, with and without spray chilling. *Journal of Food Protection,* 54(7): 514–518.

Dickson, J.S., Cutter, C.G.N. & Siragusa, G.R. 1994. Antimicrobial effects of trisodium phosphate against bacteria attached to beef tissue. *Journal of Food Protection,* 57(11): 952–955.

Dickson, J.S. & Siragusa, G.R. 1994. Survival of *Salmonella* Typhimurium, *Escherichia coli* O157:H7 and *Listeria monocytogenes* during storage on beef sanitized with organic acids. *Journal of Food Safety,* 14(4): 313–327.

Dong, P.C., Zhu, L.X., Mao, Y.W., Liang, R.R., Niu, L.B., Zhang, Y.M., Li, K. & Luo, X. 2014. Prevalence and profile of *Salmonella* from samples along the production line in Chinese beef processing plants. *Food Control,* 38: 54–60.

Dorsa, W.J., Cutter, C.N. & Siragusa, G.R. 1998a. Bacterial profile of ground beef made from carcass tissue experimentally contaminated with pathogenic and spoilage bacteria before being washed with hot water, alkaline solution, or organic acid and then stored at 4 or 12°C. *Journal of Food Protection,* 61(9): 1109–1118.

Dorsa, W.J., Cutter, C.N. & Siragusa, G.R. 1998b. Long-term bacterial profile of refrigerated ground beef made from carcass tissue, experimentally contaminated with pathogens and spoilage bacteria after hot water, alkaline, or organic acid washes. *Journal of Food Protection,* 61(12): 1615–1622.

Echeverry, A., Brooks, J.C., Miller, M.F., Collins, J.A., Loneragan, G.H. & Brashears, M.M. 2009. Validation of intervention strategies to control *Escherichia coli* O157:H7 and *Salmonella* Typhimurium DT 104 in mechanically tenderized and brine-enhanced beef. *Journal of Food Protection,* 72(8): 1616–1623.

Echeverry, A., Brooks, J.C., Miller, M.F., Collins, J.A., Loneragan, G.H. & Brashears, M.M. 2010. Validation of lactic acid bacteria, lactic acid, and acidified sodium chlorite as decontaminating interventions to control *Escherichia coli* O157:H7 and *Salmonella* Typhimurium DT 104 in mechanically tenderized and brine-enhanced (non-intact) beef at the purveyor. *Journal of Food Protection,* 73(12): 2169–2179.

Ellebracht, E.A., Castillo, A., Lucia, L.M., Miller, R.K. & Acuff, G.R. 1999. Reduction of pathogens using hot water and lactic acid on beef trimmings. *Journal of Food Science,* 64(6): 1094–1099.

Ellebracht, J.W., King, D.A., Castillo, A., Lucia, L.M., Acuff, G.R., Harris, K.B. & Savell, J.W. 2005. Evaluation of peroxyacetic acid as a potential pre-grinding treatment for control of *Escherichia coli* O157:H7 and *Salmonella* Typhimurium on beef trimmings. *Meat Science,* 70(1): 197–203.

Faith, N.G., Garcia, G., Skebba, V.P., Gandhi, N.R. & Czuprynski, C.J. 2015. Use of a commercial mixture of volatile compounds from the fungus muscodor to inhibit *Salmonella* in ground turkey and beef. *Food Control,* 47: 628–633.

Fegan, N., Vanderlinde, P., Higgs, G. & Desmarchelier, P. 2005. A study of the prevalence and enumeration of *Salmonella enterica* in cattle and on carcasses during processing. *Journal of Food Protection,* 68(6): 1147–1153.

Fouladkhah, A., Geornaras, I., Yang, H., Belk, K.E., Nightingale, K.K., Woerner, D.R., Smith, G.C. & Sofos, J.N. 2012. Sensitivity of Shiga toxin-producing *Escherichia coli*, multidrugresistant *Salmonella*, and antibiotic-susceptible *Salmonella* to lactic acid on inoculated beef trimmings. *Journal of Food Protection,* 75(10): 1751–1758.

Geornaras, I., Yang, H., Moschonas, G., Nunnelly, M.C., Belk, K.E., Nightingale, K.K., Woerner, D.R., Smith, G.C. & Sofos, J.N. 2012. Efficacy of chemical interventions against *Escherichia coli* O157:H7 and multidrug-resistant and antibiotic-susceptible *Salmonella* on inoculated beef trimmings. *Journal of Food Protection,* 75(11): 1960–1967.

Gill, C.O. & DeLacy, K.M. 1991. Growth of *Escherichia coli* and *Salmonella* Typhimurium on high-pH beef packed under vacuum or carbon dioxide. *International Journal of Food Microbiology*, 13(1): 21–30.

Gomólka-Pawlicka, M., Uradzinski, J., 2003. Antagonistic effect of chosen lactic acid bacteria strains on *Salmonella* species in meat and fermented sausages. *Polish Journal of Veterinary Sciences*, 6(1): 29–39.

Graves Delmore, L.R., Sofos, J.N., Reagan, J.O. & Smith, G.C. 1997. Hot-water rinsing and trimming/washing of beef carcasses to reduce physical and microbiological contamination. *Journal of Food Science,* 62(2): 373–376.

Greer, G.G. & Dilts, B.D. 1992. Factors affecting the susceptibility of meatborne pathogens and spoilage bacteria to organic acids. *Food Research International,* 25(5): 355–364.

Hajmeer, M.N., Marsden, J.L., Crozier-Dodson, B., Basheer, I.A. & Higgins, J.J. 1999. Reduction of microbial counts at a commercial beef koshering facility. *Journal of Food Science,* 64(4): 719–723.

Hardin, M.D., Acuff, G.R., Lucia, L.M., Oman, J.S. & Savell, J.W. 1995. Comparison of methods for decontamination from beef carcass surfaces. *Journal of Food Protection,* 58(4): 368–374.

Harris, D., Brashears, M.M., Garmyn, A.J., Brooks, J.C. & Miller, M.F. 2012. Microbiological and organoleptic characteristics of beef trim and ground beef treated with acetic acid, lactic acid, acidified sodium chlorite, or sterile water in a simulated commercial processing environment to reduce *Escherichia coli* O157:H7 and *Salmonella*. *Meat Science,* 90(3): 783–788.

Harris, K., Miller, M.F., Loneragan, G.H. & Brashears, M.M. 2006. Validation of the use of organic acids and acidified sodium chlorite to reduce *Escherichia coli* O157 and *Salmonella* Typhimurium in beef trim and ground beef in a simulated processing environment. *Journal of Food Protection,* 69(8): 1802–1807.

Hayouni, E., Chraief, I., Abedrabba, M., Bouix, M., Leveau, J.Y., Mohammed, H. & Hamdi, M. 2008. Tunisian *Salvia officinalis* L. and *Schinus molle* L. essential oils: Their chemical compositions and their preservative effects against *Salmonella* inoculated in minced beef meat. *International Journal of Food Microbiology*, 125(3): 242–251.

Hoyle, A.R., Brooks, J.C., Thompson, L.D., Palmore, W., Stephens, T.P. & Brashears, M.M. 2009. Spoilage and safety characteristics of ground beef treated with lactic acid bacteria. *Journal of Food Protection,* 72(11): 2278–2283.

Hughes, M.K., Yanamala, S., San Francisco, M., Loneragan, G.H., Miller, M.F. & Brashears, M.M. 2010. Reduction of multidrug-resistant and drug-susceptible *Salmonella* in ground beef and freshly harvested beef briskets after exposure to commonly used industry antimicrobial interventions. *Journal of Food Protection,* 73(7): 1231–1237.

Jadeja, R. & Hung, Y.C. 2014. Efficacy of near neutral and alkaline pH electrolyzed oxidizing waters to control *Escherichia coli* O157: H7 and *Salmonella* Typhimurium DT 104 from beef hides. *Food Control,* 41: 17–20.

Jensen, J.L., Saxena, A.D. & Keener, K.M. 2009. Evaluation of treatment methods for reducing bacteria in textured beef. p. 6439, *in:* Vol. 10 of Proceedings of a meeting held 21-24 June 2009, Reno, Nevada, USA. American Society of Agricultural and Biological Engineers Annual International Meeting 2009.

Kalchayanand, N., Arthur, T.M., Bosilevac, J.M., Brichta-Harhay, D., Guerini, M.N., Shackelford, S.D., Wheeler, T.L. & Koohmaraie, M. 2009. Effectiveness of 1,3-dibromo-5,5 dimethylhydantoin on reduction of *Escherichia coli* O157:H7- and *Salmonella*-inoculated fresh meat. *Journal of Food Protection,* 72(1): 151–156.

Kim, J.W. & Slavik, M.F. 1994. Trisodium phosphate (TSP) treatment of beef surfaces to reduce *Escherichia coli* O157:H7 and *Salmonella* Typhimurium. *Journal of Food Science,* 59(1): 20–22.

King, D.A., Lucia, L.M., Castillo, A., Acuff, G.R., Harris, K.B. & Savell, J.W. 2005. Evaluation of peroxyacetic acid as a post-chilling intervention for control of *Escherichia coli* O157:H7 and *Salmonella* Typhimurium on beef carcass surfaces. *Meat Science,* 69(3): 401–407.

Koohmaraie, M., Scanga, J.A., Zerda, D.L., Koohmaraie, B., Tapay, L., Beskhlebnaya, V., Mai, T., Greeson, K. & Samadpour, M. 2012. Tracking the sources of *Salmonella* in ground beef produced from non-fed cattle. *Journal of Food Protection,* 75(8): 1464–1468.

Kundu, D., Gill, A., ChenYuan, L., Goswami, N. & Holley, R. 2014. Use of low dose e-beam irradiation to reduce *E. coli* O157:H7, non-O157 (VTEC) *E. coli* and *Salmonella* viability on meat surfaces. *Meat Science,* 96(1): 413–418.

Laury, A.M., Alvarado, M.V., Nace, G., Alvarado, C.Z., Brooks, J.C., Echeverry, A. & Brashears, M.M. 2009. Validation of a lactic acid- and citric acid-based antimicrobial product for the reduction of *Escherichia coli* O157:H7 and *Salmonella* on beef tips and whole chicken carcasses. *Journal of Food Protection,* 72(10): 2208–2211.

Li, S., Kundu, D. & Holley, R.A. 2015. Use of lactic acid with electron beam irradiation for control of *Escherichia coli* O157: H7, non-O157 VTEC *E. coli,* and *Salmonella* serovars on fresh and frozen beef. *Food Microbiology,* 46: 34–39.

McCann, M.S., McGovern, A.C., McDowell, D.A., Blair, I.S. & Sheridan, J.J. 2006. Surface decontamination of beef inoculated with *Salmonella* Typhimurium DT104 or *Escherichia coli* O157:H7 using dry air in a novel heat treatment apparatus. *Journal of Applied Microbiology,* 101(5): 1177–1187.

McDaniel, J.A., Pohlman, F.W., Brown, A.H., Ricke, S.C., Morse, P.N.D., Mehall, L.N., Milillo, S.R. & Mohan, A. 2012. Evaluation of product safety enhancement through antimicrobial electrostatic spray applications on *longissimus lumborum* at the subprimal level and its impact on meat color characteristics. Arkansas Animal Science Department Report 2011. *AAES Research Series*, No. 597: 86–88.

Mehall, L.N., Pohlman, F.W., Brown, A.H., Dias-Morse, P., McKenzie, L.M. & Mohan, A. 2012a. The impact of cetylpyridinium chloride, trisodium phosphate, potassium lactate, sodium metasilicate, or water as antimicrobial interventions on microbiological characteristics of beef biceps femoris muscles. Arkansas Animal Science Department Report 2011. *AAES Research Series,* No. 597: 101–103.

Mehall, L.N., Pohlman, F.W., McDaniel, J.A., Dias-Morse, P., Rojas, T.N., 2012b. The effects of hydrochloric/citric acid mixture, cetylpyridinium chloride, trisodium phosphate, or sodium metasilicate as antimicrobial sequential treatment combinations on microbiological properties of beef longissimus lumborum muscle. Arkansas Animal Science Department Report 2011. *AAES Research Series,* No. 597: 104–107.

Mies, P.D., Covington, B.R., Harris, K.B., Lucia, L.M., Acuff, G.R. & Savell, J.W. 2004. Decontamination of cattle hides prior to slaughter using washes with and without antimicrobial agents. *Journal of Food Protection,* 67(3): 579–582.

Miya, S., Takahashi, H., Hashimoto, M., Nakazawa, M., Kuda, T., Koiso, H. & Kimura, B. 2014. Development of a controlling method for *Escherichia coli* O157:H7 and *Salmonella* spp. in fresh market beef by using polylysine and modified atmosphere packaging. *Food Control,* 37: 62–67.

Mohan, A., Pohlman, F.W., McDaniel, J.A. & Hunt, M.C. 2012. Role of peroxyacetic acid, octanoic acid, malic acid, and potassium lactate on the microbiological and instrumental color characteristics of ground beef. *Journal of Food Science,* 77(4) M188–M193.

Muluneh, G. & Kibret, M. 2015. *Salmonella* spp. and risk factors for the contamination of slaughtered cattle carcass from a slaughterhouse of Bahir Dar Town, Ethiopia. *Asian Pacific Journal of Tropical Disease,* 5: 130.

Narváez-Bravo, C., Rodas-González, A., Fuenmayor, Y., Flores-Rondon, C., Carruyo, G., Moreno, M., Perozo-Mena, A. & Hoet, A.E. 2013. *Salmonella* on faeces hides and carcasses in beef slaughter facilities in Venezuela. *International Journal of Food Microbiology,* 166(2): 226–230.

Niebuhr, S.E., Laury, A., Acuff, G.R. & Dickson, J.S. 2008. Evaluation of non-pathogenic surrogate bacteria as process validation indicators for *Salmonella enterica* for selected antimicrobial treatments, cold storage, and fermentation in meat. *Journal of Food Protection,* 71(4): 714–718.

Nissen, H., Alvseike, O., Bredholt, S., Holck, A. & Nesbakken, T. 2000. Comparison between the growth of *Yersinia enterocolitica, Listeria monocytogenes, Escherichia coli* O157:H7 and *Salmonella* spp. in ground beef packed by three commercially used packaging techniques. *International Journal of Food Microbiology*, 59(3): 211–220.

Njongmeta, N.L.A., Benli, H., Dunkley, K.D., Dunkley, C.S., Miller, D.R., Anderson, R.C., O'Bryan, C.A., Keeton, J.T., Nisbet, D.J., Crandall, P.G. & Ricke, S.C. 2011. Application of acidic calcium sulfate and epsilon-polylysine to pre-rigor beef rounds for reduction of pathogens. *Journal of Food Safety,* 31(3): 395:400.

Nou, X., Rivera-Betancourt, M., Bosilevac, J.M., Wheeler, T.L., Shackelford, S.D., Gwartney, B.L., Reagan, J.O. & Koohmaraie, M. 2003. Effect of chemical de-hairing on the prevalence of *Escherichia coli* O157:H7 and the levels of aerobic bacteria and enterobacteriaceae on carcasses in a commercial beef processing plant. *Journal of Food Protection,* 66(11): 2005–2009.

Nutsch, A.L., Phebus, R.K., Riemann, M.J., Schafer, D.E., Boyer Jr., J.E., Wilson, R.C., Leising, J.D. & Kastner, C.L. 1997. Evaluation of a steam pasteurization process in a commercial beef processing facility. *Journal of Food Protection,* 60(5): 485–492.

Olaoye, O.A. & Onilude, A.A. 2010. Investigation on the potential application of biological agents in the extension of shelf life of fresh beef in Nigeria. *World Journal of Microbiology & Biotechnology*, 26(8): 1445–1454.

Ozdemir, H., Yldrm, Y., Kuplulu, O., Koluman, A., Goncuoglu, M. & Inat, G. 2006. Effects of lactic acid and hot water treatments on *Salmonella* Typhimurium and *Listeria monocytogenes* on beef. *Food Control,* 17(4): 299–303.

Patel, J., Macarisin, D., Sanglay, G. & Murphy, C. 2012. Inactivation and injury of pathogens on intact beef treated with hydrodynamic pressure. *Innovative Food Science & Emerging Technologies*, 14: 38–45.

Patel, J.R. & Solomon, M.B. 2005. Attachment of *Escherichia coli* O157:H7, *Salmonella* Typhimurium, and *Listeria monocytogenes* to beef and inactivation using hydrodynamic pressure processing. pp. E1–E6, *in:* SPIE Proceedings Vol. 5996, Optical Sensors and Sensing Systems for Natural Resources and Food Safety and Quality.

Pearce, R. & Bolton, D.J. 2008. The anti-microbial effect of a dairy extract (LactiSAL (R)) on *Salmonella enterica* Typhimurium and *Escherichia coli* O157:H7 on different beef carcass surfaces. *Food Control*, 19(5): 449–453.

Phebus, R.K., Nutsch, A.L., Schafer, D.E., Wilson, R.C., Riemann, M.J., Leising, J.D., Kastner, C.L., Wolf, J.R. & Prasai, R.K. 1997. Comparison of steam pasteurization and other methods for reduction of pathogens on surfaces of freshly slaughtered beef. *Journal of Food Protection,* 60(5): 476–484.

Podolak, R.K., Zayas, J.F., Kastner, C.L. & Fung, D.Y. 1995. Reduction of *Listeria monocytogenes, Escherichia coli* O157:H7 and *Salmonella* Typhimurium during storage on beef sanitized with fumaric, acetic, and lactic acids. *Journal of Food Safety,* 15(3): 283–290.

Pohlman, F., Dias-Morse, P. & Pinidiya, D. 2014. Product safety and color characteristics of ground beef processed from beef trimmings treated with peroxyacetic acid alone or followed by novel organic acids. *Journal of Microbiology, Biotechnology and Food Sciences,* 4(2): 93–101.

Pohlman, F.W., Dias-Morse, P., Quilo, S.A., Brown, A.H., Crandall, P.G., Baublits, R.T., Story, R.P., Bokina, C. & Rajaratnam, G. 2009. Microbial, instrumental color and sensory characteristics of ground beef processed from beef trimmings treated with potassium lactate, sodium metasilicate, peroxyacetic acid or acidified sodium chlorite as single antimicrobial interventions. *Journal of Muscle Foods,* 20(1): 54–69.

Pohlman, F.W., Stivarius, M.R., McElyea, K.S., Johnson, Z.B. & Johnson, M.G. 2002a. The effects of ozone, chlorine dioxide, cetylpyridinium chloride and trisodium phosphate as multiple antimicrobial interventions on microbiological, instrumental color, and sensory color and odor characteristics of ground beef. *Meat Science,* 61(3): 307–313.

Pohlman, F.W., Stivarius, M.R., McElyea, K.S., Johnson, Z.B. & Johnson, M.G. 2002b. Reduction of microorganisms in ground beef using multiple intervention technology. *Meat Science,* 61(3): 315–322.

Prasai, R.K., Acuff, G.R., Lucia, L.M., Hale, D.S., Savell, J.W. & Morgan, J.B. 1991. Microbiological effects of acid decontamination of beef carcasses at various locations in processing. *Journal of Food Protection,* 54(11): 868–872.

Prasai, R.K., Kastner, C.L., Kenney, P.B., Kropf, D.H., Fung, D.Y.C., Mease, L.E., Vogt, L.R. & Johnson, D.E. 1997. Microbiological quality of beef subprimals as affected by lactic acid sprays applied at various points during vacuum storage. *Journal of Food Protection,* 60(7): 795–798.

Qiu, X.J. & Wu, V.C.H. 2007. Evaluation of *Escherichia coli* O157:H7, *Listeria monocytogenes, Salmonella* Typhimurium and *Staphylococcus aureus* in ground beef with cranberry concentrate by thin agar layer method. *Journal of Rapid Methods and Automation in Microbiology,* 15(3): 282–294.

Quilo, S.A., Pohlman, F.W., Dias-Morse, P., Brown, A.H., Crandall, P.G. & Story, R.P. 2010. Microbial, instrumental color and sensory characteristics of inoculated ground beef produced using potassium lactate, sodium metasilicate or peroxyacetic acid as multiple antimicrobial interventions. *Meat Science,* 84(3): 470–476.

Reagan, J.O., Acuff, G.R., Buege, D.R., Buyck, M.J., Dickson, J.S., Kastner, C.L., Marsden, J.L., Morgan, J.B., Nickelson II, R., Smith, G.C. & Sofos, J.N. 1996. Trimming and washing of beef carcasses as a method of improving the microbiological quality of meat. *Journal of Food Protection,* 59(7): 751–756.

Retzlaff, D., Phebus, R., Nutsch, A., Riemann, J., Kastner, C. & Marsden, J. 2004. Effectiveness of a laboratory-scale vertical tower static chamber steam pasteurization unit against *Escherichia coli* O157:H7, *Salmonella* Typhimurium, and *Listeria innocua* on pre-rigor beef tissue. *Journal of Food Protection,* 67(8): 1630–1633.

Reyes, J.E., Aguirre, C., Vargas, C., Petzold, G., Valencia, C. & Luarte, F. 2003. Efficacy of washing by aspersion with chlorine dioxide, ozone and KilolReg. to reduce bacterial contamination on surfaces of samples of cattle meat. *Alimentaria,* 40: 17.

Rivera-Betancourt, M., Shackelford, S.D., Arthur, T.M., Westmoreland, K.E., Bellinger, G., Rossman, M., Reagan, J.O. & Koohmaraie, M. 2004. Prevalence of *Escherichia coli* O157:H7, *Listeria monocytogenes*, and *Salmonella* in two geographically distant commercial beef processing plants in the United States. *Journal of Food Protection,* 67(2): 295–302.

Ruby, J.R. & Ingham, S.C. 2009. Evaluation of potential for inhibition of growth of *Escherichia coli* O157:H7 and multidrug-resistant *Salmonella* serovars in raw beef by addition of a presumptive *Lactobacillus sakei* ground beef isolate. *Journal of Food Protection,* 72(2): 251–259.

Ruby, J.R., Zhu, J. & Ingham, S.C. 2007. Using indicator bacteria and *Salmonella* test results from three large-scale beef abattoirs over an 18-month period to evaluate intervention system efficacy and plan carcass testing for *Salmonella. Journal of Food Protection,* 70(12): 2732–2740.

Ryu, S.H. & Fung, D.Y.C. 2010. Antimicrobial effect of buffered sodium citrate (BSC) on foodborne pathogens in liquid media and ground beef. *Preventive Nutrition and Food Science,* 15(3): 239–243.

Sawyer, J.E., Greiner, S.T., Acuff, G.R., Lucia, L.M., Cabrera-Diaz, E. & Hale, D.S. 2008. Effect of xylitol on adhesion of *Salmonella* Typhimurium and *Escherichia coli* O157:H7 to beef carcass surfaces. *Journal of Food Protection,* 71(2): 405–410.

Scanga, J.A., Buschow, A.W., Kauk, J.L., Burk, T.E., Koohmaraie, B., Zerda, D.L., Motlagh, A.M., Samadpour, M. & Koohmaraie, M. 2011. Localized chemical decontamination of cattle hides to reduce microbial loads and prevalence of foodborne pathogens. *Food Protection Trends,* 31(9): 569–574.

Schmidt, J.W., Bosilevac, J.M., Kalchayanand, N., Wang, R., Wheeler, T.L., Koohmaraie, M., 2014. Immersion in antimicrobial solutions reduces *Salmonella enterica* and Shiga toxin-producing *Escherichia coli* on beef cheek meat. *Journal of Food Protection,* 77(4): 538–548.

Schmidt, J.W., Wang, R., Kalchayanand, N., Wheeler, T.L. & Koohmaraie, M. 2012. Efficacy of hypobromous acid as a hide-on carcass antimicrobial intervention. *Journal of Food Protection,* 75(5): 955–958.

Schnell, T.D., Sofos, J.N., Littlefield, V.G., Morgan, J.B., Gorman, B.M., Clayton, R.P. & Smith, G.C. 1995. Effects of postexsanguination de-hairing on the microbial load and visual cleanliness of beef carcasses. *Journal of Food Protection,* 58(12): 1297–1302.

Shalaby, A.M. 2011. Antibacterial effects of mustard, garlic and acetic acid against *Salmonella* Typhimurium in ground beef. *Assiut Veterinary Medical Journal,* 57: 118–136.

Skandamis, P., Tsigarida, E. & Nychas, G-J E. 2002. The effect of oregano essential oil on survival/death of *Salmonella* Typhimurium in meat stored at 5°C under aerobic, VP/MAP conditions. *Food Microbiology,* 19(1): 97–103.

Smith, L., Mann, J.E., Harris, K., Miller, M.F. & Brashears, M.M. 2005. Reduction of *Escherichia coli* O157:H7 and *Salmonella* in ground beef using lactic acid bacteria and the impact on sensory properties. *Journal of Food Protection,* 68(8): 1587–1592.

Smith, M.G. 1992. Destruction of bacteria on fresh meat by hot water. *Epidemiology and Infection,* 109(3): 491–496.

Sofos, J.N., Kochevar, S.L., Reagan, J.O. & Smith, G.C. 1999. Incidence of *Salmonella* on beef carcasses relating to the U.S. meat and poultry inspection regulations. *Journal of Food Protection,* 6(5): 467–473.

Stelzleni, A.M., Ponrajan, A. & Harrison, M.A. 2013. Effects of buffered vinegar and sodium dodecyl sulfate plus levulinic acid on *Salmonella* Typhimurium survival, shelf-life, and sensory characteristics of ground beef patties. *Meat Science,* 95(1): 1–7.

Stivarius, M.R., Pohlman, F.W., McElyea, K.S. & Apple, J.K. 2002a. The effects of acetic acid, gluconic acid and trisodium citrate treatment of beef trimmings on microbial, color and odor characteristics of ground beef through simulated retail display. *Meat Science,* 60(3): 245–252.

Stivarius, M.R., Pohlman, F.W., McElyea, K.S. & Apple, J.K. 2002b. Microbial, instrumental color and sensory color and odor characteristics of ground beef produced from beef trimmings treated with ozone or chlorine dioxide. *Meat Science,* 60(3): 299305.

Stivarius, M.R., Pohlman, F.W., McElyea, K.S. & Waldroup, A.L. 2002c. Effects of hot water and lactic acid treatment of beef trimmings prior to grinding on microbial, instrumental color and sensory properties of ground beef during display. *Meat Science,* 60(4): 327–334.

Stopforth, J.D., Ashton, L.V., Skandamis, P.N., Scanga, J.A., Smith, G.C., Sofos, J.N. & Belk, K.E. 2005. Single and sequential treatment of beef tissue with lactic acid, ammonium hydroxide, sodium metasilicate, and acidic and basic oxidized water to reduce numbers of inoculated Escherichia coli 0157:H7 and Salmonella Typhimurium. *Food Protection Trends*, 25(1): 14–22.

Stopforth, J.D., Lopes, M., Shultz, J.E., Miksch, R.R. & Samadpour, M. 2006. Location of bung bagging during beef slaughter influences the potential for spreading pathogen contamination on beef carcasses. *Journal of Food Protection*, 69(6): 1452–1455.

Tango, C.N., Mansur, A.R., Kim, G.H. & Oh, D.H. 2014. Synergetic effect of combined fumaric acid and slightly acidic electrolysed water on the inactivation of food-borne pathogens and extending the shelf life of fresh beef. *Journal of Applied Microbiology*, 117(6): 1709–1720.

Tayel, A.A., El-Tras, W., Moussa, S.H. & El-Sabbagh, S. 2012. Surface decontamination and quality enhancement in meat steaks using plant extracts as natural biopreservatives. *Foodborne Pathogens and Disease*, 9(8): 755–761.

Tinney, K.S., Miller, M.F., Ramsey, C.B., Thompson, L.D. & Carr, M.A. 1997. Reduction of microorganisms on beef surfaces with electricity and acetic acid. *Journal of Food Protection*, 60(6): 625–628.

Tittor, A.W., Tittor, M.G., Brashears, M.M., Brooks, J.C., Garmyn, A.J. & Miller, M.F. 2011. Effects of simulated dry and wet chilling and aging of beef fat and lean tissues on the reduction of *Escherichia coli* O157:H7 and *Salmonella*. *Journal of Food Protection*, 74(2): 289–293.

Trairatapiwan, T., Lertpatarakomol, R. & Mitchaothai, J. 2011. Evaluation of *Salmonella* contamination and serovar isolation from slaughtering and cutting processes of Thai indigenous beef cattle. pp. 829–833, *in:* Proceedings of the 3rd International Conference on Sustainable Animal Agriculture for Developing Countries. Nakhon Ratchasima, Thailand.

Trivedi, S., Reynolds, A.E. & Chen, J. 2007. Use of a commercial household steam cleaning system to decontaminate beef and hog carcasses processed by four small or very small meat processing plants in Georgia. *Journal of Food Protection*, 70(3): 635–640.

Tu, L. & Mustapha, A. 2002. Reduction of *Brochothrix thermosphacta* and *Salmonella* serotype Typhimurium on vacuum-packaged fresh beef treated with nisin and nisin combined with EDTA. *Journal of Food Science*, 67(1): 302–306.

Turgis, M., Han, J., Borsa, J. & Lacroix, M. 2008. Combined effect of natural essential oils, modified atmosphere packaging, and gamma radiation on the microbial growth on ground beef. *Journal of Food Protection*, 71(6): 1237–1243.

Uhart, M., Maks, N. & Ravishankar, S. 2006. Effect of spices on growth and survival of *Salmonella* Typhimurium DT 104 in ground beef stored at 4 and 8C. *Journal of Food Safety*, 26(2): 115–125.

USDA [United States Department of Agriculture]. 2015. Update: Nationwide survey on beef and veal carcass baseline. Food Safety and Inspection Service (FSIS) Constituent Update 18–38. Available at: http://www.fsis.usda.gov/wps/portal/fsis/newsroom/meetings/newsletters/constituent-updates/archive/2015/ConstUpdate070215

Wheeler, T.L., Kalchayanand, N. & Bosilevac, J.M. 2014. Pre- and post-harvest interventions to reduce pathogen contamination in the U.S. beef industry. *Meat Science,* 98(3; special issue: 372–382.

Wilhelm, B.J., Rajic, A., Greig, J.D., Waddell, L. & Harris, J. 2011. The effect of hazard analysis critical control point programmes on microbial contamination of carcasses in abattoirs: A systematic review of published data. *Foodborne Pathogens and Disease,* 8(9): 949–960.

Wright, K.D. 2009. Validation of hot water and lactic acid sprays for the reduction of enteric pathogens on the surface of beef carcasses. Masters thesis. Texas A&M University.

Wolf, M.J., Miller, M.F., Parks, A.R., Loneragan, G.H., Garmyn, A.J., Thompson, L.D., Echeverry, A. & Brashears, M.M. 2012. Validation comparing the effectiveness of a lactic acid dip with a lactic acid spray for reducing *Escherichia coli* O157:H7, *Salmonella,* and non-O157 Shiga toxigenic *Escherichia coli* on beef trim and ground beef. *Journal of Food Protection,* 75(11): 1968–1973.

Yoder, S.F., Henning, W.R., Mills, E.W., Doores, S., Ostiguy, N. & Cutter, C.N. 2010. Investigation of water washes suitable for very small meat plants to reduce pathogens on beef surfaces. *Journal of Food Protection,* 73(5): 907–915.

Yoder, S.F., Henning, W.R., Mills, E.W., Doores, S., Ostiguy, N. & Cutter, C.N. 2012. Investigation of chemical rinses suitable for very small meat plants to reduce pathogens on beef surfaces. *Journal of Food Protection,* 75(1): 14–21.

Zhao, T., Zhao, P., Chen, D., Jadeja, R., Hung, Y.C. & Doyle, M.P. 2014. Reductions of Shiga toxin-producing *Escherichia coli* and *Salmonella* Typhimurium on beef trim by lactic acid, levulinic acid, and sodium dodecyl sulfate treatments. *Journal of Food Protection,* 77(4): 528–537.

RAPID SYSTEMATIC REVIEW

Appendix A: Search Strategy Details

This Appendix A gives the full search algorithms used for the search of peer-reviewed literature

SALMONELLA INTERVENTIONS IN BEEF

DATE	11 FEBRUARY 2015
Performed by	Ian Young
Platform/Interface	Ovid
Databases	CAB Abstracts (1910-2015 week 5)
Institution	FAO
Search string:	*(salmonella OR salmonellae) AND (biosecur* OR vaccin* OR immunis* OR immuniz* OR antibiotic* OR antimicrobial* OR antibacterial* OR phage* OR bacteriophage* OR "sodium chlorate" OR probiotic* OR "competitive exclusion" OR lactob* OR bifidobac* OR "lactic acid bacteria" OR dehid* OR dehair* OR skin* OR dress* OR vaccum* OR eviserat* OR wash* OR rins* OR steam OR pasteuriz* OR irradiat* OR spray* OR "hot water" OR chlorine OR chill* OR cool* OR debon* OR boning OR cut* OR fabricat* OR trim* OR grind* OR mitigat* OR disinfect* OR inactiv* OR reduce* OR reducing OR reduction OR intervention* OR decontaminat* OR contamination OR control OR decreas* OR treatment* OR hygiene OR trial OR efficacy OR storage) AND (beef OR veal OR cattle OR cow OR cows OR steer OR steers OR heifer* OR bull OR bulls OR calf OR calves)* in Article title OR in Abstract OR in Subject headings
Hits	2985
Limits	None

DATE	11 FEBRUARY 2015
Performed by	Ian Young
Platform/Interface	Scopus
Databases	Scopus (1823-2015)
Institution	FAO
Search string:	*(salmonella OR salmonellae) AND (biosecur* OR vaccin* OR immunis* OR immuniz* OR antibiotic* OR antimicrobial* OR antibacterial* OR phage* OR bacteriophage* OR "sodium chlorate" OR probiotic* OR "competitive exclusion" OR lactob* OR bifidobac* OR "lactic acid bacteria" OR dehid* OR dehair* OR skin* OR dress* OR vaccum* OR eviserat* OR wash* OR rins* OR steam OR pasteuriz* OR irradiat* OR spray* OR "hot water" OR chlorine OR chill* OR cool* OR debon* OR boning OR cut* OR fabricat* OR trim* OR grind* OR mitigat* OR disinfect* OR inactiv* OR reduce* OR reducing OR reduction OR intervention* OR decontaminat* OR contamination OR control OR decreas* OR treatment* OR hygiene OR trial OR efficacy OR storage) AND (beef OR veal OR cattle OR cow OR cows OR steer OR steers OR heifer* OR bull OR bulls OR calf OR calves)* in Title OR in Abstract OR Key words
Hits	3478
Limits	None

SALMONELLA INTERVENTIONS IN PORK

DATE	11 FEBRUARY 2015
Performed by	Ian Young
Platform/Interface	Ovid
Databases	CAB Abstracts (1910-2015 week 5)
Institution	FAO
Search string:	*(salmonella OR salmonellae) AND (biosecur* OR feed OR vaccin* OR immunis* OR immuniz* OR antibiotic* OR antimicrobial* OR antibacterial* OR phage* OR bacteriophage* OR "sodium chlorate" OR probiotic* OR "competitive exclusion" OR lactob* OR bifidobac* OR "lactic acid bacteria" OR eviserat* OR wash* OR rins* OR steam OR pasteuriz* OR irradiat* OR spray* OR "hot water" OR chlorine OR scald* OR chill* OR cool* OR mitigat* OR disinfect* OR inactiv* OR reduce* OR reducing OR reduction OR intervention* OR decontaminat* OR contamination OR control OR decreas* OR treatment* OR hygiene OR trial OR efficacy OR storage) AND (swine OR pig or pigs OR piglet OR piglets OR gilt OR gilts OR sow OR sows OR hog OR hogs OR weaner OR weaners OR feeder OR feeders OR finisher OR finishers OR market-weight OR porcine OR pork)* in Article title OR in Abstract OR in Subject headings
Hits	1100
Limits	Published since 2009 (update of previous systematic review)

Date	11 February 2015
Performed by	Ian Young
Platform/Interface	Scopus
Databases	Scopus (1823-2015)
Institution	FAO

Search string:	*(salmonella OR salmonellae) AND (biosecur* OR feed OR vaccin* OR immunis* OR immuniz* OR antibiotic* OR antimicrobial* OR antibacterial* OR phage* OR bacteriophage* OR "sodium chlorate" OR probiotic* OR "competitive exclusion" OR lactob* OR bifidobac* OR "lactic acid bacteria" OR eviserat* OR wash* OR rins* OR steam OR pasteuriz* OR irradiat* OR spray* OR "hot water" OR chlorine OR scald* OR chill* OR cool* OR mitigat* OR disinfect* OR inactiv* OR reduce* OR reducing OR reduction OR intervention* OR decontaminat* OR contamination OR control OR decreas* OR treatment* OR hygiene OR trial OR efficacy OR storage) AND (swine OR pig or pigs OR piglet OR piglets OR gilt OR gilts OR sow OR sows OR hog OR hogs OR weaner OR weaners OR feeder OR feeders OR finisher OR finishers OR market-weight OR porcine OR pork)* in Title OR in Abstract OR Key words
Hits	1097
Limits	Published since 2009 (update of previous systematic review)

CONFERENCE PROCEEDINGS SEARCHED FOR ADDITIONAL RELEVANT CITATIONS

TOPIC AREA	CONFERENCE	RELEVANT DATES
Pork on-farm	Safe Pork (International Conference on the Epidemiology and Control of Biological, Chemical and Physical Hazards in Pigs and Pork)	All
	Proceedings of the International Society for Veterinary Epidemiology and Economics	Since 2000
	Centre de developpement du porc du Quebec Canadian Swine Health Board	Current
Beef on-farm	Proceedings of the American Association of Bovine Practitioners/Bovine Practitioner	Since 2000
	Proceedings of International Conference on Animal Health Surveillance	Since 2000
Pork farm to abattoir	Safe Pork (International Conference on the Epidemiology and Control of Biological, Chemical and Physical Hazards in Pigs and Pork)	All
	Reciprocal Meat Science Conference	Since 2000
	Centre de developpement du porc du Quebec	Current
Beef farm to abattoir	Reciprocal Meat Science Conference	Since 2000
Pork processing	Safe Pork (International Conference on the Epidemiology and Control of Biological, Chemical and Physical Hazards in Pigs and Pork)	All
	Reciprocal Meat Science Conference ICoMST (International Congress of Meat Science and Technology)	Since 2000
Beef processing	Reciprocal Meat Science Conference ICoMST (International Congress of Meat Science and Technology)	Since 2000

DETAILS OF INTERNET SEARCHES FOR ADDITIONAL RELEVANT CITATIONS

Google search strings:

Pork

i. On-farm: run consecutively (i.e. one major intervention at a time)

Population term (swine OR pig or pigs OR piglet OR piglets OR gilt OR gilts OR sow OR sows OR hog OR hogs OR weaner OR weaners OR feeder OR feeders OR finisher OR finishers OR market-weight OR porcine) AND bacterial term (salmonella OR salmonellae) AND Intervention term:

(antibiotic* OR antimicrobial* OR antibacterial* or decontaminat* OR hygiene) OR
(acid*) OR
(biosecur* OR disinfect*) OR
(phage* OR bacteriophage) OR
("sodium chlorate") OR
(probiotic* OR "competitive exclusion" OR lactob* OR bifidobac* OR "lactic acid bacteria") OR
(vaccin* OR immunis* OR immuniz)

ii. Post-farm: run consecutively

Population term (finishers OR market-weight OR porcine OR pork) AND bacterial term ((salmonella OR salmonellae) AND intervention term

(eviserat*) OR
(wash* OR rins*) OR
(steam OR pasteuriz*) OR
(irradiat*)
(spray* OR scald* OR "hot water") OR
(chlorine*) OR
(chill* OR cool*)

Beef

i. On-farm: run consecutively (i.e. one major intervention at a time)

Population term (beef OR veal OR cattle OR cow OR cows OR steer OR steers OR heifer* OR bull OR bulls OR calf OR calves) AND bacterial term (salmonella OR salmonellae) AND Intervention term

(antibiotic* OR antimicrobial* OR antibacterial* or decontaminat* OR hygiene) OR
(acid*) OR

(biosecur* OR disinfect*) OR
(phage* OR bacteriophage) OR
("sodium chlorate") OR
(probiotic* OR "competitive exclusion" OR lactob* OR bifidobac* OR "lactic acid bacteria") OR
(vaccin* OR immunis* OR immuniz)

ii. Post-farm: run consecutively

Population term (beef OR veal OR cattle OR cow OR cows OR steer OR steers OR heifer* OR bull OR bulls) AND bacterial term (salmonella OR salmonellae) AND intervention term

(eviserat*) OR
(dehid* OR dehair* OR skin*)
(wash* OR rins*) OR
(steam OR pasteuriz*) OR
(irradiat*)
(spray* OR scald* OR "hot water") OR
(chlorine*) OR
(chill* OR cool*)

LIST OF SEARCH VERIFICATION ARTICLES WHOSE REFERENCE LISTS WERE HAND-SEARCHED

Primary research: pigs on-farm

Ahmed, S.T., Hoon, J., Mun, H.S. & Yang, C.J. 2014. Evaluation of *Lactobacillus* and Bacillus-based probiotics as alternatives to antibiotics in enteric microbial challenged weaned piglets. *African Journal of Microbiology Research,* 8(1): 96–104.

Bode, K., Baier, S. & Blaha, T. 2007. Specific cleaning and disinfection procedures for *Salmonella* infected pig herds. pp. 500–506, *in: Animal health, animal welfare and biosecurity.* Proceedings of 13th International Congress in Animal Hygiene. Tartu, Estonia.

Correia-Gomes, C., Economou, T., Mendonca, D., Vieira-Pinto, M. & Niza-Ribeiro, J. 2012. Assessing risk profiles for *Salmonella* serotypes in breeding pig operations in Portugal using a Bayesian hierarchical model. *BMC Veterinary Research,* 8: Article Number 226

De Ridder, L., Maes, D., Dewulf, J., Butaye, P., Pasmans, F., Boyen, F., Haesebrouck, F. & Van der Stede, Y. 2014. Use of a live attenuated *Salmonella enterica* serovar Typhimurium vaccine on farrow-to-finish pig farms. *Veterinary Journal,* 202: 303–308.

Goldbach, S.G. & Alban, L. 2006. A cost-benefit analysis of *Salmonella*-control strategies in Danish pork production. *Preventive Veterinary Medicine,* 77(1-2): 1–14.

Hotes, S., Kemper, N., Traulsen, I., Rave, G. & Krieter, J. 2010. Risk factors for Salmonella infection in fattening pigs – an evaluation of blood and meat juice samples. *Zoonoses and Public Health,* 57(1): 30–38.

Martin-Pelaez, S., Costabile, A., Hoyles, L., Rastall, R.A., Gibson, G.R., Ragione, R.M., Woodward, M.J., Mateu, E. & Martin-Orue, S. 2010. Evaluation of the inclusion of a mixture of organic acids or lactulose into the feed of pigs experimentally challenged with *Salmonella* Typhimurium. *Veterinary Microbiology,* 142(3-4): 337–345.

Smith, R.P., Clough, H.E. & Cook, A.J.C. 2010. Analysis of meat juice ELISA results and questionnaire data to investigate farm-level risk factors for Salmonella infection in UK pigs. *Zoonoses and Public Health,* 57(Suppl. 1): 39–48.

Upadrasta, A., O'Sullivan, L., O'Sullivan, O., Sexton, N., Lawlor, P.G., Hill, C., Fitzgerald, G.F., Stanton, C. & Ross., R.P. 2013. The effect of dietary supplementation with spent cider yeast on the swine distal gut microbiome. *PLoS ONE,* 8: Article e75714

Lo Fo Wong, D.M.A., Dahl, J., Stege, H., Van Der Wolf, P.J., Leontides, L., Von Altrock, A. & Thorberg, B.M. 2004. Herd-level risk factors for subclinical *Salmonella* infection in European finishing-pig herds. *Preventive Veterinary Medicine,* 62(4): 253–266.

Primary research: pork post-farm

Bae, Y.Y., Choi, Y.M., Kim, M.J., Kim, K.H., Kim, B.C. & Rhee, M.S. 2011a. Application of supercritical carbon dioxide for microorganism reductions in fresh pork. *Journal of Food Safety,* 31(4): 511–517.

Carpenter, C.E., Smith, J.V. & Broadbent, J.R., 2011. Efficacy of washing meat surfaces with 2% levulinic, acetic, or lactic acid for pathogen decontamination and residual growth inhibition. *Meat Science,* 88(2): 256–260.

De Busser, E.V., Maes, D., Houf, K., Dewulf, J., Imberechts, H., Bertrand, S. & De Zutter, L. 2011. Detection and characterization of Salmonella in lairage, on pig carcasses and intestines in five slaughterhouses. *International Journal of Food Microbiology,* 145(1): 279–286.

Fabrizio, K.A. & Cutter, C.N., 2004. Comparison of electrolyzed oxidizing water with other antimicrobial interventions to reduce pathogens on fresh pork. *Meat Science,* 68(3): 463–468.

Gonzales-Barron, U., Cadavez, V., Sheridan, J.J. & Butler, F. 2013. Modelling the effect of chilling on the occurrence of Salmonella on pig carcasses at study, abattoir and batch levels by meta-analysis. *International Journal of Food Microbiology,* 163(2): 101–113.

Keenliside, J., Gensler, G., McFall, M. & Goonewardene, L. 2005. Prevalence and relatedness of *Salmonella* spp. in a Canadian abattoir. p. 38, *in:* Sixth International Symposium on the Epidemiology and Control of Foodborne Pathogens in Pork. Rohnert Park, CA, USA.

Larsen, S.T., McKean, J.D., Hurd, H.S., Rostagno, M.H., Griffith, R.W. & Wesley, I.V. 2003. Impact of commercial preharvest transportation and holding on the prevalence of *Salmonella enterica* in cull sows. *Journal of Food Protection,* 66(7): 1134–1138.

Mannion, C., Fanning, J., McLernon, J., Lendrum, L., Gutierrez, M., Duggan, S. & Egan, J. 2012. The role of transport, lairage and slaughter processes in the dissemination of *Salmonella* spp. in pigs in Ireland. *Food Research International,* 45(2; Special Issue): 871–879.

O'Connor, A.M., Wang, B., Denagamage, T. & McKean, J. 2012. Process mapping the prevalence of *Salmonella* contamination on pork carcass from slaughter to chilling: A systematic review approach. *Foodborne Pathogens and Disease,* 9(5): 386–395.

Trivedi, S., Reynolds, A.E. & Chen, J. 2007. Use of a commercial household steam cleaning system to decontaminate beef and hog carcasses processed by four small or very small meat processing plants in Georgia. *Journal of Food Protection,* 70(3): 635–640.

Wilhelm, B.J., Rajic, A., Greig, J.D., Waddell, L. & Harris, J. 2011. The effect of hazard analysis critical control point programmes on microbial contamination of carcasses in abattoirs: A systematic review of published data. *Foodborne Pathogens and Disease,* 8(9): 949–960.

Review articles: pork chain

Arguello, H., Alvarez-Ordonez, A., Carvajal, A., Rubio, P. & Prieto, M. 2013. Role of slaughtering in *Salmonella* spreading and control in pork production. *Journal of Food Protection*, 76(5): 899–911.

Baer, A.A., Miller, M.J. & Dilger, A.C. 2013. Pathogens of interest to the pork industry: A review of research on interventions to assure food safety. *Comprehensive Reviews in Food Science and Food Safety*, 12(2): 183–217.

Berge, A.C. & Wierup, M. 2012. Nutritional strategies to combat *Salmonella* in monogastric food animal production. *Animal*, 6(4): 557–564.

Binter, C., Straver, J.M., Haggblom, P.,Bruggeman, G., Lindqvist, P.A., Zentek, J. & Andersson, M.G. 2011. Transmission and control of *Salmonella* in the pig feed chain: A conceptual model. *International Journal of Food Microbiology*, 145(Suppl. 1): S7–S17.

Buncic, S. & Sofos, J. 2012. Interventions to control *Salmonella* contamination during poultry, cattle and pig slaughter. *Food Research International*, 45(2; Special Issue): 641–655.

Dahl, J. 2013. Controlling *Salmonella* in live pigs and at slaughter: The Danish experience. Vet. J. 197(3): 529–530.

Rodriguez, D.M. & Suarez, M.C. 2014. *Salmonella* spp. in the pork supply chain: A risk approach. *Revista Colombiana de Ciencias Pecuarias*, 27(2): 65–75.

Rostagno, M.H. & Callaway, T.R. 2012. Pre-harvest risk factors for *Salmonella enterica* in pork production. *Food Research International*, 45(2; Special Issue): 634–640.

Wales, A.D., Cook, A.J.C. & Davies, R.H. 2011. Producing *Salmonella*-free pigs: a review focusing on interventions at weaning. *Veterinary Record*, 168(10): 267–276.

Primary research: beef on-farm

Beach, J.C., Murano, E.A. & Acuff, G.R. 2002. Prevalence of *Salmonella* and *Campylobacter* in beef cattle from transport to slaughter. *Journal of Food Protection*, 65(11): 1687–1693.

Davison, H.C., Sayers, A.R., Smith, R.P., Pascoe, S.J.S., Davies, R.H., Weaver, J.P. & Evans, S.J. 2006. Risk factors associated with the *Salmonella* status of dairy farms in England and Wales. *Veterinary Record*, 159(26): 871-880.

Edrington, T.S., Looper, M.L., Duke, S.E., Callaway, T.R., Genovese, K.J., Anderson, R.C. & Nisbet, D.J. 2006. Effect of ionophore supplementation on the incidence of *Escherichia coli* O157:H7 and *Salmonella* and antimicrobial susceptibility of faecal coliforms in stocker cattle. *Foodborne Pathogens and Disease*, 3(3): 284–291.

Fecteau, M.E., House, J.K., Kotarski, S.F., Tankersley, N.S., Ontiveros, M.M., Alcantar, C.R. & Smith, B.P. 2003. Efficacy of ceftiofur for treatment of experimental salmonellosis in neonatal calves. *American Journal of Veterinary Research*, 64(7): 918–925.

Green, A.L., Dargatz, D.A., Wagner, B.A., Fedorka-Cray, P., Ladely, S.R. & Kopral, C.A. 2010. Analysis of risk factors associated with *Salmonella* spp. isolated from U.S. feedlot cattle. *Foodborne Pathogens and Disease*, 7(7): 825–833.

Kabagambe, E.K., Wells, S.J., Garber, L.P., Salman, M.D., Wagner, B. & Fedorka-Cray, P. 2000. Risk factors for faecal shedding of *Salmonella* in 91 US dairy herds in 1996. *Preventive Veterinary Medicine*, 43(3): 177–194.

Mizuno, T., McLennan, M. & Trott, D. 2008. Intramuscular vaccination of young calves with a Salmonella Dublin metabolic-drift mutant provides superior protection to oral delivery. *Veterinary Research*, 39(3): 26–42.

Nielsen, T.D., Vesterbaek, I.L., Kudahl, A.B., Borup, K.J. & Nielsen, L.R., 2012. Effect of management on prevention of Salmonella Dublin exposure of calves during a one-year control programme in 84 Danish dairy herds. *Preventive Veterinary Medicine,* 105(1-2): 101–109.

Smith, G.W., Alley, M.L., Foster, D.M., Smith, F. & Wileman, B.W. 2014. Passive immunity stimulated by vaccination of dry cows with a *Salmonella* bacterial extract. *Journal of Veterinary Internal Medicine*], 28(5): 1602–1605

Vanselow, B.A., Hornitzky, M.A., Walker, K.H., Eamens, G.J., Bailey, G.D., Gill, P.A., Coates, K., Corney, B., Cronin, J.P. & Renilson, S. 2007. *Salmonella* and on-farm risk factors in healthy slaughter-age cattle and sheep in eastern Australia. *Australian Veterinary Journal,* 85(12): 498–502.

Primary research: beef post-farm

Arthur, T.M., Bosilevac, J.M., Brichta-Harhay, D., Kalchayanand, N., King, D.A., Shackelford, S.D., Wheeler, T.L. & Koohmaraie, M. 2008. Source tracking of *Escherichia coli* O157:H7 and *Salmonella* contamination in the lairage environment at commercial U.S. beef processing plants and identification of an effective intervention. *Journal of Food Protection,* 71(9): 1752–1760.

Barkocy-Gallagher, G., Arthur, T.M., Rivera-Betancourt, M., Nou, X., Shackelford, S.D., Wheeler, T.L. & Koohmaraie, M. 2003. Seasonal prevalence of Shiga toxin-producing *Escherichia coli,* including O157:H7 and non-O157 serotypes, and *Salmonella* in commercial beef processing plants. *Journal of Food Protection,* 66(11): 1978–1986.

Castillo, A., Lucia, L.M., Goodson, K.J., Savell, J.W. & Acuff, G.R. 1998b. Comparison of water wash, trimming, and combined hot water and lactic acid treatments for reducing bacteria of faecal origin on beef carcasses. *Journal of Food Protection,* 61(7): 823–828.

Echeverry, A., Brooks, J.C., Miller, M.F., Collins, J.A., Loneragan, G.H. & Brashears, M.M. 2010. Validation of lactic acid bacteria, lactic acid, and acidified sodium chlorite as decontaminating interventions to control *Escherichia coli* O157:H7 and *Salmonella* Typhimurium DT 104 in mechanically tenderized and brine-enhanced (non-intact) beef at the purveyor. *Journal of Food Protection,* 73(12): 2169–2179.

Gill, C.O. & DeLacy, K.M. 1991. Growth of Escherichia coli and Salmonella Typhimurium on high-pH beef packed under vacuum or carbon dioxide. *International Journal of Food Microbiology,* 13(1): 21–30.

Hoyle, A.R., Brooks, J.C., Thompson, L.D., Palmore, W., Stephens, T.P., Brashears, M.M., 2009. Spoilage and safety characteristics of ground beef treated with lactic acid bacteria. *Journal of Food Protection,* 72(11): 2278–2283.

Kundu, D., Gill, A., ChenYuan, L., Goswami, N. & Holley, R. 2014. Use of low dose e-beam irradiation to reduce E. coli O157:H7, non-O157 (VTEC) E. coli and Salmonella viability on meat surfaces. *Meat Science,* 96(1): 413–418.

Nutsch, A.L., Phebus, R.K., Riemann, M.J., Schafer, D.E., Boyer Jr., J.E., Wilson, R.C., Leising, J.D. & Kastner, C.L. 1997. Evaluation of a steam pasteurization process in a commercial beef processing facility. *Journal of Food Protection,* 60(5): 485–492.

Scanga, J.A., Buschow, A.W., Kauk, J.L., Burk, T.E., Koohmaraie, B., Zerda, D.L., Motlagh, A.M., Samadpour, M. & Koohmaraie, M. 2011. Localized chemical decontamination of cattle hides to reduce microbial loads and prevalence of foodborne pathogens. *Food Protection Trends,* 31(9): 569–574.

Stopforth, J.D., Lopes, M., Shultz, J.E., Miksch, R.R., Samadpour, M., 2006. Location of bung bagging during beef slaughter influences the potential for spreading pathogen contamination on beef carcasses. *Journal of Food Protection,* 69(6): 1452–1455.

Review articles: beef chain

Andersen, H.J., Oksbjerg, N. & Therkildsen, M. 2005. Potential quality control tools in the production of fresh pork, beef and lamb demanded by the European society. *Livestock Production Science,* 94(1-2): 105–124.

Buncic, S. & Sofos, J. 2012. Interventions to control *Salmonella* contamination during poultry, cattle and pig slaughter. *Food Research International,* 45(2): 641–655.

Dorsa, W.J. 1997. New and established carcass decontamination procedures commonly used in the beef-processing industry. *Journal of Food Protection,* 60(9): 1146–1151.

Ellis-Iversen, J. & Watson, E. 2008. A 7-point plan for control of VTEC O157, *Campylobacter jejuni/coli* and *Salmonella* serovars in young cattle. *Cattle Practice,* 16: 103–106 Part 2..

Koohmaraie, M., Arthur, T.M., Bosilevac, J.M., Guerini, M., Shackelford, S.D. & Wheeler, T.L. 2005. Post-harvest interventions to reduce/eliminate pathogens in beef. *Meat Science,* 71(1): 79–91.

McAllister, A., Beauchemin, K, Alazzeh, A., J. Baah, J., Teather, R.M. & Stanford, K. 2011. The use of direct fed microbials to mitigate pathogens and enhance production in cattle. *Canadian Journal of Animal Science,* 91(2): 193–211.

Wheeler, L., Kalchayanand, N. & Bosilevac, J.M. 2014. Pre- and post-harvest interventions to reduce pathogen contamination in the U.S. beef industry. *Meat Science,* 98(3; special issue): 372–382.

RAPID SYSTEMATIC REVIEW

Appendix B: Relevance Screening Form

QUESTION	OPTIONS	KEY DEFINITIONS
1. Does this citation describe research evaluating the efficacy and/ or effectiveness (including costs or practically of implementation) of interventions to control *Salmonella* in pork or beef at any stage from the primary production to consumption? **Selections 1-3 will pass the citation to the next review stage and the article will be procured.**	1. Yes, primary research 2. Yes, systematic review/ meta-analysis 3. Yes, risk assessment, risk profile, or other risk-based tool (e.g. cost-benefit analysis) 4. No (exclude)	**Primary research** is collection of new data in a single study. **Risk assessment** is a scientifically based process consisting of the following steps (i) hazard identification, (ii) hazard characterization, (iii) exposure assessment, and (iv) risk characterization. A **risk profile** presents the current state of knowledge related to a food safety issue, describes potential options that have been identified to date (if any), and the food safety policy context that will influence further possible actions. Other risk-based tools could include cost-benefit analyses, risk ranking, or risk prioritizations. **Systematic review** is a structured review of a clearly defined question with a transparent search strategy, relevance screening process, data extraction, risk-of-bias assessment and synthesis of results. **Meta-analysis** is a statistical technique that can be used on data collected in a systematic review. **Exclude** research on feral animals (e.g. feral pigs not produced for human consumption), and *in vitro* lab experiments.
2. If no to the above, is the article a potentially relevant narrative literature review on the subject? *(To be used for possible search verification)*	☐ Yes (check box if relevant)	**Narrative reviews** are an expert-based review not using or reporting a structured or systematic approach (often will not have a methods section).

RAPID SYSTEMATIC REVIEW

Appendix C: Relevance Confirmation Form

<table>
<tr><th>QUESTION</th><th>OPTIONS</th><th>KEY DEFINITIONS</th></tr>
<tr><td colspan="3">Relevance confirmation</td></tr>
<tr><td>Does this article investigate the efficacy and/or effectiveness of interventions to control Salmonella in beef and pork at any stage from the primary production stage to consumption and meet the PICOS eligibility criteria?</td><td>☐ Yes, proceed
☐ No, exclude
• Only investigates tertiary processed meat
• Measures irrelevant outcomes (e.g. milk samples only)
• In vitro study
• Review article
• Language other than English, French or Spanish
• No extractable data
• Other, specify:________</td><td><u>Tertiary processed meat</u> = e.g. cured, fermented or dried sausages, salamis, etc.</td></tr>
<tr><td colspan="3">Key article characteristics</td></tr>
<tr><td>What type of document is this article?</td><td>☐ Journal article
☐ Conference proceedings
☐ Government or research report
☐ Thesis
☐ Book or book chapter
☐ Other, please specify:_____</td><td></td></tr>
<tr><td>In what region and country was the study conducted?</td><td>☐ North America:__________
☐ Europe:_____________
☐ Australasia:_____
☐ Central and South America/ Caribbean:________
☐ Asia:__________________
☐ Africa:_________________
☐ Not stated</td><td>Specify country name only (not sub-regions, states, provinces, etc.)
North America: Canada, United States of America and Mexico
South America/ Caribbean: Caribbean, and all of south America.
Europe: includes, Belarus, Latvia, Ukraine, Estonia, Cyprus & west (inc. Iceland and Greenland)
Asia: Russia, Turkey, middle eastern countries and east
Australasia is limited to Australia, New Guinea, New Zealand, New Caledonia, and neighbouring islands, including the Indonesian islands</td></tr>
</table>

Specify study design	☐ Experimental research: • Randomized controlled trial • Non-randomized controlled trial • Challenge trial • Quasi-experiment ☐ Observational research • Cohort study • Case-control study • Cross-sectional study • Other ☐ Systematic review-meta-analysis ☐ Risk assessment, risk profile, cost-benefit analysis, or other risk-based tool	**Observational study:** Assignment of subjects into a treated group versus a control group is outside the control of the investigator. **Cross-sectional:** Examines the relationship of a risk factor and outcome (disease) at a point in time on representative samples of the target population. **Cohort study:** subjects with differing exposures to a suspected risk factor are observed through time for occurrence of an outcome. **Case-control study:** compares exposure to the risk factor in subjects who have an outcome (the 'cases') with subjects who do not have the outcome, but are otherwise similar (the 'controls') and drawn from the same sampling frame. **Other:** Hybrid or other observational designs but not experiments evaluating the effectiveness of interventions. **Experimental study:** Each subject is assigned to a treated group or a control group before the start of the treatment **Controlled trial:** subjects are allocated to intervention/ comparison groups and evaluated for outcomes. **Challenge trial:** Subjects are artificially challenged or exposed to the disease agent and then allocated to the intervention groups for evaluation of the outcome. **Quasi-experiment:** Observations are made on a population before and after receiving an intervention.
In what setting was the study carried out?	☐ Commercial/field conditions ☐ Research farm/pilot plant ☐ Smallholder farm/abattoir conditions ☐ Laboratory conditions ☐ Not reported	

<table>
<tr><td>What commodity is investigated?</td><td>☐ Swine/pork production
☐ Cattle/beef production</td><td></td></tr>
<tr><td>What point in the food chain and category of intervention(s) are investigated in this article?</td><td>☐ Farm:
• Biosecurity/management practices
• Vaccination
• Antimicrobials
• Competitive exclusion/ probiotics
• Feed/water acidification
• Feed characteristics/ management
• Other (e.g. bacteriophages)
☐ Transport to slaughter
☐ Processing:
• Segregated/logistic slaughter
• Cleaning/disinfection of equipment/environments
• Carcass/product washes, rinses, sprays
• Standard processing procedures/good hygienic practices (GHP)
• Irradiation
• Modified packaging
• Other
☐ Post-processing to consumer</td><td>Antimicrobials: Examples include: Fluroquinolones, cephalosporins, gentamicin, ampicillin, tetracyclines, spectinomycin, ciprofloxacin, ceftriaxone. These may be administered via feed.
Biosecurity: includes, but is not limited to, sanitation, biosafety, disinfection, hygiene and hygiene barriers, all-in-all-out production, depopulation, staff and the environment, litter testing and treatment, pest control, etc.
Competitive exclusion: May also be referred to as probiotics, prebiotics, synbiotics. May include Lactobacillus spp., bacteroides, Bifidobacterium spp., Enterococcus faecium, Aspergillus oryzae, and Saccharomyces spp. (S. cerevisiae, S. boulardii). May be caecal contents or other materials from animals or the environment that contain many different or unknown bacterial species.
Feed/water acidification: Addition of organic acids such as lactic acid, to feed or water. Would include 'nutraceuticals' such as copper, chromium, zinc, betaine or carnitine.
Feed management: E.g. comparisons of coarse/finely ground feed, fermented feed, or liquid feed.
Segregated/logistic slaughter = slaughtering/ processing of more highly contaminated lots after less contaminated lots.
Standard processing procedures/ good hygienic practices (GHP) refers to steps such as singeing, de-hiding, cooling, chilling, etc.</td></tr>
</table>

Did the study investigate outcomes other than *Salmonella*?	☐ Yes, *E. coli* (generic and/or pathogenic strains) ☐ Yes, other bacteria ☐ No	
What intervention efficacy/ effectiveness measures are investigated?	☐ Efficacy ☐ Cost ☐ Practicality ☐ Consumer acceptability ☐ Other contextual factor, specify:______________	
Does the article report any extractable data about intervention efficacy/ effectiveness that could be used for possible meta-analysis?	☐ Yes ☐ No, reason:__________	

RAPID SYSTEMATIC REVIEW

Appendix D: Data Extraction Form

Multiple forms were submitted for each unique trial (i.e. intervention/population/outcome combination) reported in a study

Question	Options	Excel code
Specify intervention category being extracted and specify points in the food chain where intervention is applied	☐ Farm:________ • Biosecurity/management practices • Vaccination • Antimicrobials • Competitive exclusion/ probiotics • Feed/water acidification • Feed characteristics/ management • Other (e.g. bacteriophages) ☐ Transport to slaughter ☐ Processing:_______ • Segregated/logistic slaughter • Cleaning/disinfection of equipment/environments • Carcass/product washes, rinses, sprays • Standard processing procedures/GHPs • Irradiation • Modified packaging/ preservation techniques Other ☐ Post-processing to consumer:_______	Int_point_in_chain Int_point_specific (text boxes) Int_category
If conducted at processing, specify point in chain where the intervention is applied?		Process_level
If conducted at farm level, did the study measure outcomes in market-weight animals or animals at slaughter?	☐ Yes (proceed with DE) ☐ No (exclude/do not proceed)	Market_weight

Intervention description (including dose, application method, etc.)	☐ ____________	Int_description
At what level was the intervention applied?	☐ Individual ☐ Group (e.g. pen, farm):____________	Int_level
If this was a challenge trial, describe the challenge details (e.g. dose, serovars, application method)	☐ ____________	ChT_description
Specify target population/sample to which intervention is applied (indicate animal age, other contextual details in text box)	☐ Swine/pork: • Breeders • Suckling swine • Nursery swine • Growers • Finishers • Swine at slaughter • Carcass • Final product ☐ Cattle beef: • Cow-calf production • Feedlot cattle • Dairy cattle • Cattle at slaughter • Carcass • Final product ☐ Other:________	Pop_category Pop_specific Pop_details
What type of sample was measured for *Salmonella*?	☐ Blood/serum ☐ Faeces ☐ Carcass muscle ☐ Organs ☐ Swab ☐ Carcass wash ☐ Other	Sample_type
Which diagnostic method(s) were used to determine the prevalence or level of the *Salmonella*? (copy and paste key test details, e.g. media, incubation time/temp, etc.)	☐ Culture ☐ ELISA ☐ PCR ☐ Other:____________	Diag_method Diag_details

What *Salmonella* serovar(s) are measured?	☐ _____________	Salm_serovar
Does this trial include indirect measurement of the intervention, population, comparison, and/or outcome of interest?	☐ Yes ☐ No **A study may indirectly address the question of interest if:** • e.g. interventions we wish to compare are measured independently in two separate trials compared with controls. • e.g. the population, intervention, comparisons or outcomes were not exactly what we are trying to draw conclusions for. (e.g., surrogate outcomes = e.g. might measure seroprevalence when outcome of interest is shedding).	Indirect_ measurement
Is there reason to believe that due to the population studied, the magnitude of effect of the intervention is likely to be underestimated?	☐ Yes ☐ No ***ONLY answer yes if there is good reason to think that the study underestimated the potential association or effect of an intervention due to the population that was sampled. (I.e. all plausible biases working to underestimate apparent intervention effect).*** *E.g. intervention was tested only on animals with a very low prevalence of Salmonella and it is likely that a stronger effect would be shown in most populations in the field.*	Magnitude_ underestimated
Was a dose-response gradient detected for the intervention effect?	☐ Yes ☐ No **If a dose response gradient is demonstrated in some or all of the studies, this increases our confidence in the findings of the study and thus we can consider upgrading the evidence.**	Dose_response
At what level are the data reported?	☐ Individual ☐ Group (e.g. pen, farm):_____________	Data_level
What outcome data were measured?	☐ Prevalence (presence/absence) ☐ Concentration (e.g. CFU, MPN) ☐ Ordinal data	Outcome_type

Extract quantitative outcome data in text boxes for each relevant category	**Dichotomous/ordinal data options**	
	☐ Raw 2×2 data	
	☐ Number positive group 1	Num_pos_group1
	☐ Number negative group 1	Num_neg_group1
	☐ Number positive group 2	Num_pos_group2
	☐ Number negative group 2	Num_neg_group2
	☐ Define group 1	Define_group1
	☐ Define group 2	Define_group2
	☐ For ordinal data, specify above for additional response categories as appropriate:	Ordinal_data
	☐ Computed effect size / measure of association (e.g. OR):	Measure_association Measure_specify
	☐ Measure of association value	
	☐ Specify measure (e.g. OR, RR, etc.)	
	☐ N in group 1	N_group1
	☐ N in group 2	N_group2
	☐ Define group 1	Define_group1
	☐ Define group 2	Define_group2
	☐ SE	SE
	☐ Variance	Variance
	☐ Lower CI	CI_low
	☐ Higher CI	CI_high
	☐ Was the outcome adjusted for other variables? If yes, check this box and specify:______	Adjusted
	Continuous data options	
	☐ Raw continuous data in each group (final outcome measure):	
	☐ Counts in group 1	Mean_group1
	☐ SD in group 1	SD_group1
	☐ N in group 1	N_group1
	☐ Counts in group 2	Mean_group2
	☐ SD in group 2	SD_group2
	☐ N in group 2	N_group2
	☐ Define group 1	Define_group1
	☐ Define group 2	Define_group2
	☐ P-value (exact Ps only)	P_value
	☐ T-value	T_value
	☐ Outcome units	Units
	☐ Difference in means:	
	☐ Difference in means (value)	Mean_difference
	☐ N group 1	N_group1
	☐ N group 2	N_group2
	☐ Common SD	Common_SD
	☐ SE	SE
	☐ Variance	Variance
	☐ Lower CI	CI_low
	☐ Higher CI	CI_high
	☐ P-value (exact Ps only)	P_value
	☐ T-value	T_value
	☐ Outcome units	
	☐ Was the outcome adjusted for other variables? If yes, check this box and specify:______	Units Adjusted
	☐ Other:________________	Other_outcomes

RAPID SYSTEMATIC REVIEW

Appendix E: Risk-of-Bias Tool for Primary Research Studies

Bias domain/ question	Risk of bias	Definitions/additional NOTES
Random sequence generation *Was the allocation sequence adequately generated?*	• Low • High • Unclear • N/A (quasi-experiments/ observational studies)	Describe the method used to generate the allocation sequence (for allocating subjects/samples into treatment groups) in sufficient detail to allow an assessment of whether it should produce comparable groups. **Low:** a random component in the sequence generation process is described (e.g. referring to a random number table or computer random number generator) OR lack of randomization unlikely to bias the results (e.g. lab-based study conducted on comparable meat samples). **High:** a non-random method is used (e.g. performed by date of birth, by preference, or convenience). **Unclear:** insufficient information provided to permit judgement (includes reported randomization with no description of randomization process).
Allocation concealment *Was the allocation sequence adequately concealed from the participants and the researcher?*	• Low • High • Unclear • N/A (quasi-experiments/ observational studies)	Describe the method used to conceal the allocation sequence in sufficient detail to determine whether intervention allocations could have been foreseen in advance of, or during, enrolment. **Low:** allocation was unlikely to influence enrolment (e.g. each farm received both treatment and untreated groups); or if the unit of allocation was by participant (e.g. individual farm received only one group), there was some form of centralised randomisation scheme such as an on-site computer system or sealed opaque envelopes were used. **High:** a non-random method is used (e.g. performed by date of birth, by preference, or convenience). **Unclear:** insufficient information provided to permit judgement.
Blinding of participants and personnel *Were participants and personnel blinded from knowledge of intervention/ control group status?*	• Low • High • Unclear • N/A (quasi-experiments/ observational studies)	Describe all measures used, if any, to blind study participants (e.g. swine/cattle farmers) and personnel from knowledge of which intervention a subject/sample received. **Low:** no blinding of outcome assessment, but the review authors judge that the outcome measurement is not likely to be influenced by lack of blinding. Blinding of outcome assessment ensured, and unlikely that the blinding could have been broken. **High:** no blinding of outcome assessment, and the outcome measurement is likely to be influenced by lack of blinding. Blinding reported, but likely that the blinding could have been broken, and the outcome measurement is likely to be influenced by lack of blinding. **Unclear:** insufficient information provided to permit judgement.

Bias domain/ question	Risk of bias	Definitions/additional NOTES
Similarity of study and target populations <u>Cohort studies:</u> *Was the level of exposure representative of exposure in the population of interest?* <u>Cross-sectional:</u> *Were the study subjects/samples selected randomly so the sample reflects disease and exposure in the population of interest?* <u>Case-control:</u> *Were the controls selected from the same source population as the cases? (case control)*	• Low • High • Unclear • N/A (experimental studies/quasi-experiments)	Note whether this is a possible **selection bias**: systematic differences between sample and target population or for case control studies between the groups being compared and an appropriate range of clinical severity. **Low:** sample and target populations are similar. **High:** • Level of exposure not representative of exposure in the population of interest (cohort studies) • Non-random subject/sample collection (cross-sectional) • Controls not selected from the same source population as cases (case-control) **Unclear:** insufficient information provided to permit judgement.
Independence of intervention effect from confounding bias *Are there any concerns that confounders have not been appropriately identified and accounted for?*	• Low • High • Unclear	Describe whether or not the intervention occurred independently of other changes over time and whether or not the outcomes may have been influenced by other confounding variables/historic events during the study period. *<u>Important confounders could include</u>:* animal age and weight; housing/feed characteristics; farm/abattoir characteristics; sampling details (e.g. time of day when sampling is conducted), etc. **Low:** there are compelling arguments that the intervention occurred independently of other changes over time and the outcome was not influenced by other confounding variables and/or historic events during study period. If randomization conducted (e.g. RCT), indicate LOW. **High:** intervention was likely not independent of other changes in time. **Unclear:** insufficient information provided to permit judgement.

Bias domain/ question	Risk of bias	Definitions/additional NOTES
Incomplete outcome data *Were losses to follow-up (attrition) and exclusions from analysis reported and comparable in both groups?*	• Low • High • Unclear • Separate risk for different outcomes, specify:	Describe the completeness of outcome data for each main outcome, including attrition and exclusions from the analysis. State whether attrition and exclusions were reported, the numbers in each intervention group (compared with total subjects/samples), whether reasons for attrition/exclusions where reported, and any re-inclusions in analyses. **Low:** missing outcome measures were unlikely to bias the results (e.g. the proportion of missing data was similar in the intervention and control groups or the proportion of missing data was less than the effect size - i.e. unlikely to overturn the study result). **High:** missing outcome data likely to bias the results. **Unclear:** not specified in the paper (Do not assume 100% follow up unless stated explicitly). ***Assessments should be made for each main outcome (or class of outcomes), as appropriate.*** *If more than one answer please specify which outcomes are associated with each answer.*
Selective reporting *Did the authors report all intended outcomes?*	• Low • High • Unclear	State how the possibility of selective outcome reporting was examined by the review authors, and what was found. **Low:** there is no evidence that outcomes were selectively reported (e.g. all relevant outcomes in the methods section are reported in the results section). **High:** some important outcomes are omitted from the results. **Unclear:** insufficient information provided to permit judgement.
Other *Was the study free of other problems that could put it at a high risk of bias?*	• Low • High • Unclear	State any important concerns about bias not addressed in the other domains in the tool (e.g. study funded by industry with concerns about sponsor involvement, inadequate control of hierarchical data structure/clustering, or important differences in baseline outcome measurements, i.e. *Salmonella* status, were present and not adjusted for in analysis) **Low:** there is no risk of other biases (please specify details). **High:** there is a risk of other biases (please specify details). **Unclear:** possible risk of other biases but insufficient information provided to permit judgement (please specify details).
Overall risk-of-bias for each outcome (within-study summary assessment)	• Low • High • Unclear • Separate risk for different outcomes, specify:	**Low:** plausible bias unlikely to seriously alter the results. Low risk of bias for key domains. **High:** plausible bias that seriously weakens confidence in the results. High risk of bias for key domains. **Unclear:** plausible bias that raises some doubt about the results. Unclear risk of bias for key domains. ***Assessments should be made for each main outcome (or class of outcomes), as appropriate.*** *If more than one answer please specify which outcomes are associated with each answer.*

RAPID SYSTEMATIC REVIEW

Appendix F: AMSTAR Quality Assessment Tool for Systematic Reviews

Question	Options
1. **Was an 'a priori' design provided?** The research question and inclusion criteria should be established before the conduct of the review. *Note: Need to refer to a protocol, ethics approval, or pre-determined/a priori published research objectives to score a "yes."*	☐ Yes ☐ No ☐ Can't answer ☐ Not applicable
2. **Was there duplicate study selection and data extraction?** There should be at least two independent data extractors and a consensus procedure for disagreements should be in place. *Note: 2 people do study selection, 2 people do data extraction, consensus process or one person checks the other's work.*	☐ Yes ☐ No ☐ Can't answer ☐ Not applicable
3. **Was a comprehensive literature search performed?** At least two electronic sources should be searched. The report must include years and databases used (e.g., Central, EMBASE, and MEDLINE). Key words and/or MESH terms must be stated and where feasible the search strategy should be provided. All searches should be supplemented by consulting current contents, reviews, textbooks, specialized registers, or experts in the particular field of study, and by reviewing the references in the studies found. *Note: If at least 2 sources + one supplementary strategy used, select "yes" (a grey literature search counts as supplementary).*	☐ Yes ☐ No ☐ Can't answer ☐ Not applicable
4. **Was the status of publication (i.e. grey literature) used as an inclusion criterion?** The authors should state that they searched for reports regardless of their publication type. The authors should state whether or not they excluded any reports (from the systematic review), based on their publication status, language etc. *Note: If review indicates that there was a search for "grey literature" or "unpublished literature," indicate "yes." SINGLE database, dissertations, conference proceedings, and trial registries are all considered grey for this purpose. If searching a source that contains both grey and non-grey, must specify that they were searching for grey/unpublished lit.*	☐ Yes ☐ No ☐ Can't answer ☐ Not applicable
5. **Was a list of studies (included and excluded) provided?** A list of included and excluded studies should be provided. *Note: Acceptable if the excluded studies are referenced. If there is an electronic link to the list but the link is dead, select "no."*	☐ Yes ☐ No ☐ Can't answer ☐ Not applicable

6. Were the characteristics of the included studies provided? In an aggregated form such as a Table, data from the original studies should be provided on the participants, interventions and outcomes. The ranges of characteristics in all the studies analyzed e.g. age, race, sex, relevant socioeconomic data, disease status, duration, severity, or other diseases should be reported. *Note: Acceptable if not in Table format as long as they are described as above.*	☐ Yes ☐ No ☐ Can't answer ☐ Not applicable
7. Was the scientific quality of the included studies assessed and documented? 'A priori' methods of assessment should be provided (e.g., for effectiveness studies if the author(s) chose to include only randomized, double-blind, placebo controlled studies, or allocation concealment as inclusion criteria); for other types of studies alternative items will be relevant. *Note: Can include use of a quality scoring tool or checklist, e.g., Jadad scale, risk of bias, sensitivity analysis, etc., or a description of quality items, with some kind of result for EACH study ("low" or "high" is fine, as long as it is clear which studies scored "low" and which scored "high"; a summary score/range for all studies is not Acceptable).*	☐ Yes ☐ No ☐ Can't answer ☐ Not applicable
8. Was the scientific quality of the included studies used appropriately in formulating conclusions? The results of the methodological rigor and scientific quality should be considered in the analysis and the conclusions of the review, and explicitly stated in formulating recommendations. *Note: Might say something such as "the results should be interpreted with caution due to poor quality of included studies." Cannot score "yes" for this question if scored "no" for question 7.*	☐ Yes ☐ No ☐ Can't answer ☐ Not applicable
9. Were the methods used to combine the findings of studies appropriate? For the pooled results, a test should be done to ensure the studies were combinable, to assess their homogeneity (i.e., Chi-squared test for homogeneity, I^2). If heterogeneity exists a random effects model should be used and/or the clinical appropriateness of combining should be taken into consideration (i.e., is it sensible to combine?). *Note: Indicate "yes" if they mention or describe heterogeneity, i.e., if they explain that they cannot pool because of heterogeneity/variability between interventions.*	☐ Yes ☐ No ☐ Can't answer ☐ Not applicable
10. Was the likelihood of publication bias assessed? An assessment of publication bias should include a combination of graphical aids (e.g., funnel plot, other available tests) and/or statistical tests (e.g., Egger regression test, Hedges-Olken). *Note: If no test values or funnel plot included, score "no". Score "yes" if mentions that publication bias could not be assessed because there were fewer than 10 included studies.*	☐ Yes ☐ No ☐ Can't answer ☐ Not applicable
11. Was the conflict of interest stated? Potential sources of support should be clearly acknowledged in both the systematic review and the included studies. *Note: To get a "yes," must indicate source of funding or support for the systematic review AND for each of the included studies.*	☐ Yes ☐ No ☐ Can't answer ☐ Not applicable

RAPID SYSTEMATIC REVIEW

Appendix G: Reporting Assessment Tool for Risk Assessments

QUALITY DOMAIN / DESCRIPTION	OPTIONS FOR EACH CRITERION
A. Internal validity	
1. Timeliness/scope i. Is the scope of the study clearly defined (e.g. regarding species/ time/place/sample matrices)? ii. Is scope/level of detail of RA justified by authors in relation to direction from end-users/stakeholders?	1= Justified 0= Not justified 99= Not applicable
2. Quality and treatment of the data i. Are current, relevant and sufficient data presented (e.g. are published relevant studies excluded from dataset?) ii. Are sources of data justified and quality of data assessed? iii. Is expert opinion elicited via structured transparent process, to augment existing data or fill data gaps? iv. Where expert opinion is substituted for existing primary research, do the authors justify this decision? v. Is rationale for inclusion/exclusion of data provided?	1= Yes 0= No 2= Can't answer 99= Not applicable
3. Assumptions, expert opinion, scientific support i. Are assumptions explicitly stated and justified? ii. Are premises or underlying assumptions logical, in terms of theoretical arguments or empirical results?	1= Yes 0= No 2= Can't answer 99= Not applicable
4. Inference of probability i. Are probabilities transparently estimated? ii. Is the choice of distributions transparent? iii. If the model is stochastic, is the number of model iterations reported? iv. In semi-quantitative or qualitative assessments, is there a clear definition of the meaning of descriptors/elements of risk matrices such as negligible, low, medium, high? v. Is the logic used in combining semi-quantitative values sound? vi. Is the description of how conclusions are derived and magnitudes of probability and severity are assigned, clear?	1= Yes 0= No 2= Can't answer 99= Not applicable
5. Internal consistency i. Are the correct units used for parameters throughout the model? ii. Does the RA methodology demonstrate sound logic and inference, or if the reader cannot assess this, at least no circular logic? iii. Does the RA demonstrate systematic reasoning in assigning values for risk ranking and other non-mathematical evaluations? iv. Are RA outcomes consistent with observed data (although not necessarily in agreement should be within plausible limits and account for all observations)? v. Was sensitivity analysis performed and reported? vi. Was the potential effect of the most influential assumption(s) discussed?	1= Yes 0= No 2= Can't answer 99= Not applicable

QUALITY DOMAIN / DESCRIPTION	OPTIONS FOR EACH CRITERION
B. External validity	
6. Target populations i. Is the scope/target population/target scenario(s) for the RA explicitly stated? ii. Are the data sources appropriate for making inferences beyond this group? iii. Are the assumptions appropriate for making inferences beyond this group?	1= Yes 0= No 2= Can't answer 99= Not applicable
7. Uncertainty i. Was uncertainty analysis performed and reported? ii. Was the potential effect of the sources of greatest uncertainty discussed? iii. Is uncertainty investigated at each appropriate level (e.g. model uncertainty/path uncertainty/parameter uncertainty)?	1= Yes 0= No 2= Can't answer 99= Not applicable
C. Reporting, peer review, and communication	
8. Transparency with regards to: i. Do the authors present systematic development/description of the risk assessment steps and structure? ii. Is the use of alternatives to close data gaps (e.g. expert knowledge, surrogate data, assumptions) clearly described? iv. Are all sources of included data referenced? v. Is code for the model(s) provided?	1= Yes 0= No 2= Can't answer 99= Not applicable
D. Overall: Does the reader have confidence in the risk characterization and options presented, given the attendant uncertainties reported?	1= Yes 0= No 2= Can't answer

RAPID SYSTEMATIC REVIEW

Appendix H: GRADE Approach

CRITERION	GRADE POINTS	EXPLANATION
Downgrading criterion		
Individual study risk-of-bias rating **Serious** = >50% of trials had an unclear or high overall risk-of-bias rating **Very serious** = the risk of bias may have a strong influence on the estimate of effect and study limitations are present in the majority of studies	a) = -1 b) = -2	Rating from risk-of-bias tool
Inconsistency of findings among studies **Serious** = some inconsistency is noted (point estimates vary widely across studies and/or confidence intervals show minimal overlap, or significant heterogeneity is noted in meta-analysis). **Very serious** = studies show widely different directions of effect (some positive, some negative).	a) = -1 b) = -2	Heterogeneity in the results is measured by I^2 and the Q test (only if meta-analysis is possible).
Imprecision of effect estimates The total number of samples is less than that required by a conventional sample size calculation for a single adequately powered controlled trial	a) = -1	For prevalence outcomes, assume at least a 10% reduction in risk (for common outcomes, ~750-800 samples, for rare outcomes, ~200-300). (Check median control group risk in sample size calculator). For concentration outcomes, total sample size is >400. For rare events, can be less. (Check median treatment and control group means and SDs in sample size calculator).
Indirectness of individual study parameter as representative of target parameter **Serious** = >50% of trials indirectly measure the intervention, population, comparison, or outcome **Very serious** = >50% of trials measure two or more of the above parameters indirectly	a) = -1 b) = -2	Indirectness indicates studies do not directly measure the target parameter of interest to the review question. E.g. studies only measuring surrogate/intermediate outcomes can be rated down. E.g. studies conducted under non-commercial conditions (e.g. low density farm housing, or laboratory conditions) to be rated down as not directly generalizable.

CRITERION	GRADE POINTS	EXPLANATION
Publication bias Detected or suspected in data subset	a) = -1	If meta-analysis is possible, publication bias detected in dataset. If meta-analysis is not possible, publication bias suspected by other means (e.g. lack of small studies showing no treatment effect).
Upgrading criterion		
Large magnitude of effect Large effect in the absence of plausible confounders and major threats to validity	a) = +1	Large effect considered at least a 2-fold reduction in risk.
Direction of plausible bias Results may have been underestimated due to the study design (e.g. population sampled)	a) = +1	E.g. only sicker patients receive an experimental intervention, yet they still fare better, and the actual effect of the intervention is likely to be even greater than the data suggest.
Dose-response gradient Majority of trials identified a dose-response relationship	a) = +1	

RAPID SYSTEMATIC REVIEW

Appendix I: Summary of factors that could influence the applicability and effectiveness of interventions to control *Salmonella* in beef and pork

This document serves as a companion to the Summary Cards that describe the efficacy of various interventions for *Salmonella* control in beef and pork. The systematic review identified many options for control of *Salmonella* in beef and pork chains, from farm production to consumption. Many interventions were shown to be effective in research study settings, but for various reasons, there was frequently very low to low confidence, so one could expect a similar intervention effect in practice due to various complex and dynamic factors. Some of these factors were related to the studies themselves (e.g. small sample sizes, possible risks of bias, and methodological differences), while others are related to variability in intervention effects across study settings. This purpose of this document is to outline some of the latter factors that might influence the effect of interventions across different contexts and settings. Some of these factors are within the control of various stakeholders (e.g. producers or processors), and some may be outside of their control but are important to consider in deciding on the appropriateness of a particular control strategy.

SALMONELLA BURDEN AND CARRIAGE IN CATTLE AND PIGS

- Prevalence and concentration of *Salmonella* in cattle and pigs on individual farms, within herds, entering abattoirs, and on processing lines is dynamic and is not always predictable. Interventions aimed at controlling *Salmonella* may be affected by these baseline contamination levels. Multiple host factors influence these levels in animals, such as species, age, health status and co-infections, stress, reproduction, and high shedding events.
- At the farm level, units may have status varying from negligible risk of infection, to endemic. Farms negative for *Salmonella* detection will more appropriately be interested in control measures to prevent infection from entering the farm, while those that are positive will have greater interest in measures to limit spread and reduce prevalence.
- The burden of *Salmonella* can increase between farm and abattoirs due to the introduction of contamination during transport, lairage and marketing, deriving from sources such as other animals, transport vehicles, holding yards and environments. If animals with undetectable levels of *Salmonella* on-farm,

particularly pigs, are exposed to salmonellae after leaving the farm they can develop infection rapidly and the prevalence at slaughter can be higher than on-farm.
- At processing, cross-contamination and possible re-introduction of *Salmonella* can mitigate the impact of pathogen-reduction steps and decontamination interventions.
- *Salmonella* are carried in lymphoid tissues of cattle and pigs where they can be protected from exposure to decontamination agents, possibly reducing the efficacy of processing interventions.

SALMONELLA STRAINS

- *Salmonella* strains vary in their virulence, ecology, environmental persistence and ability to elicit responses to stress. Interventions have to be effective on the prevailing strains where the intervention is implemented.
- *Salmonella* can adapt and become resistant to control measures over time; regular monitoring of strain types and the effectiveness of interventions is required and corrective actions undertaken when necessary.

ENVIRONMENT

- At the farm level, *Salmonella* is widespread in farm environments and its presence is influenced by a complexity of non-host as well as host-related factors.
- The non-host factors that can interact in a complex manner on the presence of *Salmonella* in farm environments can include geography and location, weather, season, wildlife, and natural flora and fauna.
- An assessment of the risks of an existing or new setting for a production system is required to understand factors that need to be controlled either before or at the same time as the on-farm interventions, and whether an intervention can be transferred and adopted effectively.

PRODUCTION SYSTEMS AND INFRASTRUCTURE

- Different farm types and associated animal groups (e.g. mixed-species farms, dairy, feedlot, pasture, breeder and finishing farms) vary in their size, resources, animal husbandry and biosecurity practices, and the level of exposure to *Salmonella* and could affect intervention suitability and effectiveness.
- Fully integrated production systems under single management may offer greater opportunity for control than fragmented systems with multiple inde-

pendent managers; closer proximity of the production to the processing facilities will reduce the risk of an increase in the pathogen burden.

- At processing, establishment size, operating facilities, resources, and availability of services need to be considered in assessing the appropriateness and cost-effectiveness of interventions.
- Management commitment and staff capability and training can influence the implementation, effectiveness, and sustainability of interventions, and they are essential prerequisites to *Salmonella* control.

SOCIO-BEHAVIOURAL, POLITICAL AND REGULATORY FACTORS

- Various other contextual factors will influence implementation and uptake of interventions: acceptable costs and perceived benefits for those implementing the intervention; maintenance of market access and opportunity for new international markets; meeting customer specifications and economic incentives; consumer acceptance; regulation; and prevailing political and economic conditions.

IMPLICATIONS

- Reports of interventions are described in research studies in terms of *Salmonella* prevalence or concentration changes at one or more points in the food chain, and not on impacts on the final consumer risk of salmonellosis.
- A risk-based approach should be taken based on relevant country data and the unique context to identify the most appropriate points for *Salmonella* control. Controls in different sectors may be co-dependent and a combination of controls along the food chain will probably be required.

Appendix J: Detailed GRADE Assessment Results

					Downgrading Criteria					Upgrading Criteria			
Commodity	Point in chain	Study design	Intervention subgroup	GRADE starting level	Risk of bias	Inconsistency	Imprecision	Indirectness	Pub. bias	Large effect	Direction of bias	Dose response	GRADE final level
Pork	On-farm	ChT	Antimicrobials - phage therapy - lymph nodes	3	-1	0	-1	-1	0	0	0	0	1
Pork	On-farm	ChT	Antimicrobials - phage therapy - faecal	3	-1	0	-1	-1	0	0	0	0	1
Pork	On-farm	CT	Biosecurity - strategic movement	4	-1	0	-1	0	0	0	0	0	2
Pork	On-farm	CT	Feed/water acidification- faecal	4	-1	-1	0	0	0	0	0	0	2
Pork	On-farm	CT	Feed/water acidification- serum ELISA	4	-1	-1	0	0	0	0	0	0	2
Pork	On-farm	CT	Feed/water acidification- lymph node	4	-1	0	0	0	0	0	0	0	3
Pork	On-farm	CT	Feed/water acidification- rectal content	4	-1	-1	0	0	0	0	0	0	2
Pork	On-farm	CT	Feed management - meal - ELISA - finishers	4	-1	-1	0	0	0	0	0	0	2
Pork	On-farm	CT	Feed management - coarse vs fine - faecal	4	-1	0	0	0	0	0	0	0	3
Pork	On-farm	CT	Vaccination - individual faecal	4	-1	0	0	0	0	0	0	0	3
Pork	On-farm	CT	Vaccination - pen faecal	4	-1	-1	-1	0	0	0	0	0	1

					Downgrading Criteria					Upgrading Criteria			
Commodity	**Point in chain**	**Study design**	**Intervention subgroup**	**GRADE starting level**	**Risk of bias**	**Inconsistency**	**Imprecision**	**Indirectness**	**Pub. bias**	**Large effect**	**Direction of bias**	**Dose response**	**GRADE final level**
Pork	On-farm	CT	Vaccination - carcass lymph nodes	4	-1	0	0	0	0	0	0	0	3
Pork	On-farm	CT	Vaccination - carcass caecal content	4	-1	0	0	0	0	0	0	0	3
Pork	Transport	CT	No lairage - caecal contents	4	-1	-1	0	0	0	0	0	0	2
Pork	Transport	CT	No lairage - lymph nodes	4	-1	-1	0	0	0	0	0	0	2
Pork	Processing	Quasi	Scalding	3	0	-1	0	0	0	0	0	0	2
Pork	Processing	Quasi	Singeing	3	0	0	0	0	0	0	0	0	3
Pork	Processing	ChT	Singeing	3	0	0	-1	0	0	0	0	0	2
Pork	Processing	CT	Pre-chill - thermal wash	4	0	0	0	0	0	0	0	0	4
Pork	Processing	CT	Pre-chill - acid wash vs no treatment	4	0	0	0	0	0	0	0	0	4
Pork	Processing	CT	Pre-chill - acid wash vs water	4	-1	0	-1	0	0	0	0	0	2
Pork	Processing	CT	Pre-chill - acid wash vs no treatment - 24 hours post measure	4	-1	0	-1	0	0	0	0	0	2
Pork	Processing	CT	Pre-chill - other wash	4	-1	0	-1	0	0	0	0	0	2
Pork	Processing	CT	Pre-chill – steam vacuum vs trim	4	-1	0	-1	0	0	0	0	0	2
Pork	Processing	CT	Pre-chill - multiple	4	0	0	-1	-1	0	0	0	0	2
Pork	Processing	Quasi	Pre-chill - water wash	3	0	0	-1	-1	0	0	0	0	1
Pork	Processing	Quasi	Pre-chill - organic wash	3	0	0	-1	0	0	0	0	0	2

					Downgrading Criteria					Upgrading Criteria			
Commodity	Point in chain	Study design	Intervention subgroup	GRADE starting level	Risk of bias	Inconsistency	Imprecision	Indirectness	Pub. bias	Large effect	Direction of bias	Dose response	GRADE final level
Pork	Processing	Quasi	Pre-chill - unspecified wash	3	-1	0	0	0	0	0	0	0	2
Pork	Processing	Quasi	Pre-chill - multiple interventions	3	0	0	0	0	0	0	0	0	3
Pork	Processing	ChT	Pre-chill - water wash	3	0	-1	-1	-1	0	0	0	0	1
Pork	Processing	ChT	Pre-chill - thermal wash	3	0	-1	-1	0	0	0	0	0	1
Pork	Processing	ChT	Pre-chill - organic wash vs water	3	0	-1	-1	-1	0	0	0	0	1
Pork	Processing	ChT	Pre-chill - organic wash vs no treatment	3	0	-1	-1	-1	0	0	0	0	1
Pork	Processing	ChT	Pre-chill - other wash vs water	3	0	0	-1	0	0	0	0	0	2
Pork	Processing	ChT	Pre-chill - other wash vs no treatment	3	0	-1	-1	0	0	0	0	0	1
Pork	Processing	ChT	Pre-chill - multiple interventions vs chilling	3	0	0	-1	0	0	0	0	0	2
Pork	Processing	ChT	Pre-chill - multiple interventions vs pre-treatment	3	-1	0	-1	0	0	0	0	0	1
Pork	Processing	SR	Chilling	3	-1	-1	0	0	-1	0	0	0	1
Pork	Processing	CT	Chilling - spray vs conventional	4	-1	0	-1	0	0	0	0	0	2
Pork	Processing	SR	Multiple stages – HACCP	3	0	-1	0	0	0	0	0	0	2
Pork	Processing	Quasi	Cleaning/ disinfection	3	0	0	0	-1	0	0	0	0	2
Pork	Processing	Quasi	Multiple stages - post-exsanguination	3	-1	0	-1	0	0	0	0	0	1

					Downgrading Criteria					Upgrading Criteria			
Commodity	**Point in chain**	**Study design**	**Intervention subgroup**	**GRADE starting level**	**Risk of bias**	**Inconsistency**	**Imprecision**	**Indirectness**	**Pub. bias**	**Large effect**	**Direction of bias**	**Dose response**	**GRADE final level**
Pork	Processing	Quasi	Multiple stages - pre-evisceration	3	-1	-1	0	0	0	0	0	0	1
Pork	Processing	Quasi	Multiple stages - post-evisceration	3	-1	-1	0	0	0	0	0	0	1
Pork	Processing	ChT	Non-thermal - UV - CFU/cm^2	3	0	0	-1	-1	0	0	0	0	1
Pork	Processing	ChT	Non-thermal - UV - CFU/g	3	0	0	-1	-1	0	0	0	1	2
Pork	Processing	ChT	Non-thermal - plasma	3	0	0	-1	-1	0	0	0	1	2
Beef	On-farm	CT	Antimicrobials - ionophores	4	-1	0	0	0	0	0	0	0	3
Beef	On-farm	CT	Gut manipulation - probiotics	4	-1	0	-1	0	0	0	0	0	2
Beef	On-farm	CT	Vaccination - dairy cow - shedding	4	-1	-1	0	-1	0	0	0	0	1
Beef	Processing	Quasi	Live animal - water wash	3	-1	0	-1	0	0	0	0	0	1
Beef	Processing	Quasi	Post-exsanguination - water wash	3	0	-1	-1	0	0	0	0	0	1
Beef	Processing	Quasi	Live animal - acid wash	3	-1	0	-1	0	0	0	0	0	1
Beef	Processing	Quasi	Post-exsanguination - acid wash	3	0	0	-1	0	0	0	0	0	2
Beef	Processing	Quasi	Live animal - other wash	3	-1	0	-1	0	0	0	0	0	1
Beef	Processing	Quasi	Post-exsanguination - other wash	3	0	0	0	0	0	0	0	0	3

					Downgrading Criteria					Upgrading Criteria			
Commodity	Point in chain	Study design	Intervention subgroup	GRADE starting level	Risk of bias	Inconsistency	Imprecision	Indirectness	Pub. bias	Large effect	Direction of bias	Dose response	GRADE final level
Beef	Processing	ChT	Post-exsanguination - acid wash	3	0	-1	-1	0	0	0	0	0	1
Beef	Processing	ChT	Post-exsanguination - other wash	3	0	-1	-1	0	0	0	0	0	1
Beef	Processing	ChT	Post-exsanguination – de-hairing vs no treatment	3	0	0	-1	-1	0	0	0	0	1
Beef	Processing	ChT	Post-exsanguination – de-hairing vs water	3	0	0	-1	0	0	0	0	0	1
Beef	Processing	CT	Bunging	4	-1	0	-1	0	0	0	0	0	2
Beef	Processing	Quasi	Post-evisceration - water wash	3	-1	0	0	0	0	0	0	0	2
Beef	Processing	Quasi	Post-evisceration - thermal wash	3	0	0	0	0	0	0	0	0	3
Beef	Processing	Quasi	Post-evisceration - multiple treatments	3	-1	0	0	0	0	0	0	0	2
Beef	Processing	CT	Post-evisceration - thermal	4	0	0	0	0	0	0	0	0	4
Beef	Processing	CT	Post-evisceration - Other interventions	4	-1	0	-1	-1	0	0	0	0	1
Beef	Processing	ChT	Post-evisceration - water wash - carcass sample	3	0	-1	-1	0	0	0	0	0	1
Beef	Processing	ChT	Post-evisceration - water wash - ground beef sample	3	0	0	-1	0	0	0	0	0	2
Beef	Processing	ChT	Post-evisceration - thermal vs water wash - carcass sample	3	0	-1	-1	0	0	0	0	0	1

					Downgrading Criteria					Upgrading Criteria			
Commodity	**Point in chain**	**Study design**	**Intervention subgroup**	**GRADE starting level**	**Risk of bias**	**Inconsistency**	**Imprecision**	**Indirectness**	**Pub. bias**	**Large effect**	**Direction of bias**	**Dose response**	**GRADE final level**
Beef	Processing	ChT	Post-evisceration - thermal vs no wash - carcass sample	3	0	-1	-1	0	0	0	0	0	1
Beef	Processing	ChT	Post-evisceration - thermal vs water wash - ground beef sample	3	0	0	-1	0	0	0	0	0	2
Beef	Processing	ChT	Post-evisceration - thermal vs no wash - ground beef sample	3	0	0	-1	0	0	0	0	0	2
Beef	Processing	ChT	Post-evisceration - acid vs water wash - carcass sample	3	0	-1	0	0	0	0	0	0	2
Beef	Processing	ChT	Post-evisceration - acid vs no wash - carcass sample	3	0	-1	-1	0	0	0	0	0	1
Beef	Processing	ChT	Post-evisceration - acid vs water wash - ground beef sample	3	0	0	-1	0	0	0	0	0	2
Beef	Processing	ChT	Post-evisceration - acid vs no wash - ground beef sample	3	0	0	-1	0	0	0	0	0	2
Beef	Processing	ChT	Post-evisceration - other vs water wash - carcass sample	3	0	-1	0	0	0	0	0	0	2
Beef	Processing	ChT	Post-evisceration - other vs no wash - carcass sample	3	0	-1	-1	0	0	0	0	0	1
Beef	Processing	ChT	Post-evisceration - other vs water wash - ground beef sample	3	0	0	-1	0	0	0	0	0	2
Beef	Processing	ChT	Post-evisceration - other vs no wash - ground beef sample	3	0	0	-1	0	0	0	0	0	2

					Downgrading Criteria					Upgrading Criteria			
Commodity	**Point in chain**	**Study design**	**Intervention subgroup**	**GRADE starting level**	**Risk of bias**	**Inconsistency**	**Imprecision**	**Indirectness**	**Pub. bias**	**Large effect**	**Direction of bias**	**Dose response**	**GRADE final level**
Beef	Processing	ChT	Post-evisceration - dry heat vs water wash - ground beef sample	3	0	0	-1	0	0	0	0	0	2
Beef	Processing	ChT	Post-evisceration - electricity	3	0	0	-1	-1	0	0	0	0	1
Beef	Processing	ChT	Post-evisceration - extracts vs no treatment	3	0	-1	-1	-1	0	0	0	0	1
Beef	Processing	ChT	Post-evisceration - extracts vs water	3	0	0	-1	-1	0	0	0	0	1
Beef	Processing	ChT	Post-evisceration - trimming	3	0	0	-1	0	0	0	0	0	2
Beef	Processing	ChT	Post-evisceration - multiple treatments vs water wash	3	0	-1	-1	0	0	0	0	0	1
Beef	Processing	ChT	Post-evisceration - multiple treatments vs no treatment	3	0	-1	-1	0	0	0	0	0	1
Beef	Processing	Quasi	Chilling	3	-1	0	0	0	0	0	0	0	2
Beef	Processing	ChT	Chilling - spray	3	0	0	-1	-1	0	0	0	0	1
Beef	Processing	ChT	Chilling - acid	3	0	-1	-1	-1	0	0	0	0	1
Beef	Processing	ChT	Chilling - multiple	3	0	0	-1	-1	0	0	0	0	1
Beef	Processing	Quasi	Post-chill - acid wash	3	-1	0	-1	0	0	0	0	0	1
Beef	Processing	ChT	Post-chill - thermal wash	3	0	0	-1	0	0	0	0	0	2
Beef	Processing	ChT	Post-chill - other wash	3	0	0	-1	0	0	0	0	0	2
Beef	Processing	ChT	Post-chill - multiple treatments	3	0	0	-1	0	0	0	0	0	2
Beef	Processing	ChT	Post-fabrication - water wash - ground beef sample	3	0	-1	-1	0	0	0	0	0	1

					Downgrading Criteria					Upgrading Criteria			
Commodity	Point in chain	Study design	Intervention subgroup	GRADE starting level	Risk of bias	Inconsistency	Imprecision	Indirectness	Pub. bias	Large effect	Direction of bias	Dose response	GRADE final level
Beef	Processing	ChT	Post-fabrication - water wash - steak sample	3	0	0	-1	0	0	0	0	0	2
Beef	Processing	ChT	Post-fabrication - water wash - cheek meat	3	-1	0	-1	-1	0	0	0	0	1
Beef	Processing	ChT	Post-fabrication - thermal wash - ground beef sample	3	0	-1	-1	-1	0	0	0	0	1
Beef	Processing	ChT	Post-fabrication - thermal wash - steak/cut sample	3	0	0	-1	-1	0	0	0	0	1
Beef	Processing	ChT	Post-fabrication - thermal wash - cheek meat	3	-1	0	-1	-1	0	0	0	0	1
Beef	Processing	ChT	Post-fabrication - acid vs water wash - ground beef sample	3	-1	0	-1	0	0	0	0	0	1
Beef	Processing	ChT	Post-fabrication - acid vs no wash - ground beef sample	3	0	-1	-1	-1	0	0	0	0	1
Beef	Processing	ChT	Post-fabrication - acid vs water wash - steak/cuts sample	3	0	-1	-1	0	0	0	0	0	1
Beef	Processing	ChT	Post-fabrication - acid vs no wash - steak/cuts sample	3	0	-1	-1	-1	0	0	0	0	1
Beef	Processing	ChT	Post-fabrication - acid wash - cheek meat	3	-1	0	-1	-1	0	0	0	0	1
Beef	Processing	ChT	Post-fabrication - other vs water wash - ground beef sample	3	0	-1	-1	0	0	0	0	0	1
Beef	Processing	ChT	Post-fabrication - other vs no wash - ground beef sample	3	0	-1	-1	-1	0	0	0	0	1

					Downgrading Criteria					Upgrading Criteria			
Commodity	Point in chain	Study design	Intervention subgroup	GRADE starting level	Risk of bias	Inconsistency	Imprecision	Indirectness	Pub. bias	Large effect	Direction of bias	Dose response	GRADE final level
Beef	Processing	ChT	Post-fabrication - other vs water wash - steak/cut sample	3	0	-1	-1	0	0	0	0	0	1
Beef	Processing	ChT	Post-fabrication - other vs no wash - steak/cut sample	3	0	-1	-1	-1	0	0	0	0	1
Beef	Processing	ChT	Post-fabrication - other wash - cheek meat	3	-1	0	-1	-1	0	0	0	0	1
Beef	Processing	ChT	Post-fabrication - irradiation	3	0	-1	-1	0	0	0	0	0	1
Beef	Processing	ChT	Post-fabrication - lactic acid bacteria	3	0	-1	-1	0	0	0	0	0	1
Beef	Processing	ChT	Post-fabrication - multiple treatments vs water wash - ground beef sample	3	0	0	-1	0	0	0	0	0	2
Beef	Processing	ChT	Post-fabrication - multiple treatments vs no treatment - ground beef sample	3	0	-1	-1	-1	0	0	0	0	1
Beef	Processing	ChT	Post-fabrication - multiple treatments vs water wash - steak/cut sample	3	0	-1	-1	-1	0	0	0	0	1
Beef	Processing	ChT	Post-fabrication - multiple treatments vs no treatment - steak/cut sample	3	0	-1	-1	-1	0	0	0	0	1
Beef	Processing	SR	Multiple stages - HACCP	3	0	0	0	0	0	0	0	0	3
Beef	Processing	Quasi	Multiple stages - multiple interventions	3	-1	0	0	0	0	0	0	0	2
Beef	Processing	ChT	Multiple stages	3	0	-1	-1	-1	0	0	0	0	1

NOTES: CT = controlled trial; ChT = challenge trial; Quasi = quasi-experiment; SR = systematic review.

Annex 2

Steps in the beef and pork food chains as outlined in the Codex Committee on Food Hygiene Draft Guidelines

A generic flow diagram of the basic beef and pork production processes is shown below, as presented in the CCFH "Proposed Draft Guidelines for the Control of Non-typhoidal *Salmonella* spp. in Beef and Pork Meat". This was the flow diagram which was available at the time of the expert meeting. Subsequent to the expert meeting some small updates were made to the diagram, in particular the inclusion of an additional step before Chilling, entitled Pre-chill Treatment. This diagram and the revised Codex guidance which reflects many of the recommendations made in this report can be found in Appendix II of the report of the 47th Session of the Codex Committee on Food Hygiene[1].

Beef chain steps:

1. **Primary Production**
2. **Transport to Slaughter**
3. **Receive and Unload**
4. **Lairage**
5. **Stunning**
6. **Shackling**
7. **Sticking/Bleeding**
8. **De-hiding**

(continued)

[1] Available at: http://www.fao.org/fao-who-codexalimentarius/sh-proxy/en/?lnk=1&url=https%253A%252F%252Fworkspace.fao.org%252Fsites%252Fcodex%252FMeetings%252FCX-712-47%252FReport%252FREP16_FHe.pdf

(continued from previous page)

9. **Head Removal/Head Washing**
10. **Bunging**
11. **Brisket Opening**
12. **Rodding/Tying the Weasand**
13. **Evisceration**
14. **Splitting**
15. **Post Mortem Inspection**
16. **Chilling**
17. **Carcass Fabrication**
18. **Trim/Grinding**
19. **Packaging Finished Product**
20. **Transport to Distribution Channels**
21. **Cold Storage/Aging**
22. **Receiving at Purveyor**
23. **Finished Product Fabrication**
24. **Mechanical Tenderization**
25. **Distribution/Retail**
26. **Consumer**

Pork chain steps:

1. Primary Production
2. Transport to Slaughter
3. Receive and Unload
4. Lairage
5. Stunning
6. Sticking/Bleeding
7. Scalding
8. De-hairing
9. Gambrelling
10. Singeing
11. Polishing
12. Bunging
13. Midline Opening
14. Evisceration
15. Splitting
16. Head Dropping/Removal
17. Post Mortem inspection
18. Chilling

(continued)

(continued from previous page)

19. Carcass Fabrication
20. Mechanical Tenderization/Mincing
21. Packing Product
22. Transport to Distribution Channels
23. Cold Storage
24. Distribution/Retail
25. Consumer

FAO/WHO Microbiological Risk Assessment Series

1 Risk assessments of *Salmonella* in eggs and broiler chickens: Interpretative Summary, 2002

2 Risk assessments of *Salmonella* in eggs and broiler chickens, 2002

3 Hazard characterization for pathogens in food and water: Guidelines, 2003

4 Risk assessment of *Listeria monocytogenes* in ready-to-eat foods: Interpretative Summary, 2004

5 Risk assessment of *Listeria monocytogenes* in ready-to-eat foods: Technical Report, 2004

6 *Enterobacter sakazakii* and microorganisms in powdered infant formula: Meeting Report, 2004

7 Exposure assessment of microbiological hazards in food: Guidelines, 2008

8 Risk assessment of *Vibrio vulnificus* in raw oysters: Interpretative Summary and Technical Report, 2005

9 Risk assessment of choleragenic *Vibrio cholerae* 01 and 0139 in warm-water shrimp in international trade: Interpretative Summary and Technical Report, 2005

10 *Enterobacter sakazakii* and *Salmonella* in powdered infant formula: Meeting Report, 2006

11 Risk assessment of *Campylobacter* spp. in broiler chickens: Interpretative Summary, 2008

12 Risk assessment of *Campylobacter* spp. in broiler chickens: Technical Report, 2008

13 Viruses in food: Scientific Advice to Support Risk Management Activities: Meeting Report, 2008

14 Microbiological hazards in fresh leafy vegetables and herbs: Meeting Report, 2008

15 *Enterobacter sakazakii* (*Cronobacter* spp.) in powdered follow-up formula: Meeting Report, 2008

16 Risk assessment of *Vibrio parahaemolyticus* in seafood: Interpretative Summary and Technical Report, 2009

17 Risk characterization of microbiological hazards in food: Guidelines, 2009.

18 Enterohaemorragic *Escherichia coli* in meat and meat products: Meeting Report, 2010

19 *Salmonella* and *Campylobacter* in chicken meat: Meeting Report, 2009

20 Risk assessment tools for *Vibrio parahaemolyticus* and *Vibrio vulnificus* associated with seafood: Meeting Report and Follow-up, In press

21 *Salmonella* spp. In bivalve molluscs: Risk Assessment and Meeting Report, In press

22 Selection and application of methods for the detection and enumeration of human pathogenic *Vibrio* spp. in seafood: Guidance, 2016

23 Multicriteria-based ranking for risk management of food-borne parasites, 2014

24 Statistical aspects of microbiolgical criteria related to foods: A risk managers guide, 2016

25 A risk based approach for the control of *Trichinella* in pigs and *Taenia saginata* in beef: Meeting Report, In press

26 Ranking of low moisture foods in support of microbiological risk management: Meeting Report and Systematic Review, In press

27 Microbiological hazards associated with spices and dried aromatic herbs: Meeting Report, In press

28 Microbial Safety of lipid based ready-to-use foods for the management of moderate acute and severe acute malnutrition: First meeting report, 2016

29 Microbial Safety of lipid based ready-to-use foods for the management of moderate acute and severe acute malnutrition: Second meeting report, In press

30 Interventions for the Control of Non-typhoidal *Salmonella* spp. in Beef and Pork: Meeting Report and Systematic Review, 2016